Lecture Notes in Computer Scie

Commenced Publication in 1973
Founding and Former Series Editors:
Gerhard Goos, Juris Hartmanis, and Jan van Leeuwen

Editorial Board

José Luiz Fiadeiro   Peter D. Mosses
Fernando Orejas (Eds.)

Recent Trends
in Algebraic
Development Techniques

17th International Workshop, WADT 2004
Barcelona, Spain, March 27-29, 2004
Revised Selected Papers

 Springer

Volume Editors

José Luiz Fiadeiro
University of Leicester, Department of Computer Science
Leicester LE1 7RH, UK
E-mail: jose@fiadeiro.org

Peter D. Mosses
University of Wales Swansea, Department of Computer Science
Singleton Park, Swansea SA2 8PP, UK
E-mail: P.D.Mosses@swan.ac.uk

Fernando Orejas
Universitat Politècnica de Catalunya
Departament de Llenguatges i Sistemes Informàtics
Campus Nord C5, Jordi Girona 1-3, 08034 Barcelona, Spain
E-mail: orejas@lsi.upc.es

Library of Congress Control Number: 2005922176

CR Subject Classification (1998): F.3.1, F.4, D.2.1, I.1

ISSN 0302-9743
ISBN 3-540-25327-0 Springer Berlin Heidelberg New York

Springer is a part of Springer Science+Business Media

springeronline.com

© Springer-Verlag Berlin Heidelberg 2005
Printed in Germany

Typesetting: Camera-ready by author, data conversion by Scientific Publishing Services, Chennai, India
Printed on acid-free paper   SPIN: 11407355   06/3142   5 4 3 2 1 0

Preface

This volume contains selected papers from WADT 2004, the 17th International Workshop on Algebraic Development Techniques. Like its predecessors, WADT 2004 focussed on the algebraic approach to the specification and development of systems, an area that was born around the algebraic specification of abstract data types and encompasses today the formal design of software systems, new specification frameworks and a wide range of application areas.

WADT 2004 took place at the Technical University of Catalonia (UPC), Barcelona, Spain, on 27–29 March 2004, and was organized by Fernando Orejas and Jordi Cortadella.

The program consisted of invited talks by Luís Caires (Universidade Nova de Lisboa, Portugal) and Reiko Heckel (University of Paderborn, Germany), and 33 presentations describing ongoing research on main topics of the workshop: formal methods for system development, specification languages and methods, systems and techniques for reasoning about specifications, specification development systems, methods and techniques for concurrent, distributed and mobile systems, and algebraic and co-algebraic foundations.

The Steering Committee of WADT, consisting of Michel Bidoit, José Fiadeiro, Hans-Jöerg Kreowski, Peter Mosses, Fernando Orejas, Francesco Parisi-Presicce, and Andrzej Tarlecki, with the additional help of Christine Choppy and Till Mossakowski, selected several presentations and invited their authors to submit a full paper for possible inclusion in this volume. All submissions underwent a careful refereeing process. We are extremely grateful to all the referees who helped in reviewing the submissions: H. Baumeister, L. Caires, A. Cherchago, R. Heckel, R. Hennicker, F. Jacquemard, R. Klempien-Hinrichs, C. Lüth, S. Merz, W. Pawlowski, and L. Schröder.

This volume contains the final versions of the 14 contributions that were accepted. It contains also the invited paper of Reiko Heckel, co-authored with Sebastian Thöne.

The workshop was jointly organized with IFIP WG 1.3 (Foundations of System Specification), and received generous sponsorship from the following organizations:

- Spanish Ministry of Science and Technology (MCYT)
- Catalan Department for University, Research and Information Society (DURSI)
- Technical University of Catalonia (UPC)

David Banyeres, Robert Clariso, Kyller Costa, Nilesh Modi, Jiangtao Meng, Nikos Mylonakis, Sonia Perez, Edelmira Pasarella, and Elvira Pino provided invaluable help throughout the preparation and organization of the workshop. We are grateful to Springer for its helpful collaboration and quick publication.

Finally, we would like to announce that, starting in 2005, WADT will join forces and reputations with CMCS, the International Workshop on Coalgebraic Methods in Computer Science, to create a new high-level biennial international event: CALCO, the Conference on Algebra and Coalgebra in Computer Science.

December 2004   José Fiadeiro, Peter Mosses, Fernando Orejas

Table of Contents

Invited Technical Paper

Contributed Papers

Behavior-Preserving Refinement Relations Between Dynamic Software Architectures

Reiko Heckel[1] and Sebastian Thöne[2]

[1] Department of Computer Science
[2] International Graduate School Dynamic Intelligent Systems,
University of Paderborn, Germany
{reiko, seb}@upb.de

Abstract. In this paper, we address the refinement of abstract architectural models into more platform-specific representations. For each level of abstraction, we employ an architectural style covering structural restrictions on component configurations as well as supported communication and reconfiguration operations. Architectural styles are formalized as graph transformation systems with graph transformation rules defining the available operations. Architectural models are given as graphs to which one can directly apply the transformation rules in order to simulate operations and their effects.

In addition to previous work, we include process descriptions into our architectural models in order to control the communication and reconfiguration behavior of the components. The execution semantics of these processes is also covered by graph transformation systems.

We propose a notion of refinement which requires the preservation of both structure and behavior at the lower level of abstraction. Based on formal refinement relationships between abstract and platform-specific styles, we can use model checking techniques to verify that abstract scenarios can also be realized in the platform-specific architecture.

1 Introduction

In the development of complex software systems, a model of the *software architecture* [30] allows for early reasoning on the system at a high level of abstraction. An architectural model covers the involved run-time *configuration* of system components, the *communication* between these components, and possible *reconfiguration* operations that enable the system to react to upcoming requirements and events. Such *dynamic* architectures gain increasing attention in the context of e-business, self-healing, and mobile systems.

Since software architectures[1] are intended to bridge the gap between system requirements and implementation, they have to conform to both business-driven requirements as well as restrictions and mechanisms imposed by the chosen run-time infrastructure. In order to integrate both aspects, we propose a stepwise refinement approach starting with an abstract, *business-level* architecture which

---

[1] We use the term software architecture as a synonym for the model of an architecture.

J.L. Fiadeiro, P. Mosses, and F. Orejas (Eds.): WADT 2004, LNCS 3423, pp. 1–27, 2005.
© Springer-Verlag Berlin Heidelberg 2005

can be derived from user and business requirements. This business-level architecture is then refined into a more concrete description which also integrates *platform-specific* aspects like supported reconfiguration operations and communication mechanisms.

A recent example of this general principle of model refinement is the *Model-Driven Architecture (MDA)* [26] put forward by the OMG. Here, platform-specific details are initially ignored at the model-level to allow for maximum portability. Then, these platform-independent models are refined by adding details required to map to a given target platform. At each refinement level, more assumptions on the resources, constraints, and services of the chosen platform are incorporated into the model.

Similarly, as described in a previous paper [3], we use *architectural styles* [2], formalized as graph transformation systems, for defining the assumptions on a certain level of platform abstraction, i.e., the vocabulary, structural constraints, and available communication and reconfiguration mechanisms. Then, an architecture at a certain level of abstraction has to conform to the corresponding architectural style.

In our previous work, we applied the available style-specific reconfiguration and communication operations to an architecture without further control. In this paper (see Section 4), we provide an extension which allows the definition of *processes* and their operational semantics. These processes control the order in which available operations are invoked by the individual software components. This leads to a more detailed picture of the architectural behavior.

We do not consider architectural refinement as the internal decomposition of components into subcomponents, as done by other authors, but rather focus on porting an abstract architecture to a more platform-specific level which usually requires additional platform-related entities and resources. For this purpose, we define different architectural styles for different levels of platform abstraction, namely a generic, platform-independent style for business-level architectures and a more specific style for architectural models at the platform-specific level.

When refining software architectures from the abstract to the concrete level, we have to preserve both structural and behavioral properties. This leads to the following two requirements:

1. **Architectural consistency:** After being ported to the lower level of abstraction, the concrete architecture has to satisfy the same functional requirements as the abstract architecture. Therefore, we have to refine configurations of components, connections, and other resources in a way that all business-relevant entities of the abstract architecture are also preserved at the concrete level.

2. **Behavior preservation:** Similarly, the concrete architecture has to preserve the abstract communication and reconfiguration behavior. In particular, we require that all business-relevant scenarios of the abstract architecture are also realizable in the concrete architecture.

While porting the abstract behavior to the platform-specific level we have to respect the capabilities of the chosen target platform according to its reconfiguration and communication mechanisms. In many cases, depending on the current situation where an operation is to be applied and the effects of preceding actions, the refinement of an abstract action varies from other situations and cannot be decided locally. Thus, we believe that behavior preservation cannot be solved by a fixed syntactic mapping between abstract and concrete operations but has to be dealt with at a semantic level.

Further requirements include a high degree of *reusability* which means that the refinement relationship between certain levels of platform abstraction should not only apply for one specific system, but should be reusable for other architectures as well.

Since refinement is not an easy task and thus error-prone and cost-intensive if done by hand, we are also aiming at *tool support*. However, it is difficult to automate the *construction* of refined architectures, because this is a creative process, and computers cannot invent details for the concrete level that are not existent at the abstract level. Nevertheless, we intend to investigate tool support for *checking* if a concrete architecture satisfies the formal refinement relationship we prescribe for the refinement from abstract to concrete level. Combined with user interaction for modifying invalid concrete models, we achieve a semi-automated approach for creating refined architectural models.

The refinement relationship, as already proposed in [5], is *style-based* meaning that it is defined between two architectural styles rather than between individual architectures. Since this relationship can be applied to any instances of the styles, we achieve the desired degree of reusability.

To check for architectural consistency, we have to compare the business-relevant entities of the abstract and the concrete model. For this purpose, we use an *abstraction function* which lifts concrete models to the abstract style. To check for behavior preservation, we have to prove that all states of an abstract scenario are also reachable in a corresponding concrete scenario, preferably with the help of model checking techniques. For this purpose, we employ a contravariant *translation function* which transforms abstract states into requirements for states at the platform-specific level. A model checker can then search for concrete states satisfying these requirements.

The rest of this paper is organized as follows. We survey related work in Section 2. In Section 3, we revisit the modeling of architectural styles based on graph theory, and in Section 4 we extend the proposed architecture description technique by processes for controlling architectural behavior. In Section 5, we use this formal framework to define our notion of refinement under the obligation of architectural consistency, and Section 6 covers the problem of behavior preservation by a semantic requirement that can be checked by model checking tools. Section 7 concludes the paper.

2   Related Work

Refinement is a long-known design principle in software engineering. First ideas in the context of program development go back to Wirth [34]. In the sense of a systematic top-down methodology, he argued for the expansion of high-level program instructions to lower level macros and procedures.

While Wirth mainly investigated sequential programs, the refinement of concurrent systems became popular as *action refinement* in the context of process algebras (cf. [17] for a survey on this topic). This field considers the refinement of abstract actions into sequences of concrete actions, also called *processes*, and the potential interleaving of multiple concurrent processes.

Our approach is different from this work for two reasons. First, we want to avoid a fixed, sometimes even syntactically defined substitution of an abstract action by a concrete process wherever the abstract action occurs. Instead, we are aiming at a more flexible notion of refinement which also allows for alternate refinements of an action depending on the context where the action occurs. Second, we also want to enable refinement in those cases where the two levels of abstraction are so different that it becomes hard to relate the corresponding actions with each other.

Apart from action refinement, we also have to mention the different notions of refinement in the field of software architecture. For instance, Batory et. al. [6] consider *feature refinement* which is modifying models, code, and other artifacts in order to integrate additional features with every refinement step. Different to this work, Canal et. al. [9] consider refinement as the decomposition of a software component into subcomponents and the specialization of components under certain compatibility conditions.

In our case, we neither want to add any extra-functionality to the architecture nor to look into the internals of the components, but we rather want to port a business-level architecture to a more platform-specific level considering all the restrictions and mechanisms of the chosen target platform.

Refinement of architectures in this sense has first been discussed by Moriconi et al. in [25]. Building a formalization in first-order logic, the authors describe a general approach of rule-based refinement replacing a structural pattern in an abstract style by its realization in the concrete style. The approach is related to ours, but focuses on refinement of the structure only and does not take reconfiguration and communication behavior into account. Also, applying the logic-based theory to concrete architecture description languages is not trivial. The general idea of rule-based refinement, however, is applicable in our context, too.

Garlan [16] stresses the fact that it is more powerful to have rules operating on architectural styles rather than on style instances. He formalizes refinements as abstraction functions from the concrete to the abstract style. We use a similar approach to define refinement relationships (see Section 5). Also, he argues that no single definition of refinement can be provided, but that one should state what properties are preserved. In our case, we concentrate on the preservation of architectural consistency and the dynamic semantics of reconfiguration and communication scenarios.

Other proposals on architecture refinement like [1, 12] concentrate on structural refinements only, which is complementary to our work. The only formal approach we are aware of that considers refinement of dynamic reconfiguration can be found in [8]. But, the paper only sketches the ideas without any concrete definition. Moreover, the approach is targeted on the translation from one Architecture Description Language to another rather than on the refinement between architectural styles that represent different levels of platform abstraction.

Since we use graph transformation systems as the underlying formalism to describe dynamic software architectures, which is in the tradition of [21, 22, 24, 31, 33], it is also worth to look at existing work on refinement of graph transformation systems. The general idea is to relate the transformation rules and, thus, the behavior of an abstract graph transformation system to the rules of a more concrete transformation system. One can judge these refinement relationships along a continuum from syntactical relationships to more semantical ones.

Große-Rhode et. al. [18], for instance, propose a refinement relationship between abstract and concrete rules that can be checked syntactically. One of the conditions requires that, e.g., the abstract rule and its refinement must have the same pre- and post-conditions except for retyping. Based on this very restrictive definition they can prove that the application of the concrete rule expression yields the same behavior as the corresponding abstract rule. The draw-back of this approach is that it cannot handle those cases where the refining rule expression should have additional effects on platform-specific elements that do not occur in the abstract rule. And, similar to action refinement, the approach does not allow alternate refinements for the same abstract rule.

Similarly, the work by Heckel et. al. [20] is based on a syntactical relationship between two graph transformation systems. Although this approach is less restrictive as it allows additional (platform-specific) elements at the concrete level, it is still difficult to apply if there are no direct correspondences between abstract and concrete rules. Moreover, their objective is to project any given concrete transformation behavior to the abstract level and not vice versa.

In our work, we propose a more flexible, semantic-based notion of refinement. We do not define a fixed mapping between the various transformation rules but only between the structural parts of the graph transformation system. Then, we check whether all system states of an abstract model are also reachable at the concrete level, no matter by which order of transformation rules. By avoiding the functional refinement mapping between transformation rules, we can also relate transformation systems with completely different behavior, and we are flexible enough to cope with alternate refinements.

3   Graph Transformation Systems as Architectural Styles

As already introduced in [3], we use *architectural styles* as conceptual platform models. Such a platform model has to define the vocabulary of elements to be considered, to restrict the possible relationships among those elements, and to specify communication as well as reconfiguration mechanisms supported by the platform. We use different styles for different levels of platform abstraction.

In this section, we present the formal definition of architectural styles as *typed graph transformation systems* [10] together with two exemplary styles, namely an abstract style for business-level architectures and a platform-specific style for service-oriented architectures. In Section 5, we explain how a refinement relationship between these styles can be used to refine business-level architectures, which abstract from platform-specific vocabulary and restrictions, to service-oriented architectures.

Informally, a typed graph transformation system consists of (1) a *type graph* to define the vocabulary of architectural elements, (2) a set of *constraints* to further restrict the valid models, and (3) a set of *graph transformation rules* for communication and reconfiguration operations. A system architecture that conforms to a given style is represented as an *instance graph* of the type graph.

Definition 1 (Graph and Graph Morphism). *A graph is a tuple $G = (N, E, src, tar)$ with a set N of nodes, a set E of edges, and functions $src, tar : E \rightarrow N$ that assign source and target nodes to each edge. A graph morphism $f = (f_N, f_E) : G \rightarrow G'$ is a pair of functions $f_N : N \rightarrow N'$ and $f_E : E \rightarrow E'$ preserving source and target ($src' \circ f_E = f_N \circ src$ and $tar' \circ f_E = f_N \circ tar$).*

Definition 2 (Typed Graph). *Given a graph TG, a TG-typed graph $\langle G, tp_G \rangle$ is a graph G equipped with a structure-preserving graph morphism $tp_G : G \rightarrow TG$. We call TG type graph and $\langle G, tp_G \rangle$ instance graph over TG. The category of TG-typed instance graphs is called $\mathbf{Graph}_{TG}$.*

The graphs we use are directed and unlabeled; for the sake of clarity, nodes (and edges) can be named by unique identifiers. Type graphs can be represented by *UML class diagrams* and instance graphs by *UML object diagrams* [19]. The typing morphism tp_G is depicted by referencing the type names. As an example, Figure 1(a) shows the type graph of the business-level style we have defined in [4]. Figure 1(b) shows a corresponding instance graph.

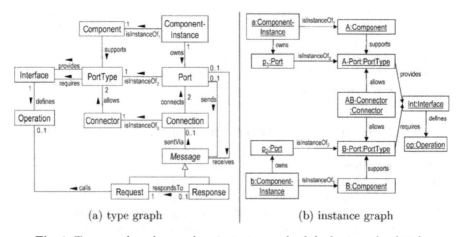

(a) type graph   (b) instance graph

Fig. 1. Type graph and exemplary instance graph of the business-level style

According to this type graph, architectures consist of ComponentInstances which externalize their functionalities through Ports. They can interact with each other through a Connection between their Ports. The state of a communication is encoded by Request and Response message nodes.

Besides the elements for run-time configurations, the type graph also defines nodes for the application-specific types of these elements. For example, Component, PortType, and Connector nodes can be used to describe certain types of components, ports, or connections; PortTypes are characterized by provided and required Interfaces. This way, a corresponding instance graph incorporates both the actual configuration at a certain run-time state as well as application-specific type information about the involved entities.

For example, the instance graph in Fig. 1(b) defines a system that consists of an instance a of component A and an instance b of component B. Both component instances own a port of type A-Port and B-Port respectively, which could be connected by an instance of the AB-Connector. The A-Port provides the interface Int with the operation op, while the B-Port requires this interface.

Along with the type graph comes a set C of *constraints* that further restricts the set of valid instance graphs. Simple constraints already included in the class diagram are cardinalities that restrict the multiplicity of links between the elements (omitted cardinality means 0..n by default). More complex restrictions can be defined, e.g., using expressions of the *Object Constraint Language (OCL)*, which is part of the UML.

Graph transformation. Graph transformation rules [13] are used to define rewriting operations on graphs. Since our instance graphs represent system configurations, transformation rules nicely fit to define *reconfiguration operations* provided by the platform. If we encode communication-related information into the graphs, as done by the Message node and its subtypes in Fig. 1(a), then transformation rules are also suitable to represent *communication mechanisms*. A certain reconfiguration and communication scenario can be modeled as a sequence of transformation rules which are applied to an initial instance graph. The set of meaningful sequences can be restricted by additional *control processes* as discussed in Section 4.

Formally, a graph transformation rule $r : L \rightsquigarrow R$ consists of a pair of TG-typed instance graphs L, R such that the intersection $L \cap R$ is well-defined (this means that, e.g., edges which appear in both L and R are connected to the same vertices in both graphs, or that vertices with the same name have to have the same type, etc.). The left-hand side L represents the pre-conditions of the rule while the right-hand side R describes the post-conditions. The left-hand side can also state negative pre-conditions (*negative application conditions, NAG*).

According to the *Double-Pushout* semantics (DPO [14]), the application of a rule r is performed in three steps, yielding a transformation step $G \Rightarrow H$:

1. Find an occurrence o_L of the left-hand side L in the current object graph G. Formally, this is a total graph morphism $o_L : L \rightarrow G$ which maps the left-hand side L to a matching subgraph in G. The occurrence is only valid, if $o_L(L)$ cannot be extended by the forbidden elements of a NAG.

2. Remove all the vertices and edges from G which are matched by $L \setminus R$. We must also be sure that the remaining structure $D := G \setminus o_L(L \setminus R)$ is still a legal graph, i.e., that no edges are left dangling because of the deletion of their source or target vertices. In this case, the *dangling condition* [14] is violated and the application of the rule is prohibited.
3. Glue D with a copy of $R \setminus L$ to obtain the derived graph H. We assume that all newly created nodes and edges get fresh identities, so that $G \cap H$ is well-defined and equal to the intermediate graph D.

As an example, consider the reconfiguration rule connect depicted in Fig. 2. According to the left-hand side, the rule can be applied if there are two component instances with free ports whose types can be connected by a connector. According to the right-hand side, an application of this rule, e.g., to the graph in Fig. 1(b), results in the creation of a new connection between the two ports.

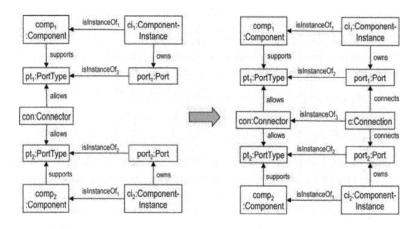

Fig. 2. Reconfiguration rule connect

Definition 3 (Typed graph transformation system). *A typed graph transformation system* $\mathcal{G} = \langle TG, C, R \rangle$ *consists of a type graph* TG, *a set of structural constraints* C *over* TG, *and a set* R *of transformation rules* $r : L \rightsquigarrow R$ *over* TG.

A transformation sequence $s = (G_0 \overset{r_1(o_1)}{\Longrightarrow}_{\mathcal{G}} \cdots \overset{r_n(o_n)}{\Longrightarrow}_{\mathcal{G}} G_n)$ in $\mathcal{G}$, briefly $G_0 \overset{*}{\Longrightarrow}_{\mathcal{G}} G_n$, is a sequence of consecutive transformations such that all graphs $G_0, \ldots, G_n$ satisfy the constraints C. As above, we assume that fresh identifiers are given to newly created elements, i.e., ones that have not been used before in the transformation sequence. In this case, for any $i < j \leq n$ the intersection $G_i \cap G_j$ is well-defined and represents that part of the structure which has been preserved in the transformation from G_i to G_j.

Besides the rule connect, the graph transformation system for the business-level style contains about 10 transformation rules which handle, for instance, creation and deletion of ports and connections as well as sending and receiving of messages. The complete specification of the style can be found in [4], where we also define a platform-specific style for *service-oriented architectures* as follows.

A SOA-specific architectural style. In service-oriented architectures (SOA), software components expose their functionality as *services* over a network to service requesters. The objective of SOA is to enable *dynamic service discovery* at runtime, even if service providers and requesters do not know each other in advance. For this purpose, the service provider has to deliver a detailed description of the service with all necessary information about its interface, access point, quality-of-service, and so forth. The service description is usually published to third-party *discovery agencies* where service requesters can retrieve it from. As soon as the requester finds a description that fits the requirements, it can use it to connect to the component that provides the desired service.

For the definition of the SOA-specific architectural style, we extend the type graph of the business-level style as partially depicted in Fig. 3. The new subtypes of Component can be used to define a software component as Service or, if functioning as discovery agency, as DiscoveryService. There are also special PortTypes used for communication to discovery services. A central SOA element is the ServiceDescription which describes a specific ServiceInstance. The knows relationship indicates which components have access to the description. Besides ordinary Request and Response messages, there are additional SOA message types for service discovery, namely ServicePublication, ServiceQuery, and QueryResult.

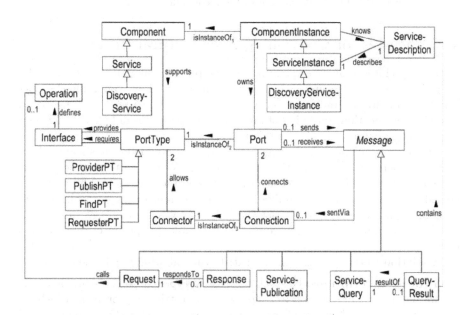

Fig. 3. Type graph of the SOA-specific style

For the creation and deletion of ports and connections, the SOA-specific style contains the same transformation rules as the business-level style, except for setting up a connection to a service which requires that the requester knows the service description beforehand. The SOA-specific variant of the connect rule, which includes this requirement as an additional precondition, is depicted in Fig. 4. To model SOA-specific platform mechanisms for dynamic service discovery, the SOA style contains additional transformation rules for publishing and querying service descriptions. Altogether, there are about 20 transformation rules which can be found in [4].

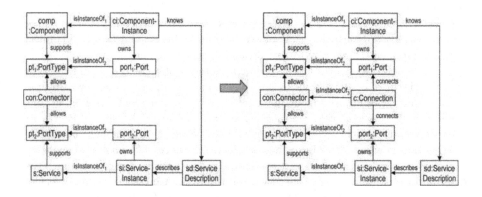

Fig. 4. SOA-specific variant of the reconfiguration rule connect

4   Processes for Controlling Architectural Behavior

With the architectural styles presented in the previous section, we can formally define software architectures as instance graphs to which we can apply the rule-based reconfiguration and communication operations. However, such a rule-based behavior specification consisting of pre- and postconditions only is not sufficient to completely express architectural behavior.

The main problem is that a rule can be applied non-deterministically whenever and wherever its precondition is satisfied. This can lead to non-meaningful sequences of operations, for instance the deletion of a connection while the connected components are still running a certain communication protocol. Instead, we would like to be able to restrict the behavior of a system, e.g., in order to coordinate reconfiguration with communication and to specify communication protocols.

A solution to this problem should satisfy the following requirements:

1. **Process descriptions:** For each component type of a software architecture, we require the description of a *process* that restricts the order in which reconfiguration and communication operations of the underlying architectural

style can be applied to any instance of the type. Thus, an individual action of a process should correspond to the invocation of a transformation rule which conforms to the current level of abstraction. We also require operational semantics for these process descriptions which smoothly suits to the existing graph transformation framework.

2. **Parameters:** Since a process specifies the behavior of all instances of a certain component type, its actions have to relate the invocation of an operation to the local context of the respective component instance. Thus, we have to allow for *input parameters* that refer to dedicated elements within the system architecture. Similarly, we have to equip actions with *output parameters* so that the result of a rule application, e.g., a newly created port or connection, can be referred to in subsequent actions.

3. **Concurrent threads:** Since an architecture may contain several run-time instances of the same component type, we have to allow for a concurrent execution of multiple independent *threads* of the same process. Moreover, one can think of component types that follow different processes simultaneously which results in a branch of concurrent threads for each component instance. Other reasons for concurrency are situations where a server has to supply a service to multiple concurrent client requests each of which is represented by its own thread.

4. **Synchronization:** Many reconfiguration and communication operations involve more than one component instance, for example, when a connection is created between two instances. In such cases, all involved components should agree on the execution of the desired operation which gives rise to *synchronization* issues: The threads which first reach the shared action have to wait until all other participating threads have also reached that action.

The process descriptions are required for models at all levels of abstraction. As a solution, we propose an extension of the architectural styles introduced in Section 3: We integrate a relatively simple meta-model for process descriptions into the type graphs which allows us to include component-specific process descriptions into the instance graphs. Furthermore, we adapt the existing graph transformation rules so that they respect the behavior restrictions imposed by the processes. Eventually, we define a few additional transformation rules that are required for managing, i.e., starting and terminating the individual threads.

Although many authors use *process algebras* [7], petri nets, or event structures to specify and reason about concurrent processes, we stick to graph transformation theory. One reason is that we can continue to apply standard graph transformation tools for executing and analyzing the *process-controlled* transformation systems. Moreover, we can save additional efforts that would be required to combine the operational semantics of graph transformations with the different process formalisms.

Another witness for the suitability of graph transformation is existing work that uses graph transformation systems for defining the semantics of process algebras like the π-calculus. In this context, graph-based approaches are especially used for algebras that support structural changes, e.g., messages that carry references to certain communication channels; of course, such structural changes play an essential role in dynamic software architectures, too.

Type graph extensions for processes. Figure 5 shows the type graph extensions related to process specification and thread management. Note that conceptually these extensions are similar to a UML-like meta-model for process descriptions. According to the left part of Fig. 5, each Component can follow one or more Processes, and each ComponentInstance runs one or more Threads as instances of a process. Each Process has a set of *Action*s ordered by the next association. A Thread has a pointer, named previous, to the most recent action it has executed. Together with the next association this determines the possible subsequent actions. A process can declare Variables, and threads can store values for these variables as References to arbitrary model elements. For this purpose, we introduce the abstract type *Element* as a supertype to all other types in the type graph.

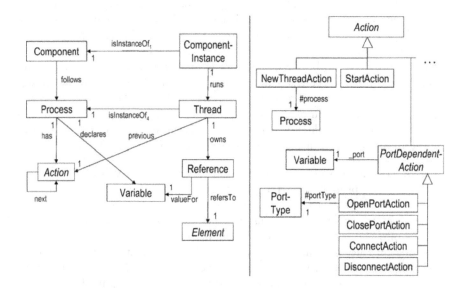

Fig. 5. Type graph extensions for processes and actions

The right part of Fig. 5 shows some of the subtypes of *Action*. Besides the NewThreadAction, which is used to create new process instances, and the Start-Action, which indicates the entry point of a process, there is a special xAction node for every transformation rule of the underlying architectural style (where x is the name of the rule). For better readability, we use additional abstract action types like *PortDependentAction* to group actions with equal parameters.

Parameters are defined by special associations outgoing from an action node. We distinguish between *constants* and *variables*. Constants, named by a preceding "#", refer to elements that define application-specific *types* like, e.g., Component, Connector, PortType, Operation, or Interface which are already known at design-time of the process. Variables, named by a preceding "_", refer to a Variable node where elements that represent *run-time instances* like, e.g., ComponentInstance, Connection, Port, or Request can be stored at execution-time of the process.

With the help of these type graph extensions, we can now include process definitions in our instance graphs in order to specify component behavior.

Transformation rule extensions for processes. The semantics of an action in a process is that the transformation rule it refers to can only be applied at that point of the process. This introduces an additional precondition that has to be checked before a rule is applied. Note that this is only a necessary but not commensurate precondition: If the other preconditions of the rule are not satisfied, then the rule cannot be applied immediately, and the process has to wait until the remaining preconditions are satisfied, too. In order to properly interpret actions in this way, we have to adapt the existing transformation rules of our graph transformation systems and to restrict their applicability. We call such adapted transformation rules *process-controlled*.

Consider, for example, the OpenPortAction in Fig. 5 which should create a new port whose type is specified by the input parameter #portType and return this port through the output variable _port. The corresponding process-controlled graph transformation rule is depicted in Fig. 6. The upper part of the rule (with gray background) indicates the original reconfiguration rule which creates a new port of the selected port type.

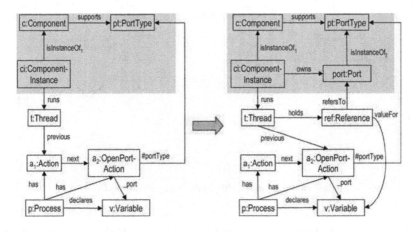

Fig. 6. The process-controlled reconfiguration rule openPort

The new, lower part of the rule restricts its application to those situations where an OpenPortAction belongs to the next actions of the running thread and where the #portType parameter selects the right PortType. According to the lower right-hand side of the rule, which defines additional process-related effects, the rule creates a Reference node for the current thread which refers to the new port node as value for the output variable _port. As another effect, all such process-controlled transformation rules update the pointer to the previous action like a program counter that is increased after completion of a command.

The concurrency requirements discussed at the beginning of this section are implicitly satisfied by the process-controlled transformation rules because the rules are non-deterministically applied wherever possible. And, since every rule application represents the execution of one of the pending actions, this corresponds to a non-deterministic interleaving of the concurrent threads. Thus, concurrency is the default behavior of graph transformation systems.

In addition, the management of threads is handled by the two special transformation rules in Fig. 7: The rule newThread presented in Fig. 7(a) creates new instances of a process whenever a NewThreadAction occurs. The input parameter #process determines which process has to be started. If the parameter refers to the action's own process, a new thread of the same process is started, e.g., if multiple threads of the process are required to serve multiple incoming requests.

A thread terminates after an action that has no more subsequent actions according to the next relation. In this case, the garbage collection rule clearThread of Fig. 7(b) can be applied in order to remove the remaining Thread and Reference nodes (crossed out elements are negative application conditions).

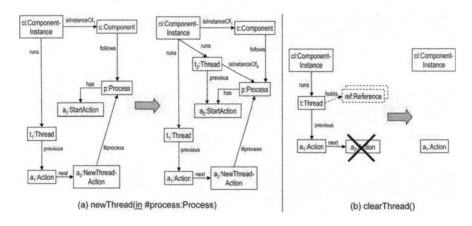

(a) newThread(in #process:Process)   (b) clearThread()

Fig. 7. Thread management rules newThread (a) and clearThread (b)

Synchronization between concurrent threads is required, if a reconfiguration or communication operation involves more than one component. Then, it becomes necessary to get agreements from all involved threads before the corresponding rule can be applied. We satisfy this requirement by *non-local rules* whose precondition comprises the current state of more than one component.

Remember, e.g., the reconfiguration rule **connect** depicted in Fig. 2. In this case, both components should agree to the creation of the new connection beforehand. Figure 8 shows how this synchronization requirement is integrated as a non-local precondition into the left-hand side of the rule. This way, the two involved threads synchronize at the virtually shared **ConnectAction**.

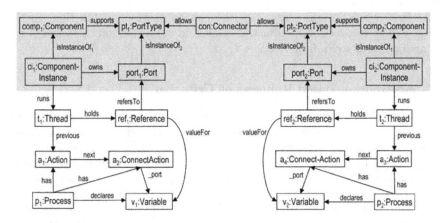

Fig. 8. The process-controlled variant of rule **connect** (left-hand side only!)

Although both the process-controlled transformation rules and the process-aware instance graphs become more complex, we benefit from the uniformity of the theory and its built-in operational semantics. Also, one can think of the formal graph-based model representation as an internal format for tools only, while software architects do not necessarily have to encode their architectures and process definitions as instance graphs. In fact, they can rather use a better concrete notation like *UML component diagrams* for structural aspects and *UML activity diagrams* or *statecharts* for process definitions, which is then internally translated by a suitable CASE tool. Moreover, one could also think of import and export interfaces to other process description standards like the *Business Process Execution Language for Web Services* (BPEL4WS [15]).

5   Refinement of Architectural Configurations

As already discussed in Section 1, refinement is an important concept for developing software architectures. This section deals with the refinement of architectural configurations while preserving the provided system functionality (architectural consistency). In an instance graph, functional elements are reflected by nodes for component and connection types and run-time instances thereof (see Section 3) as opposed to behavior-related elements for control processes (see Section 4).

We do not provide a functional refinement operator which takes an abstract architecture and returns the corresponding concrete architecture, because such a refinement is a creative, non-deterministic design process which includes several alternative options which can all lead to a valid refinement at the concrete level. However, what we do provide is a formal criterion for deciding if a given concrete architectural configuration is a valid refinement of a certain abstract configuration or not.

At first, we define this criterion based on an abstraction function. In the second part of this section, we illustrate this kind of abstraction by means of the two architectural styles from Section 3.

5.1 Abstraction-Based Refinement Criterion

Our notion of refinement is *style-based*, i.e., it is based on a relationship between an abstract architectural style $\mathcal{G} = \langle TG, C, R \rangle$, e.g., the business-level style from Section 3, and a concrete style $\mathcal{G}' = \langle TG', C', R' \rangle$, e.g., the service-oriented style from Section 3. The refinement of architectural configurations formally corresponds to the refinement of the structural parts of an abstract instance graph $G \in \mathbf{Graph}_{TG}$ into a concrete instance graph $G' \in \mathbf{Graph}_{TG'}$.

In the previous section, we extended the type graphs by behavior-related elements in order to encode process descriptions into instance graphs. However, to reason solely about structural aspects of an architecture, we have to distinguish the structure-related part TG_S of the underlying type graph TG from the process-related extensions defined in Section 4. From this distinction, we can derive a *projection function* on instance graphs which preserves structure-related elements only and neglects all behavior-related elements.

Definition 4 (Projection). *Given a type graph TG and a subgraph $TG_S \subseteq TG$ thereof. The projection of instance graphs from TG to TG_S is defined as a function $proj_{TG_S} : \mathbf{Graph}_{TG} \to \mathbf{Graph}_{TG_S}$ which returns for any $G \in \mathbf{Graph}_{TG}$ with nodes N, edges E, and typing $tp : G \to TG$ a graph $G_S \in \mathbf{Graph}_{TG_S}$ with nodes $N_S = \{n \in N | tp(n) \in TG_S\}$, edges $E_S = \{e \in E | tp(e) \in TG_S\}$, and typing $tp|_{G_S}$. Note that G_S is a subgraph of G ($G_S \subseteq G$).*

According to the requirements stated in Section 1, a refinement criterion has to respect architectural consistency, meaning that for a valid refinement the concrete architecture has to preserve the functionality of the abstract architecture. Thus, all structural entities like components and connectors of the business-level model have to be preserved in the platform-specific model.

Since the two instance graphs to be compared are expressed in terms of different architectural styles, i.e., different type graphs, one cannot simply compare them and check if the abstract graph is part of the concrete graph. Before we can do so, we rather have to express one of the graphs in terms of the other architectural style.

The canonical solution to this problem is by means of an *abstraction* function $abs : \mathbf{Graph}_{TG'} \to \mathbf{Graph}_{TG}$ which takes the concrete instance graph and, by abstracting from all platform-specific details, lifts it to the abstract level. Then, we can check if the resulting abstraction contains the same functional elements as the original abstract graph. For this purpose, we regard the original abstract graph as a *property* that has to be *satisfied* by the lifted concrete graph according to the following definition:

Definition 5 (Satisfaction). *Given a model represented as an instance graph G and a property represented as an instance graph P, both typed over the same type graph TG. Also, given a type graph distinction $TG_S \subseteq TG$ with a corresponding projection $proj_{TG_S}$.*
We say that G satisfies P, i.e., $G \models P$, iff there is a total, injective graph morphism $m : proj_{TG_S}(P) \rightarrow proj_{TG_S}(G)$. This means that the relevant part of P can be embedded into the relevant part of G.

Based on this definition, we can now formally define structural refinement based on abstraction as follows:

Definition 6 (Structural Refinement). *Given an abstract type graph TG, a concrete type graph TG', and an abstraction function $abs : \mathbf{Graph}_{TG'} \rightarrow \mathbf{Graph}_{TG}$. A concrete instance graph $G' \in \mathbf{Graph}_{TG'}$ is a structural refinement of an abstract instance graph $G \in \mathbf{Graph}_{TG}$, if $abs(G') \models G$.*

5.2   Abstraction Function Based on a Type Graph Morphism

The abstraction function *abs* is a semantic mapping, associating with each concrete configuration a corresponding abstract configuration. There is a range of possibilities for the concrete definition of *abs* depending on the characteristics of the respective architectural styles. For example, it is sufficient to base the abstraction function *abs* on a mapping between the abstract and the concrete type graph, if the abstraction of a concrete instance graph consists of adapting the types of business-relevant elements and omitting platform-specific elements.

Other cases, not discussed in detail in this paper, might require more complex transformations which map entire patterns of concrete elements to abstract elements. For instance, a combination of two unidirectional channels at the platform-specific level could be used to realize an abstract bidirectional channel. These complex mappings can be defined by graph transformation systems or even more sophisticated methods like *triple graph grammars* [29].

In the rest of this section, we take the platform-independent (pi) style of Section 3 as abstract transformation system $\mathcal{G}^{pi} = \langle TG^{pi}, C^{pi}, R^{pi} \rangle$ and the service-oriented (so) style as concrete transformation system $\mathcal{G}^{so} = \langle TG^{so}, C^{so}, R^{so} \rangle$. We define a mapping between the two type graphs TG^{so} and TG^{pi} and exemplify how to derive an appropriate abstraction function from this mapping.

Type graph mapping. A type graph mapping is a partial graph morphism $t : TG^{so} \rightarrow TG^{pi}$ which maps structure-related elements of the concrete type graph TG^{so} to structure-related elements of the abstract type graph TG^{pi} as partially shown in Fig. 9.

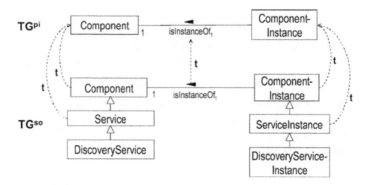

Fig. 9. Part of the type graph mapping t

The concrete definition of t is driven by semantic correspondences between the structure-related elements of the two type graphs. For instance, both Component and Service nodes in a service-oriented architecture represent what we call a Component in the business-level style. Since there is no distinction between private and published components at the abstract level, t maps both types to the abstract type Component.

Those elements that represent purely platform-specific concepts not occurring at the abstract level like, e.g., DiscoveryService, ServiceDescription, or the SOA-specific port types (see Fig. 3) are not mapped to the abstract type graph. For behavior-related elements like, e.g., Process and Action nodes (see Fig. 5), the type mapping is undefined, too, because the abstraction function only needs to lift structural aspects to the abstract level.

Abstraction function. From the type mapping t, we can now derive an abstraction function $abs_t : \mathbf{Graph}_{TG^{so}} \to \mathbf{Graph}_{TG^{pi}}$ that abstracts instance graphs typed over TG^{so} to those typed over TG^{pi}. This abstraction informally consists of (1) renaming the types of all elements whose type has an image in TG^{pi} according to the definition of t, (2) deleting all nodes and edges which, due to the partiality of t, have a type in TG^{so} but not in TG^{pi}, and (3) deleting all dangling edges and those adjacent nodes whose number of connected neighbor nodes falls below the lower bound of the relevant cardinality constraint.

Figure 10 illustrates the effect of the abstraction function abs_t for a small instance graph in the service-oriented style (shown in the upper left corner). The instance graph defines a service S that supports a port type AccessS for using the service and another port type PublishDesc for sending a service description to available discovery services. The represented run-time snapshot contains one instance si of the service which owns a port for each of the supported port types. Besides, there is a description document descr describing the service instance.

In a first step, we apply the type mapping t and rename the types of the Service and ServiceInstance nodes into Component and ComponentInstance (1).

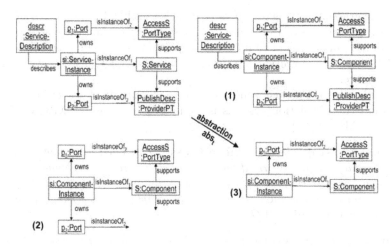

Fig. 10. Abstraction of a small instance graph

Then, we delete the ProviderPT and ServiceDescription nodes and the describes edge because they have no mapping to TG^{pi} under t (2). The deletion of the ProviderPT node leads to the deletion of the adjacent Port node in the third step, because otherwise the cardinality constraint would be violated which says that every Port requires a PortType. Eventually, all dangling edges are removed (3).

6   Behavior-Preserving Refinement

In addition to structural refinement, we also want to check if the refinement preserves the architectural behavior. In this section, we extend our refinement criterion by a corresponding semantic condition, discuss possibilities and requirements for an automated verification of this criterion, and exemplify the concepts with the two style examples from Section 3.

6.1 Extended Criterion for Behavior-Preserving Refinement

While a single instance graph represents a certain system state, the application of a transformation rule represents a transition from one state to a subsequent state. This way, the potential behavior can be represented as a *transition system* whose states are the reachable instance graphs and whose transitions are generated by rule applications. If we also encode application-specific process definitions into the instance graphs, as introduced in Section 4, then the architectural behavior represents the concurrent execution of these processes.

Given the initial state of the architecture as a start graph, one can generate and explore the transition system by continuously applying the transformation rules to previously generated states. We can reduce the state space by considering isomorphic graphs as a single state only. This is sensible, because for isomorphic graphs the same set of transformation rules is applicable, and the result of a rule application is only determined up to isomorphism in the DPO approach anyway (cf. [11]). Note that, however, in some cases the resulting transition system may still be infinite.

Recently, the automated generation and exploration of transition systems from graph transformation systems is supported by tools like GROOVE [27] and CheckVML [28, 32].

Definition 7 (Graph Isomorphism Class). *Two TG-typed graphs G and H are isomorphic, briefly $G \cong H$, if there is a bijective graph morphism $i : G \to H$, called isomorphism, which also preserves their typings, i.e., $tp_H \circ i = tp_G$.*

An isomorphism class $[G]$ is the set of all graphs that are isomorphic to G, i.e., $[G] = \{H \in \mathbf{Graph}_{TG} \mid G \cong H\}$.

Definition 8 (Architectural Behavior). *Given an architectural style represented by a typed graph transformation system $\mathcal{G} = \langle TG, C, R \rangle$ and an initial system state represented by an instance graph $G_0 \in \mathbf{Graph}_{TG}$, then the architectural behavior is defined by a transition system $TS_{G_0} = (S, \Rightarrow)$ with*

- *a set of states $S = \left\{ [G] \mid G_0 \overset{*}{\Rightarrow}_{\mathcal{G}} G \right\}$ consisting of isomorphism classes of all graphs reachable in $\mathcal{G}$, with $[G_0] \in S$,*
- *a transition relation $\Rightarrow \subseteq S \times S$ which is defined by all possible rule applications in $\mathcal{G}$, i.e., $G_i \Rightarrow G_j$ iff $G_i \Rightarrow_{\mathcal{G}} G_j$.*

For the sake of simplicity, we continue to denote the states of the transition system as graphs rather than as graph isomorphism classes. Thus, when speaking of $G \in S$, we precisely mean $[G] \in S$. Moreover, the concatenation of consecutive transitions starting from G and leading to H is called a *path* from G to H, denoted by $G \overset{*}{\Longrightarrow} H$.

For the intended refinement criterion, we again consider an abstract architecture as instance graph G of a platform-independent style $\mathcal{G} = \langle TG, C, R \rangle$ and a concrete architecture as instance graph G' of a platform-specific style $\mathcal{G}' = \langle TG', C', R' \rangle$. Besides structural refinement, we now also require that the concrete architecture preserves the behavior of the abstract architecture.

This property is expressed in terms of structural refinements for all states reachable in the abstract behavior. To be more precise, we demand, that for every path $G \Rightarrow G_1 \Rightarrow \ldots \Rightarrow G_n$ in the abstract transition system TS_G there exist paths $G' \overset{*}{\Longrightarrow} G'_1 \overset{*}{\Longrightarrow} \ldots \overset{*}{\Longrightarrow} G'_n$ in the concrete transition system $TS_{G'}$ with G'_i structurally refining G_i (that is, $abs(G'_i) \models G_i$) for all $i = 1 \ldots n$.

Since we are now dealing with isomorphism classes, we require that the abstraction function abs preserves isomorphisms: $G \cong H \implies abs(G) \cong abs(H)$. As a consequence, it is indifferent to which representative of an isomorphism class the function is applied.

In terms of software architecture, a path represents a certain scenario of communication and reconfiguration operations, and, for a behavior-preserving refinement, we want to ensure that every abstract, business-level scenario can also be realized at the platform-specific level. This criterion can be formulated as a co-inductive definition as follows:

Definition 9 (Behavior-Preserving Refinement). *Given an abstract architectural style* $\mathcal{G} = \langle TG, C, R \rangle$, *a concrete architectural style* $\mathcal{G}' = \langle TG', C', R' \rangle$, *and an abstraction function* $abs : \mathbf{Graph}_{TG'} \to \mathbf{Graph}_{TG}$ *which preserves isomorphisms. A concrete instance graph* $G' \in \mathbf{Graph}_{TG'}$ *with behavior* $TS_{G'}$ *refines an abstract instance graph* $G \in \mathbf{Graph}_{TG}$ *with behavior* TS_G, *if*

- $abs(G') \models G$
- *for every transition* $G \Rightarrow H$ *in the abstract system* TS_G *there exists a path* $G' \overset{*}{\Longrightarrow} H'$ *in the concrete system* $TS_{G'}$ *such that* H' *refines* H.

According to this definition, a single transformation *step* $G \Rightarrow_{\mathcal{G}} H$ is refined by a transformation *sequence* $G' \overset{*}{\Longrightarrow}_{\mathcal{G}'} H'$. This is because it might be necessary to perform a number of platform-specific steps in order to realize the abstract step. For example, consider an application of the reconfiguration rule connect (see Fig. 2). In a service-oriented architecture, it is not directly possible to apply the corresponding SOA-specific connect rule (see Fig. 4), because connecting to a service requires knowledge about its description beforehand. If the description is not known to the requester, other SOA-specific rules for service publication and discovery have to be applied first as shown in Fig. 11.

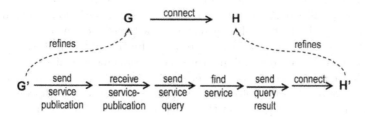

Fig. 11. Refinement of an abstract transformation step

Since the behavior-preserving refinement solely depends on the structural refinement of reachable states, we do not need to provide a fixed mapping between the transformation rules of the two involved styles. This is especially advantageous, if abstract and concrete operations are very different.

Moreover, this approach implicitly allows for alternate refinements of an abstract operation depending on the context of its application. In the above case, for instance, the service description might have been published to the discovery agency already. Then, the first two operations of the transformation sequence in Fig. 11 can be omitted. Or, the description might already be known to the requester due to some previous look-up so that even the query operations can be omitted.

6.2   Tool-Based Verification of Behavior Preservation

Based on the tools for automated state space generation mentioned above, one can apply various analysis techniques like *model checking* to the resulting transition system. We propose to express the behavioral refinement check as a *reachability problem* in the concrete transition system that can be solved by model checkers.

According to Definition 9, for a given abstract transformation step $G \Rightarrow H$ the reachability problem consists of searching a path $G' \overset{*}{\Longrightarrow} H'$ in the concrete transition system with $abs(H') \models H$. Consequently, the abstraction function abs has to be applied to every visited state in order to find an appropriate target graph H'. Since this affects the computational complexity, we would rather express the same property solely at the level of the concrete system.

For this purpose, we assume a second translation, contravariant to abstraction. A function $trans : \mathbf{Graph}_{TG} \rightarrow \mathbf{Graph}_{TG'}$ associates an abstract instance graph with a concrete one representing the reformulation of an abstract state over the concrete type system. Note that the concrete graph does not necessarily represent a complete state of the concrete architecture, but rather a minimal pattern which has to be present in order for the requirements of the abstract graph to be fulfilled. Thus, we consider a concrete instance graph G' as a valid structural refinement of an abstract graph G if it *satisfies* this pattern, formally $G' \models trans(G)$.

Since the abstraction function abs is in general not injective, there are various alternative possibilities for translating an abstract configuration to the concrete level which are all valid structural refinements. The translation function $trans$ selects for every abstract graph only one possible translation instead of returning the set of all potential translations. Thus, the definition of $trans$ already includes certain design decisions determining the specific refinement of abstract elements.

As a consequence, not every concrete configuration that is a valid refinement according to abs is a valid refinement according to $trans$, too. However, in the opposite direction we require that the translation function has to be compatible with the abstraction function so that a refinement according to $trans$ entails a refinement according to abs. This is formally expressed as a *satisfaction condition*, reminiscent of similar conditions in algebraic specification or logics.

Definition 10 (Satisfaction Condition). *Given abstract type graph TG and concrete type graph TG'. A translation function $trans : \mathbf{Graph}_{TG} \rightarrow \mathbf{Graph}_{TG'}$ is* compatible *to an abstraction function $abs : \mathbf{Graph}_{TG'} \rightarrow \mathbf{Graph}_{TG}$, if the following satisfaction condition holds for all $G \in \mathbf{Graph}_{TG}$ and all $G' \in \mathbf{Graph}_{TG'}$:*

$$G' \models trans(G) \implies abs(G') \models G$$

Note that, due to the argument above, the opposite direction of the entailment is in general not true. This reflects the fact that the translation function contains more information than the abstraction function. Thus, the left condition is more specific and not necessarily entailed by the right condition. It would be different if we had defined the translation function as returning the set of all possible translations of an abstract configuration. But then, it would be more difficult to reformulate the reachability problem solely for the concrete transition system.

Under the assumption of the satisfaction condition, we can now easily derive the following theorem which allows to express behavior-preserving refinement solely at the concrete level.

Theorem 1. *Given an abstract architectural style* $\mathcal{G} = \langle TG, C, R \rangle$, *a concrete architectural style* $\mathcal{G}' = \langle TG', C', R' \rangle$, *and compatible abstraction and translation functions abs and trans. A concrete instance graph* $G' \in \mathbf{Graph}_{TG'}$ *refines an abstract instance graph* $G \in \mathbf{Graph}_{TG}$ *according to Definition 9, if*

- $G' \models trans(G)$
- *for every transition* $G \Rightarrow H$ *in the abstract system* TS_G *there exists a path* $G' \stackrel{*}{\Longrightarrow} H'$ *in the concrete system* $TS_{G'}$ *such that* H' *refines* H.

Proof. Since *trans* is compatible to *abs*, we can conclude from $G' \models trans(G)$ that $abs(G') \models G$ holds, too. The second clause already conforms to Definition 9.

A model checker like SPIN [23] can now be used to refine an abstract transformation step $G \Rightarrow H$ by looking for a state that satisfies $trans(H)$, technically speaking a graph that contains $trans(H)$ as a subgraph. With the help of temporal logics such as *linear-time temporal logic* (LTL), we can even formulate the reachability of entire abstract sequences $G \Rightarrow G_1 \Rightarrow \ldots \Rightarrow G_n$ as:[2]

$$\Diamond(trans(G_1) \wedge \Diamond(trans(G_2) \wedge \ldots \Diamond(trans(G_n)) \ldots))$$

Since we require only one path to satisfy the above formula while an LTL formula always refers to all paths of the transition system, we have to negate the above formula and let the model checker look for a counter example. A counter example that violates the negated formula can then be used as a witness for the original formula.

Although we cannot verify the refinement of the complete abstract transition system this way, we are able to check at least the most important scenarios of business-level behavior.

6.3 Compatible Translation Function for a Type Graph-Based Abstraction Function

In order to satisfy the required compatibility, the definition of a translation function heavily depends on the definition of the contravariant abstraction function. In this subsection, we revisit the abstraction function from Section 5.2, which is based on a mapping between the concrete and the abstract type graph elements, and we show how a compatible translation function looks like for this kind of abstraction function.

---

[2] The LTL operator $\Diamond$ means "at some time in the future"

In Section 5.2, we define a semantic mapping $t : TG^{so} \rightarrow TG^{pi}$ from the concrete type graph to the abstract type graph. From this type graph morphism, we derive the abstraction function abs_t on instance graphs which consists of renaming concrete types to abstract types and, due to the partiality of t, deleting purely platform-specific elements.

For the definition of a compatible translation function, we *invert* the type mapping t. However, since t is not injective (e.g., both Component and Service are mapped to Component in Fig. 9), the resulting inverse $\bar{t}$ is a *relation* between the elements of the two type graphs, which can be expressed by a function $\bar{t} : N_{TG} \rightarrow \mathcal{P}(N_{TG'})$. If t maps a concrete node type nt' to the abstract type nt, then $nt' \in \bar{t}(nt)$. Analogously, $\bar{t}$ can be extended to edge types as well.

From the inverted type mapping $\bar{t}$, we can now derive a translation function $trans_{\bar{t}} : \textbf{Graph}_{TG} \rightarrow \textbf{Graph}_{TG'}$. For an instance graphs $G \in \textbf{Graph}_{TG}$ with typing morphism tp, it

1. deletes all nodes n whose type has no image under $\bar{t}$, i.e., $\bar{t}(tp(n)) = \emptyset$
2. changes the type of n to a certain $nt' \in \bar{t}(tp(n))$ else.

The first case is relevant for behavior-related nodes, e.g., for process-descriptions, which are also excluded from the original type graph mapping t. The second case adapts the types of the remaining elements to the vocabulary of the concrete style. Since there might be several alternatives returned by $\bar{t}$ for adapting the type of an abstract element, the translation function cannot completely be defined at the type level but requires additional user decisions at the instance level for translating individual nodes.

Technically, these user decisions can be integrated by additional node attributes that are set by the engineer to determine the desired translation option for a node. By evaluating the values of these attributes for a given instance graph, the translation function determines the intended translation to the concrete level.

What remains is to show the compatibility of $trans_{\bar{t}}$ to the original abstraction function abs_t. According to Definition 10, we have to show that

$$G' \models trans_{\bar{t}}(G) \implies abs_t(G') \models G$$

Proof sketch. For arbitrary $G \in \textbf{Graph}_{TG}$ and $G' \in \textbf{Graph}_{TG'}$ be

$$G' \models trans_{\bar{t}}(G) \tag{1}$$

Since $\bar{t}$ is the inverse of t, $trans_{\bar{t}}(G)$ contains only elements whose types are in the domain of t. These elements are preserved by abstraction on both sides of (1). Thus, the satisfaction relation still holds after application of the abstraction function:

$$abs_t(G') \models abs_t(trans_{\bar{t}}(G)) \tag{2}$$

Since the application of $trans_{\bar{t}} \circ abs_t$ is the identity for structure-related elements, we receive:

$$abs_t(G') \models G \tag{3}$$

$\square$

7 Conclusion

In this paper, we introduced a formal technique for modeling dynamic architectures as instances of graph transformation systems. Graph transformation rules were used to express available communication and reconfiguration operations in a certain architectural style. Architectural models were enriched by process descriptions with operational semantics that restrict and coordinate the architectural behavior.

We have discussed semantic conditions for the behavior-preserving refinement of architectural models under the obligation of architectural consistency across different levels of platform abstraction. Style-based abstraction and translation functions were introduced as formal refinement criteria that can be checked with the help of analysis tools.

The presented approach is very flexible because it is not based on a fixed syntactical mapping between the different operations but on a semantic relationship that also respects context-dependent alternatives for refining abstract behavior.

Future work includes further investigations on the verification of behavioral refinement by existing simulation algorithms for related transition systems. We are planning to support the approach by a coupling of CASE tools with editors and analysis for graph transformation systems, presently conducting experiments with existing model checkers.

References

[1] M. Abi-Antoun and N. Medvidovic. Enabling the refinement of a software architecture into a design. In Proc. UML 99 - The Unified Modeling Language, volume 1723 of LNCS, pages 17–31. Springer, 1999.

[2] G. D. Abowd, R. Allen, and D. Garlan. Using style to understand descriptions of software architectures. ACM Software Engineering Notes, 18(5):9–20, 1993.

[3] L. Baresi, R. Heckel, S. Thöne, and D. Varró. Modeling and validation of service-oriented architectures: Application vs. style. In Proc. European Software Engineering Conference and ACM SIGSOFT Symposium on the Foundations of Software Engineering, ESEC/FSE 03, pages 68–77. ACM Press, 2003.

[4] L. Baresi, R. Heckel, S. Thöne, and D. Varró. Style-based modeling and refinement of service-oriented architectures. Technical Report TR-RI-04-250, University of Paderborn, 2004. ftp://ftp.upb.de/doc/techreports/Informatik/tr-ri-04-250.pdf.

[5] L. Baresi, R. Heckel, S. Thöne, and D. Varró. Style-based refinement of dynamic software architectures. In Proc. 4th Working IEEE/IFIP Conference on Software Architecture, WICSA 4, pages 155–164. IEEE, 2004.

[6] D. Batory, J. N. Sarvela, and A. Rauschmayer. Scaling step-wise refinement. In Proc. ICSE 2003 - Int. Conference on Software Engineering, pages 187–197. IEEE, 2003.

[7] J.A. Bergstra, A. Ponse, and S.A. Smolka, editors. Handbook of Process Algebra. Elsevier Science, 2001.

[8] T. Bolusset and F. Oquendo. Formal refinement of software architectures based on rewriting logic. In Proc. RCS 02 Int. Workshop on Refinement of Critical Systems, 2002. www-lsr.imag.fr/zb2002/.

[9] C. Canal, E. Pimentel, and J. M. Troya. Specification and refinement of dynamic software architectures. In Proc. WICSA 1, First Working IFIP Conference on Software Architecture, volume 140 of IFIP Conference Proceedings, pages 107–126. Kluwer, 1999.

[10] A. Corradini, U. Montanari, and F. Rossi. Graph processes. Fundamenta Informaticae, 26(3,4):241–265, 1996.

[11] A. Corradini, U. Montanari, F. Rossi, H. Ehrig, R. Heckel, and M. Löwe. Algebraic approaches to graph transformation; basic concepts and double pushout approach. In G. Rozenberg, editor, Handbook of Graph Grammars and Computing by Graph Transformation, volume 1: Foundations. World Scientific, 1997.

[12] M. Denford, T. O'Neill, and J. Leaney. Architecture-based design of computer based systems. In Proc. StraW 03, Int. Workshop From Software Requirements to Architectures, 2003. se.uwaterloo.ca/~straw03/.

[13] H. Ehrig, G. Engels, H.-J. Kreowski, and G. Rozenberg, editors. Handbook of Graph Grammars and Computing by Graph Transformation, volume 2: Applications, Languages and Tools. World Scientific, 1999.

[14] H. Ehrig, M. Pfender, and H. J. Schneider. Graph grammars: an algebraic approach. In 14th Annual IEEE Symposium on Switching and Automata Theory, pages 167–180. IEEE, 1973.

[15] T. Andrews et al. Specification: Business Process Execution Language for Web Services Version 1.1. BEA, IBM, Microsoft, SAP AG and Siebel Systems, 2003. http://www-106.ibm.com/developerworks/library/ws-bpel/.

[16] D. Garlan. Style-based refinement for software architecture. In Proc. ISAW -2, 2nd Int. Software Architecture Workshop on SIGSOFT '96, pages 72–75. ACM Press, 1996.

[17] R. Gorrieri and A. Rensink. Action refinement. In J.A. Bergstra, A. Ponse, and S.A. Smolka, editors, Handbook of Process Algebra, pages 1047–1147. Elsevier, 2001.

[18] M. Große-Rhode, F. Parisi Presicce, and M. Simeoni. Spatial and temporal refinement of typed graph transformation systems. In Proc. Math. Foundations of Comp. Science 1998, volume 1450 of LNCS, pages 553–561. Springer, 1998.

[19] Object Management Group. The UML website. www.uml.org.

[20] R. Heckel, A. Corradini, H. Ehrig, and M. Löwe. Horizontal and vertical structuring of typed graph transformation systems. Math. Struct. in Computer Science, 6:613–648, 1996.

[21] D. Hirsch. Graph transformation models for software architecture styles. PhD thesis, Departamento de Computación, Universidad de Buenos Aires, 2003.

[22] D. Hirsch and U. Montanari. Synchronized hyperedge replacement with name mobility. In Proc. CONCUR 2001 - Concurrency Theory, volume 2154 of LNCS, pages 121–136. Springer, 2001.

[23] G. Holzmann. The model checker SPIN. IEEE Transactions on Software Engineering, 23(5):279–295, 1997.

[24] D. Le Métayer. Software architecture styles as graph grammars. In Proc. 4th ACM SIGSOFT Symposium on the Foundations of Software Engineering, volume 216 of ACM Software Engineering Notes, pages 15–23. ACM Press, 1996.

[25] M. Moriconi, X. Qian, and R. A. Riemenschneider. Correct architecture refinement. IEEE Transactions on Software Engineering, 21(4):356–372, 1995.

[26] Object Management Group. Model-Driven Architecture. www.omg.org/mda/.

[27] A. Rensink. The GROOVE simulator: A tool for state space generation. In M. Nagl, J. Pfalz, and B. Böhlen, editors, Proc. Application of Graph Transformations with Industrial Relevance (AGTIVE '03), volume 3062 of LNCS. Springer, 2003.

[28] Á. Schmidt and D. Varró. CheckVML: A tool for model checking visual modeling languages. In Proc. UML 2003 - The United Modeling Language, volume 2863 of LNCS, pages 92–95, 2003.

[29] A. Schürr. Specification of graph translators with triple graph grammars. In Tinhofer, editor, Proc. WG '94 Int. Workshop on Graph-Theoretic Concepts in Computer Science, volume 903 of LNCS, pages 151–163. Springer, 1994.

[30] M. Shaw and D. Garlan. Software Architecture: Perspectives on an Emerging Discipline. Prentice-Hall, 1996.

[31] G. Taentzer, M. Goedicke, and T. Meyer. Dynamic change management by distributed graph transformation: Towards configurable distributed systems. In Proc. TAGT '98 - Theory and Application of Graph Transformations, volume 1764 of LNCS, pages 179–193. Springer, 2000.

[32] D. Varró. Towards symbolic analysis of visual modeling languages. In Proc. GT-VMT 2002 - Int. Workshop on Graph Transformation and Visual Modeling Techniques, volume 72 of ENTCS, pages 57–70. Elsevier, 2002.

[33] M. Wermelinger and J. L. Fiadero. A graph transformation approach to software architecture reconfiguration. Science of Computer Programming, 44(2):133–155, 2002.

[34] N. Wirth. Program development by stepwise refinement. Communications of the ACM, 14(4):221–227, 1971.

Modelling Mobility with Petri Hypernets*

Marek A. Bednarczyk[1], Luca Bernardinello[2],
Wiesław Pawłowski[1], and Lucia Pomello[2]

[1] Institute of Computer Science, P.A.S., Gdańsk, Poland
[2] DISCO, Università degli Studi di Milano Bicocca, Milano, Italy

Abstract. Petri hypernets, a novel framework for modeling mobile agents based on **nets-within-nets** paradigm is presented. Hypernets employ a local and finitary character of interactions between agents, and provide means for a modular and hierarchical description. They are capable of modelling mobile agents tfrahat can dynamically change their hierarchy, and can communicate with each other and with the outside world by exchanging messages, i.e., other mobile agents.

1 Introduction

The idea of *mobility* has already attracted a lot of interest in the computing science community. Many existing formalisms devised to cope with specific application areas have been enriched by 'mobility-related' features. Here, examples range from extensions of Milner's CCS, like *mobile processes* in π-calculus, cf. [10], or 'programming-level' notations such as *UNITY* or specification formalisms like *UML* (see e.g., [1]). Efforts to capture the essence of mobility also resulted a new, dedicated frameworks like the calculus of *mobile ambients* [6], *Join calculus* [7] or *mobile Petri nets* [2].

Here, yet another model called *Petri hypernets* is introduced. Petri nets, cf. [11], are well-known as a general and intuitive framework in which concurrent, asynchronous and distributed systems can be modeled. Our plan is to retain these strengths in an extension based on the principle that *mobile agents* should be modeled as Petri nets, and that other mobile agents should be able to manipulate them.

In contrast, e.g., *mobile nets* aim to capture the essence of mobility by allowing *names* of places of a Petri net to be sent as tokens. This requires the mechanisms of fresh variable creation and variable binding to be used in the formalism. We consider this as a departure from one of the fundamental concepts underlying Petri nets philosophy, namely that the interaction between places and transitions should be *local* and *finitary*. This sets our model apart from the formalisms like π-calculus, Join calculus, mobile nets, etc.

* This research has been partially supported by the KBN grant No. 7 T11C 002 21, and by the CATNET and CATALYSIS projects within CNR/PAN and CNRS/PAN cooperation programmes, respectively.

We intend to model mobile agents as nets and allow that nets are manipulated by other nets. Within a Petri net framework this can be realized by assuming that some nets are 'tokens' for other nets. This **nets-within-nets** paradigm was proposed by Valk and is being developed by his students and followers cf. [12, 13, 14, 8, 9]. Petri hypernets can be seen as yet another forking of the **nets-within-nets** paradigm. Each hypernet is a collection of mobile agents, called *open nets*, together with an assignment of mobile agents as tokens to places of other mobile agents, called *hypermarking*.

Petri hypernets support a *modular* and *hierarchical* description of reality.

There is a natural hierarchy of agents that corresponds to the assumption that any mobile agent should be higher in the hierarchy than any of the tokens it manipulates.

Modularity is imposed on mobile objects by assumption that each open net is a synchronous composition of *modules*, each one responsible for manipulation of mobile objects traveling along a fixed *channel*.

Co-operation between modules, each associated to a *different* channel within an open net, is enforced by *hand-shake synchronization* of their transitions. Namely, if the same transition t occurs in several different modules of an open net N, then in order to fire t in N all these modules must be ready to participate. It is well-known that every elementary net can be obtained as such a synchronization of state machines [4]. A *state machine* is a Petri net such that every transition has exactly one precondition and exactly one postcondition, and with only one token in exactly one place initially marked. Thus, all reachable markings are one-token markings. As a result the state machines are purely *sequential*, i.e., two transitions can never fire concurrently in a reachable marking. Something similar holds for 1-safe Petri nets. Namely, given such a net N one can construct a net N' with the same behavior as the synchronization of state machines, see [3]. These observations encourage us to restrict attention only to modules which have, essentially, the structure of sequential machines. Only the assumption that there is at most one token in exactly one place is dropped.

The features of hypernets discussed above could be found in the existing ramifications of the **nets-within-nets** paradigm. There are, however, several important novelties as well.

First, we assume that a mobile agent does not make any assumptions about the structure of other mobile agents. Following Valk's original idea, also in our model an open net is allowed to synchronize the firing of its transition t with the firing of the transitions with the same name present in its tokens, if needed. Namely, each module in an open net has facilities to communicate locally, with a module of *its* kind in an adjacent open net. The nets *adjacent* to the given open net are those immediately *below* and *above* it in the hierarchy. However, unlike in Valk's proposal, in hypernets such an inter-level synchronization is achieved solely by means of exchanging *messages*. This readiness to communicate with its parent net and its token nets, explains the terminology 'open nets' used to refer to mobile agents in our formalization.

Second, messages are also just mobile agents. This assumption implies that in Petri hypernets mobile agents can migrate between the levels of the hierarchy. Thus, the hierarchy itself may change.

Third, Valk gives two different semantics for his object nets. Consider, for instance, a transition with one input and two output places in the top level net. Then one has the problem explaning what is the outcome of firing this transition when a token net enters the input place. Valk considers two versions of the semantics, both based on the assumption that the transition does not, in fact, move the real token net, but *references* to it. Our decision to use sequential machines as moduls solves this problem. In particular, firing a transition preserves the identity of agents. We can think that a transition manipulates token nets as *values*, that it moves them, not their references, from the input place to the output place. The choice of sequential machines also entails that no agent is created, and no agent is destroyed. Thus, the hypernet has a finite state space, even if the agents it comprises are truly mobile, and can change their cooperation potential dynamically.

Finally, it is worth stressing that in case of hypernets the decompositions play an important structuring rôle. We see a hypernet as a hierarchical structure of open nets, each of them decomposed into a collection of sequential modules. From this perspective, firing of a transition t of the hypernet should be seen as a complex *transaction*, which involves firing transitions t which occur in the modules of all the open nets involved.

The reason for writing the paper is to present the simple ideas, together with their slightly less simple formalization. The main result states that the formalization is well-founded. Namely, firing a transaction preserves the forest-like hierarchy of open nets within a hypernet.

2   Petri Hypernets — A Gentle Introduction

2.1   Example: Air Travels

To introduce the model and illustrate the features of Petri hypernets let us recall a simple air travel case study considered e.g., in [1]. The case study requires the modeler to cope with mobile agents of at least 3 different kinds: *airports*, *planes* and *travelers*. The task is to describe the most basic aspects of the air travel involving possible collections of mobile agents of these kinds.

2.2   Kinds and Modules in the Airport

An agent of each kind exhibits dynamic behavior which may involve manipulation of objects of other kinds. For instance, while at an airport, travelers and planes are under the rules set forth by the airport.

One of the assumptions underlying the definition of hypernets is that agents cannot directly manipulate other agents of their kind. In fact, the separation of agents into *kinds* is one of the basic features of the model. Most of the time kinds will be called *channels* to stress the communication aspect played by them within hypernets.

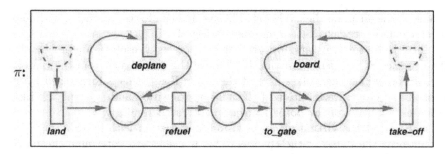

Fig. 1. Airport — plane manipulation module

An airport perspective of the activities involved in handling of airplanes is presented on Fig. 1. This is a purely sequential view. The planes *land* at a gate. While at the gate they can *deplane* the travelers, one after another. Action *refuel* then takes planes to the refueling place. Then, *to_gate* moves planes to another gate where the plane may accept new travelers, and finally *take-off*.

The graphical description of the plane handling activities in Fig. 1 is essentially standard. The exception concerns the precondition of *land* and postcondition of *take-off* transitions. Here, the open world assumption comes into play. The airport is expecting that the planes are landing in co-operation with some *higher-level* traffic control authority. Similarly, the airport authority should be able to safely assume that upon *take-off* somebody above the hierarchy ladder will take care of the plane with its passengers.

The dashed half circles which provide the input for *land* and the output for *take-off* on Fig. 1 are intended to capture the above intuition. Drawing half of a place is meant to indicate that co-operation with higher level is required to successfully conclude the operation, here *land* and *take-off*. The dashes highlight the *virtual* character of the half-places. Namely, they are a means of synchronization of transitions between adjacent levels rather than real places that can store tokens. This evident link with the concept of *zero-places* proposed by Bruni and Montanari, see [5], remains to be investigated.

Fig. 1 describes the π-module of an airport agent only. We use the prefix π to indicate that the tokens manipulated within the module are 'planes'. A τ-module for manipulating travelers at the airport is defined in Fig. 2. Recall that the π-module presented on Fig. 1 describes co-operation with the higher level only. Thus, the hypothetical higher level air traffic control agent should contain a π-module which can deliver the planes to airports for landing, and handle the

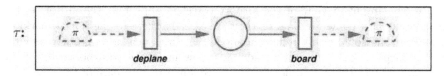

Fig. 2. Airport — traveler handling module

planes in the air upon their take-off. This mechanism for co-operation with lower level agents is present in the τ-module on Fig. 2.

Intuitively, the travelers appear in the airport as a result of a plane landing. The essence of the *deplane* action is to move the passengers from the plane to the airport which then takes care of them. On Fig. 2 this is formalized by the dashed half circle precondition of the *deplane* transition and marking it with π. The postcondition of the *board* transition is treated similarly.

Notice that the virtual places used for co-operation with lower level mobile agents bear the name of the channel. This is needed to determine the proper place which is supposed to provide/receive the manipulated mobile agent. Due to the tree-like hierarchy assumption such annotations were not necessary on Fig. 1 — simply, there is at most one agent above.

2.3   Airport — An Open Net

By synchronizing the π-module with the τ-module we obtain a net representation of an airport, see Fig. 3. The net is well-formed in the sense that the references

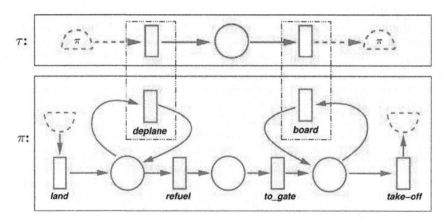

Fig. 3. Airport — traveler and plane manipulation modules

to other modules are consistent with respect to their structure. For instance, the transaction board in the τ-module assumes that the π-module also has the board transition. Thus, the passanger provided for boarding is guaranteed to have a plane ready for accepting it at the postcondition of board in the π-module, which is a local place. Such well-formed synchronized modules are called *open nets*. The terminology intends to highlight the communication capabilities of the agents.

In our simplified view of the airport travelers appear at the airport either as a result of a plane landing, or they are already there in the initial marking. However, one could easily refine the model. For instance, adding a new module to the airport could amount to adding a new means of transportation of the passangers. Clearly, τ-module should be redefined to take care of the new ways of handling travelers. In principle, though, π-module could be left unchanged,

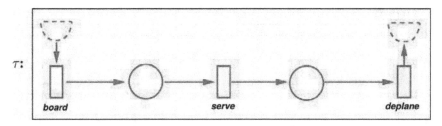

Fig. 4. An airplane

unless some ways of direct interaction between the two means of transportation need to be introduced. This seems to indicate that the modular structure of mobile agents may increase re-usability of components.

2.4   Plane

A very simple model of a plane is presented on Fig. 4. It consists of just one module, a τ-module to handle travelers. The passengers may *board* the plane as a result of co-operation with a higher level agent and, after being *served* they may *deplane*, again in co-operation with the higher level mobile agent.

2.5   Petri Hypernet

An open net is, essentially, a synchronous product of sequential nets, whose transitions are capable also of sending and receiving tokens to and from other open nets. In fact, the tokens themselves are also open nets.

Accordingly, a *Petri hypernet* is a set of open nets $\mathcal{N}$ plus a *hypermarking*:

$$m : \mathcal{N} \rightharpoonup \bigcup_{N \in \mathcal{N}} P_N$$

which describes the distribution of the elements of $\mathcal{N}$ as *token nets* in places of other open nets in $\mathcal{N}$. The assumption that m is a partial function captures the idea that each open net has at most one *level up* net. To ensure that the net N' such that $m(N) \in P_{N'}$ is indeed 'higher' we require that the hierarchy, i.e., the transitive closure of this 'one level up' relation, is irreflexive. If it is, we call the hypermarking *well-founded*.

Fig. 5 presents an open Petri hypernet which consists of one top level net, the airport, one plane and two (unspecified) traveler nets.

The hypernet is called *open*, since the top level net has a transition ready to receive and another ready to send tokens to a higher level.

Clearly, the hierarchy described by the hypermarking is well-founded.

2.6   Transactions — Firing a Transition in a Petri Hypernet

In the hypermarking presented on Fig. 5 one traveler can *board* the plane. Result of firing this transition is described on Fig. 6.

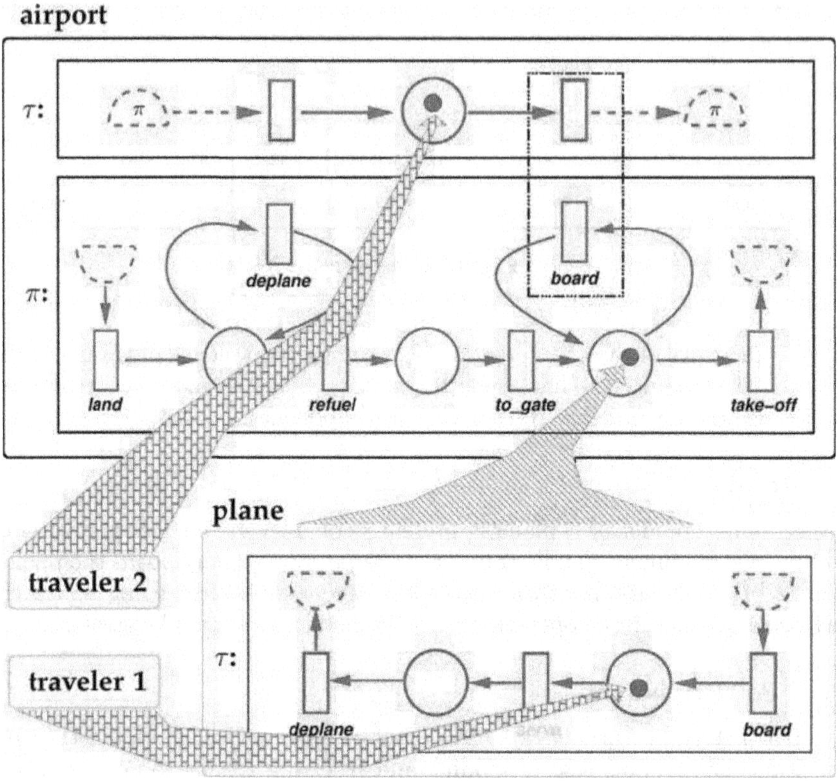

airport

Fig. 5. An open Petri hypernet

The 'traveler 2' net is sent through a 'virtual channel' connecting the airport and the airplane open nets. As we already mentioned, tokens while traveling along *virtual channels* cannot be observed. Hence, the middle picture on Fig. 6 does not really describe a valid hypernet and serves as an intuitive explanation of the firing process only.

As the result of boarding the traveler net involved disappears from the horizon of the airport net, and is moved to the plane net postcondition of *board*. Thus, the hierarchy changes.

This single *action* involves *board* transitions on two different levels: two instances of *board* in the airport net, and one instance of *board* in the plane net. Boarding also involves manipulation of two token nets: the plane and a traveler. From this perspective firing looks like a complex *transaction* which involves synchronization of activities on many different levels in the hypernet.

This transactional view of transition firing provides insight into preservation of the well-foundedness of hypermarkings. The point to notice is this. The transaction involves making connections between complementary and well-connected virtual places. In our example there are two such places: the virtual output of *board* in the τ-module of the airport net, and the virtual input of *board* in the τ-

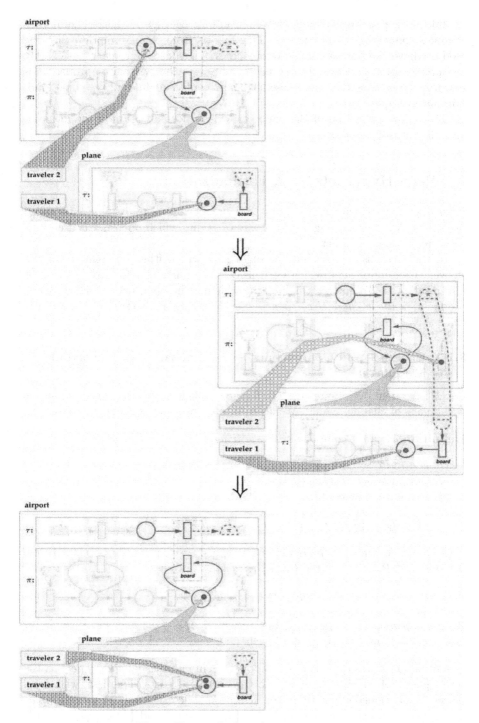

Fig. 6. Firing of a boarding transaction

module of the plane net. Gluing such connections together establishes temporary channels connecting *proper* places. This follows from the assumption that the modules have the structure of state machines. Thus, each transition has exactly one precondition and exactly one postcondition, either virtual or proper. It is also important to notice that these channels involve modules of the same kind, τ in our example.

The following section of the paper provides a formalization of the ideas informally introduced above.

3   Petri Hypernets — A Formalization

Petri hypernets which we are about to define consist of individual nets, called *open nets*. A hypermarking is just a way of saying which open net is used as a token in another open net.

In the formalism presented the open nets will neither be created nor destroyed. Cloning of nets, which we consider as a particular form of creation, is also disallowed. To enforce this policy we define the open nets as synchronous products of *single channel components*, each having a structure of a state machine.

3.1   Channels and Components — Formalization of Kinds and Modules

Let Σ, Δ range over finite sets of *channels*, taken from some fixed countable vocabulary of channel names. We let α, β, etc., range over channel names.

Definition 1. *An α-component C is a triple $C = \langle P, T, F \rangle$, where*

- *P is a finite set of local places of C,*
- *T is a finite set of transitions of C,*
- *F is a finite flow relation,*

$$F \subseteq (P \cup \{?\} \cup \{?\beta \mid \beta \neq \alpha\}) \times T \cup T \times (P \cup \{!\} \cup \{!\beta \mid \beta \neq \alpha\})$$

such that for every $t \in T$ there exists a unique $p \in P \cup \{?\} \cup \{?\beta \mid \beta \neq \alpha\}$, and a unique $q \in P \cup \{!\} \cup \{!\beta \mid \beta \neq \alpha\}$ such that

$$F(p,t) \quad and \quad F(t,q) \tag{1}$$

An α-component C is of sort Σ, notation $C{:}\Sigma$, if $\alpha \in \Sigma$, and $\beta \in \Sigma$ whenever either $F(?\beta,t)$ or $F(t,!\beta)$ hold. Clearly, $C{:}\Sigma$ and $\Sigma \subseteq \Delta$ implies $C{:}\Delta$.

Condition (1) states that each α-component is in fact a usual *state machine*, the set of places of which is divided into three disjoint parts. The first, with local places, is mentioned explicitly in the definition. The other two are *virtual inputs* and *virtual outputs*, contained in $\{?\} \cup \{?\beta \mid \beta \neq \alpha\}$ and $\{!\} \cup \{!\beta \mid \beta \neq \alpha\}$, respectively. In accord with the intuitions put forward in the previous section, input ? and output ! refer to communication with the implicit upper level net. Inputs of the form $?\beta$ and outputs $!\beta$ indicate communications with the lower level nets.

The unique places p and q, local or otherwise, such that $F(p, t)$ and $F(t, q)$ holds are called the *precondition* and *postcondition* of t in C.

3.2   Open Nets — Formalization of Mobile Agents

Suppose that for each $\alpha \in \Sigma$ an α-component of sort Σ is given. Under suitable conditions concerning the inter-component communication, such a collection of components synchronized together forms an *open Petri net* — the counterpart of a mobile agent in our formalization.

Definition 2. *An open net N is a pair $N = \langle\, \Sigma, \{\, N^\alpha \,\}_{\alpha \in \Sigma} \,\rangle$, such that*

1. *$N^\alpha \triangleq \langle\, P^\alpha, T^\alpha, F^\alpha \,\rangle$ is an α-component of sort Σ, for every $\alpha \in \Sigma$*
2. *$F^\alpha(?\beta, t)$ implies $(\exists p \in P^\beta).F^\beta(p, t)$, for every $\alpha, \beta \in \Sigma$, $t \in T_N$*
3. *$F^\alpha(t, !\beta)$ implies $(\exists q \in P^\beta).F^\beta(t, q)$, for every $\alpha, \beta \in \Sigma$, $t \in T_N$*
4. *$\alpha \neq \beta$ implies $P^\alpha \cap P^\beta = \emptyset$*

where T_N denotes the set $\bigcup_{\alpha \in \Sigma} T^\alpha$.

An open net N consists of sequential components N^α, for $\alpha \in \Sigma$. Co-operation of the components within N is enforced by taking the *(partial) synchronous product*. That is, one glues the components on identical transitions while keeping their state spaces apart. The latter is ensured by Def. 2.4.

A sequential component formalizes the notion of *module* which copes with specific activities associated with its channel. From this perspective the synchronous product of the components of an open net is a means to force cooperation of different modules, cf Fig. 3. Additionally, and this is specific to our proposal, the synchronization facilitates the communication between the open net and its tokens, cf. Fig. 6.

Thus, flow $F^\alpha(?\beta, t)$ in the α-component of N signals that the net, while executing t, intends to exchange communication with its token net traveling along channel β. Specifically, it means that transition t in the component N^α wants to read what the token net enabling t in N^β 'sends up' on channel α. Note that, by Def. 2.1, N has a β-component. Def. 2.2 states that in such a case the precondition of t in the component N^β has to be a *local place* from P^β. This, as discussed later, ensures the local character of firing a transition.

In terms of our running example, consider N^τ and N^π to be the sequential components of the airport net responsible for managing travelers and planes respectively, see Fig. 3. Then, Def. 2.2 ensures that disembarkment of passengers from a plane requires the presence of the local place in N^τ, which we could think of as an 'arrival gate'.

Similarly, $F^\alpha(t, !\beta)$ means that while performing t, the component N^α will send *its* token one level down, as a token of the token net traveling along channel β. Def. 2.3 ensures that in this case the postcondition of t in N^β is also a proper place. In our example this situation is demonstrated by the *board* transition and the 'departure gate' in the π-component of the airport.

Note also that the above discussion relies on the assumption that the sequential components are indeed state machines, i.e., that each occurrence of transition t has exactly one input, and one output place.

3.3   Hypernets

Let $N = \langle \Sigma, \{ N^\alpha \}_{\alpha \in \Sigma} \rangle$ be an open net. In the sequel the notation $P_N = \bigcup_{\alpha \in \Sigma} P^\alpha$ and $T_N = \bigcup_{\alpha \in \Sigma} T^\alpha$ is used to denote the collection of all local places and all transitions of N, respectively.

Definition 3. *A Petri hypernet H is a pair $H = \langle \mathcal{N}, m \rangle$, where*

- $\mathcal{N}$ *is a finite set of open nets.*
- $m : \mathcal{N} \rightharpoonup \bigcup_{N \in \mathcal{N}} P_N$ *is a partial function called the* hypermarking.

such that the following conditions hold.

1. *$N \neq N'$ implies $P_N \cap P_{N'} = \emptyset$, for $N, N' \in \mathcal{N}$.*
2. *The partial m-depth function $d_m : \mathcal{N} \rightharpoonup \mathbb{N}$ inductively defined by*

$$d_m(N) \triangleq \begin{cases} 0 & \text{if } m(N) \text{ is undefined} \\ d_m(N') + 1 & \text{if } m(N) \in P_{N'} \end{cases}$$

is total.

Def. 3.1 captures the idea that different mobile agents cannot share any local places. Similarly, sequential modules of an open net have disjoint local spaces.

If the value $m(N)$ is defined, then, by Def. 3.1, there exists a unique $N' \in \mathcal{N}$ such that $m(N)$ is a local place of N'. In this case N is a *token* or *token net* of N'. Thus, the clause in Def. 3.2 indeed defines d_m as a partial function.

The condition of Def. 3.2 really states that the hypermarking of a Petri hypernet is *well-founded* in the sense that its depth function stratifies $\mathcal{N}$ into a finite, forest-like hierarchy. A token net N in H is a *top-level* net iff $d_m(N) = 0$ iff $m(N)$ is undefined.

In the sequel usual conventions apply, e.g., P_H stands for $\bigcup_{N \in \mathcal{N}} P_N$, etc. We write $\overline{m}(N)$ for N' such that $m(N) \in P_{N'}$, and $\chi_m(N)$ for the unique channel α such that $m(N)$ is a place of the α-component of N'. In the sequel the use $\overline{m}(N)$ always implies that $m(N)$ is defined.

3.4   Consortia

In Petri hypernets firing a transition may involve nets at several levels in the current hierarchy. Moreover, as an effect of firing the transition the hierarchy itself may change. All this makes the situation more complex than usual. Yet, the central paradigm underlying Petri net theory is retained — transition firing has *local* and *finitary* character.

We have already argued that with the explicit decomposition of a Petri hypernet into open nets, and these in turn into components, it is appropriate to view the global firing of a transition t as a result of a complex *transaction* which involves these mobile agents and all their modules in which an occurrence of t takes part in reshuffling of tokens involved in the transition.

To start with we consider *consortia*. The idea behind the notion is to take account of all open nets involved in a transition. This involvement may take two forms: the net is being moved around, or it is involved in moving other tokens around. In fact, some token nets may well play both rôles at the same time, as demonstrated by the plane agent in our running example, see Fig.6.

Let $H = \langle \mathcal{N}, m \rangle$ be a hypernet. Consider $t \in T_H$, and a non-empty family $\mathcal{T}$ of open nets containing t, i.e., $N \in \mathcal{T}$ implies $t \in T_N$. For $\mathcal{T}' \subseteq \mathcal{T}$ let us define $\mathsf{Inp}_t(\mathcal{T}') = \{p \in P_H \mid \exists N \in \mathcal{T}'.F_N(p,t)\}$ and call it the set of t-*input-places* of $\mathcal{T}'$. We use the notation $\mathsf{Inp}_t^\alpha(\mathcal{T}')$ when we restrict attention to places in α-components of the nets from $\mathcal{T}'$ only. If $\mathcal{T}'$ is a singleton $\{N\}$ we simply write $\mathsf{Inp}_t(N)$ instead of $\mathsf{Inp}_t(\{N\})$. The set $\mathsf{Out}_t(\mathcal{T})$ of t-*output-places* for $\mathcal{T}$ is defined analogously.

A t-*consortium* selects the set of open nets $\mathcal{T}$ involved in performing the transition, as well as an input token from every input place of the transition t in any net belonging to $\mathcal{T}$. This choice is modeled by the function ξ in the definition below.

The definition of a t-consortium lists five *structural conditions* which have to be satisfied by the nets from the set $\mathcal{T}$ and the function ξ. First, it is required that the choice of token nets given by ξ agrees with the hypermarking of H, i.e., for any t-*input-place* in $\mathcal{T}$ only a token assigned to this place in the hypermarking can be selected by ξ. The remaining four conditions describe the relationship between a (component of a) *token net* and its *parent net*, which make the *inter-level exchange of tokens* possible. Conditions 2 and 3 describe the situation from the token net point of view, while 4 and 5 take the other perspective into account.

Definition 4. *A pair $\langle \mathcal{T}, \xi \rangle$, where $\xi : \mathsf{Inp}_t(\mathcal{T}) \to \mathcal{N}$, is a t-consortium in H provided the following conditions hold for $N \in \mathcal{T}$.*

1. $F_N(p,t)$ *implies* $m(\xi(p)) = p$, *for* $p \in \mathsf{Inp}_t(\mathcal{T})$.
2. $F_N^\alpha(?,t)$ *implies* $\overline{m}(N) \in \mathcal{T} \ \wedge \ F_{\overline{m}(N)}^\alpha(t, !\overline{\alpha}) \ \wedge \ \xi(\mathsf{Inp}_t^{\overline{\alpha}}(\overline{m}(N))) = N$
   where $\overline{\alpha} = \chi_m(N)$.
3. $F_N^\alpha(t, !)$ *implies* $\overline{m}(N) \in \mathcal{T} \ \wedge \ F_{\overline{m}(N)}^\alpha(?\overline{\alpha}, t) \ \wedge \ \xi(\mathsf{Inp}_t^{\overline{\alpha}}(\overline{m}(N))) = N$
   where $\overline{\alpha} = \chi_m(N)$.
4. $F_N^\alpha(?\beta, t)$ *implies* $\xi(\mathsf{Inp}_t^\beta(N)) \in \mathcal{T} \ \wedge \ F_{\xi(\mathsf{Inp}_t^\beta(N))}^\alpha(t, !)$.
5. $F_N^\alpha(t, !\beta)$ *implies* $\xi(\mathsf{Inp}_t^\beta(N)) \in \mathcal{T} \ \wedge \ F_{\xi(\mathsf{Inp}_t^\beta(N))}^\alpha(?, t)$.

An important observation concerning the notion of consortium is that we do not require $\mathcal{T}$ to contain *all* open nets which contain t among their transitions. It is easy to develop an example in which there are two disjoint t-consortia present in the hypernet. Then, clearly, their union is also a t-consortium.

From the condition 1 it follows that the function ξ is injective.

Condition 2 states that if the α-component of N expects a token from the upper level, then the upper level net $\overline{m}(N)$ exists and belongs to $\mathcal{T}$. Moreover, the α-component of $\overline{m}(N)$ must be ready to *send* something *down* to its (unique) $\overline{\alpha}$-channel such that N is the input token for t in $\overline{m}(N)^{\overline{\alpha}}$ chosen by ξ. Channel $\overline{\alpha}$ is the channel through which N travels in the upper level net, i.e., $\overline{\alpha} = \chi_m(N)$.

In terms of the airport example, an instance of this condition would say that in any **board**-*consortium* if the plane net is ready to accept a passenger, then the airport net must be ready to provide one, and for this to happen the plane has to be at the departure gate (which is a **board**-*input-place* of the plane-manipulation component of the airport).

Condition 3. is similar, but refers to the ability of *sending* a token *up*.

Above, the unique *upper channel* from both conditions 2. and 3. is slightly ambiguously denoted $\overline{\alpha}$. It is hoped that the precise meaning can always be deduced from the context.

Condition 4. states, that if the α-component of N expects a token to be provided by the token net selected by ξ in the β-precondition of t, then the token net selected for this precondition by ξ is in $\mathcal{T}$, and has an α-component in which t is ready to send something up. For example, in any **deplane**-*consortium* if the airport is ready to accept a passenger from a plane, the plane must be at the **deplane**-*input-place* (i.e., an arrival gate) and has to be ready to send the passenger up to the passenger handling τ-component of the airport.

As we know, transitions in $\mathcal{T}$ may also involve *virtual inputs/outputs*. The set containing both input-places and virtual inputs defined as

$$\mathsf{PreCon}_t(\mathcal{T}) = \{\, \langle p, N^\alpha \rangle \mid N \in \mathcal{T} \,\wedge\, \alpha \in \Sigma_N \,\wedge\, F_N^\alpha(p,t) \,\}$$

will be called the set of *t-preconditions in* $\mathcal{T}$. The set of *t-postconditions in* $\mathcal{T}$ is defined in an analogous way and denoted by $\mathsf{PostCon}_t(\mathcal{T})$.

Again, when restricting attention to transitions occurring in α-components of the nets from $\mathcal{T}$ only, we shall write $\mathsf{PreCon}_t^\alpha(\mathcal{T})$ and $\mathsf{PostCon}_t^\alpha(\mathcal{T})$.

3.5   Transactions, or Firing a *t*-Consortium

Let $H = \langle \mathcal{N}, m \rangle$ be a hypernet, $t \in T_H$, and $\mathcal{T}$ be a *t*-consortium in H. As we know, the occurrences of t in the open nets from $\mathcal{T}$ may have virtual inputs/outputs as their preconditions/postconditions. Informally speaking, in the process of firing the consortium the virtual places *along channels* are glued together and become invisible. Token nets are being moved from the *input places* they occupy (via ξ) to the corresponding *output places*. Both input places and output places are 'proper', i.e., local places from P_H.

To describe the effect of firing the transition t in H let us first recursively define a family of partial functions:

- $\mathsf{src}_t^\alpha : \mathsf{PreCon}_t^\alpha(\mathcal{T}) \cup \mathsf{PostCon}_t^\alpha(\mathcal{T}) \rightharpoonup \mathsf{Inp}_t^\alpha(\mathcal{T})$
- $\mathsf{trg}_t^\alpha : \mathsf{PreCon}_t^\alpha(\mathcal{T}) \cup \mathsf{PostCon}_t^\alpha(\mathcal{T}) \rightharpoonup \mathsf{Out}_t^\alpha(\mathcal{T})$

for all α in the alphabet of $\mathcal{T}$ (i.e. $\alpha \in \bigcup \{\, \Sigma_N \mid N \in \mathcal{T} \,\}$).

For an arbitrary *t-precondition/t-postcondition* in $\mathcal{T}$ the functions $\mathbf{src}_t^\alpha$ and $\mathbf{trg}_t^\alpha$ assign the '*source input-place*' and the '*target output-place*' for the particular occurrence of t respectively (both are elements of P_H).

$$\mathbf{src}_t^\alpha(\langle p, N^\alpha\rangle) \cong \begin{cases} p & p \in P_N^\alpha \ \wedge \ F_N^\alpha(p,t) \\ \mathbf{src}_t^\alpha(\langle p', N^\alpha\rangle) & F_N^\alpha(t,p) \ \wedge \ F_N^\alpha(p',t) \ \wedge \ p \neq p' \\ \mathbf{src}_t^\alpha(\langle !\overline{\alpha}, \overline{m}(N)^\alpha\rangle) & p = \text{``?''} \\ \mathbf{src}_t^\alpha(\langle !, \xi(q)^\alpha\rangle) & p = \text{``?}\beta\text{''} \end{cases}$$

$$\text{where } q = \mathsf{Inp}_t^\beta(N)$$

$$\mathbf{trg}_t^\alpha(\langle p, N^\alpha\rangle) \cong \begin{cases} p & p \in P_N^\alpha \ \wedge \ F_N^\alpha(t,p) \\ \mathbf{trg}_t^\alpha(\langle p', N^\alpha\rangle) & F_N^\alpha(p,t) \ \wedge \ F_N^\alpha(t,p') \ \wedge \ p \neq p' \\ \mathbf{trg}_t^\alpha(\langle ?\overline{\alpha}, \overline{m}(N)^\alpha\rangle) & p = \text{``!''} \\ \mathbf{trg}_t^\alpha(\langle ?, \xi(q)^\alpha\rangle) & p = \text{``!}\beta\text{''} \end{cases}$$

$$\text{where } q = \mathbf{src}_t^\beta(\langle \mathsf{Out}_t^\beta(N), N^\beta\rangle)$$

Let us briefly comment on the above definition. The first two clauses in the definition of $\mathbf{src}_t^\alpha$ and the definition of $\mathbf{trg}_t^\alpha$ are self-explanatory. The third clause in both cases refers to $\overline{m}(N)^\alpha$ i.e., the α-component of the 'parent net' of N (wrt. to the hypermarking m). Such a parent net exists by conditions 2 and 3 of the Definition 4 respectively.

The third clause of the definition of $\mathbf{src}_t^\alpha$ simply says that to find the *source input-place* of an occurrence of t which *reads* something '*from above*' we have to look for the source input-place of t in the corresponding component of the parent net. The third clause of the definition of $\mathbf{trg}_t^\alpha$ has an analogous interpretation.

Up to this point definitions of $\mathbf{src}_t^\alpha$ and $\mathbf{trg}_t^\alpha$ were symmetric. The last clauses brake this symmetry for obvious reasons. In the case of $\mathbf{src}_t^\alpha$ the situation is simple. We have to search for a matching !-virtual place in the token provided to t in the α-component of N. This token is in the unique local place q such that $F_N^\beta(q,t)$ holds, i.e., $q = \mathsf{Inp}_t^\beta(N)$.

A symmetric argument in case of $\mathbf{trg}_t^\alpha$ would lead to a local place $r = \mathsf{Out}_t^\beta(N)$ which is a *postcondition* of t, so ξ cannot be directly applied to it. To cope with this we have to first compute the local place which is going to end up in r, and only then apply ξ to it.

The following result is crucial to guarantee correctness of our last definition.

Lemma 1. *Let* $H = \langle \mathcal{N}, m\rangle$ *be a hypernet,* $t \in T_H$, *and* $\mathcal{T}$ *be a t-consortium in* H.
1. $\mathbf{src}_t^\alpha\langle p, N^\alpha\rangle$ *is defined for any local* $p \in \mathsf{Out}_t^\alpha(\mathcal{T})$.
2. $\mathbf{trg}_t^\alpha\langle p, N^\alpha\rangle$ *is defined for any local* $p \in \mathsf{Inp}_t^\alpha(\mathcal{T})$.

Note that the definition of $\mathbf{src}_t^\alpha$ and $\mathbf{trg}_t^\alpha$ does not directly refer to $\mathbf{src}_t^\beta/\mathbf{trg}_t^\beta$ for $\beta \neq \alpha$. Intuitively it stresses the fact that the flow of tokens within a hypernet takes place 'along channels', i.e., a token can move from an α-component of one open net to a β-component of another open net within H only if β and α are equal (of course other conditions have to be satisfied as well).

Definition 5. *The result of firing a consortium* $\langle T, \xi \rangle$ *in a hypernet* $\langle N, m \rangle$ *is a hypernet* $\langle N, m' \rangle$ *such that:*

$$m'(N) = \begin{cases} \mathbf{trg}_t^\alpha(\langle m(N), \overline{m}(N)^\alpha \rangle) & N \in \xi[\mathsf{Inp}_t^\alpha(T)], \ m(N) \in \overline{m}(N)^\alpha \\ m(N) & otherwise \end{cases}$$

i.e., the new hypermarking m' *is obtained from the original one by moving all the input tokens designated by* ξ *to their target output-places given by the appropriate instance of the function* $\mathbf{trg}_t$.

Finally, let us formulate the main result. We state it without proof which is quite involved and therefore omitted due to space limitations.

Proposition 1. *Definition 5 is correct. In particular, the new assignment of open nets as tokens which result from the firing of a t-consortium in a hypernet is a well-founded hypermarking.*

4   Conclusion

Petri hypernets, an extension of elementary Petri nets, have been introduced as a means to represent systems of interacting mobile agents. Some basic properties of the model have also been established. The main features of Petri hypernets are the following.

- Local and finitary character of interactions between mobile agents, retained from Petri nets.
  This allows to reuse the usual notions from Petri net theory. For instance, two consortia can be fired concurrently, provided their resources are disjoint, etc. In fact, this seems to apply to all ramifications of Valk's **nets-within-nets** paradigm.
- Limited expressive power.
  Clearly, the state space of a finite closed hypernet is finite. Thus, the model promises to capture some essential features without resort to powerful semantical concepts, like free variable generation and binding.
- The *channels* and the *hierarchy* as a support for modularization.
  Channels and modules allow separation of concerns with regard to agents of different kinds. The hierarchy helps structure the control flow of mobile agents.
- Flexibility of the hierarchy structure, and inter-level migration.
  The ability of open nets to move up and down in the hierarchy is one of the main differences between hypernets and the other ramifications of Valk's **nets-within-nets** paradigm.
- Semantics based on the principle of preservation of mobile agents identity.

The comparable alternative approaches to the problem of specifying systems of mobile agents can be very roughly divided in two classes: the π-calculus and the calculi inspired by it, or akin to it; the net-based models following the paradigm of **nets-within-nets**, proposed by Valk. We have chosen the latter paradigm in order to adhere to the key principles of Petri's net theory: locality of states and transitions, and finiteness of the basic model. This choice sets bounds to the expressive power of the model. In particular, each finite hypernet has a finite state space if it is closed or no interactions with the outside world are made.

A more detailed comparison of our proposal with respect to both trends has been made in the introduction. Let us finish by discussing plans for further work.

The next step we intend to undertake is to develop sound reasoning techniques for verification of properties of Petri hypernets. This should involve defining suitable logics to match the behavioral notions of the model. In particular, the logics have to take the individual character of tokens into account.

Petri hypernets provide a natural model in cases similar to the airport case study. Here, it is natural to think of places as physical locations and of tokens as *physical objects*. However, even in simple generalizations, manipulation of references seems unavoidable. For instance, if passengers were allowed to carry luggage, then the task of collecting the luggage after landing by its owner could be realized by references. The luggage should have a reference to its owner, and the owner could have a reference to its luggage. We plan to investigate if hypernets could be extended to handle also references to agents. Perhaps one of Valk's referential semantics could be adapted here. Within this context it might also be natural to consider the non-well-founded hierarchies.

References

1. P. Andrade and P. Baldan, H. Baumeister, et al. AGILE: Software architecture for mobility. In M. Wirsing, D. Pattinson, and R. Hennicker, editors, Recent Trends in Algebraic Development Techniques. 16th International Workshop WADT 2002, Frauenchiemsee, Germany, September 24-27, 2002. Revised Selected Papers, volume 2755 of LNCS, pages 48–70. Springer-Verlag, 2003.
2. A. Asperti and N. Busi. Mobile Petri nets. Technical Report UBLCS-96-10, Lab. for Computer Science, University of Bologna, Italy, 1996.
3. M. A. Bednarczyk and A. M. Borzyszkowski. On concurrent realizations of reactive systems and their morphisms. In H. Ehrig, G. Juhas, J. Padberg, and G. Rozenberg, editors, Unifying Petri nets, Advances in Petri nets, volume 2128 of LNCS, pages 346–379. Springer-Verlag, 2001.
4. L. Bernardinello. Synthesis of net systems. In M. A. Marsan, editor, Application and Theory of Petri Nets 1993, 14th International Conference, Chicago, Illinois, USA, June 21-25, 1993, Proceedings, volume 691 of LNCS, pages 89–105. Springer-Verlag, 1993.
5. R. Bruni and U. Montanari. Transactions and zero-safe nets. In H. Ehrig, G. Juhas, J. Padberg, and G. Rozenberg, editors, Unifying Petri nets, Advances in Petri nets, volume 2128 of LNCS, pages 346–379. Springer-Verlag, 2001.
6. L. Cardelli, G. Ghelli, and A. D. Gordon. Mobility types for mobile ambients. In J. Wiedermann, P. van Emde Boas, and M. Nielsen, editors, Automata, Languages and Programming: 26th International Colloquium, ICALP '99, Proceedings, volume 1644 of LNCS, pages 230–239. Springer-Verlag, 1999.

7. C. Fournet and G. Gonthier. The reflexive chemical abstract machine and the Join calculus. In Proceedings of POPL '96, pages 372–385, 1996.
8. M. Köhler, D. Moldt, and H. Rölke. Modelling mobility and mobile agents using nets within nets. In W. van der Aalst and E. Best, editors, Applications and Theory of Petri Nets 2003, Proceedings, volume 2679 of LNCS, pages 121–139. Springer-Verlag, 2003.
9. I. A. Lomazova and P. Schnoebelen. Some decidability results for nested Petri nets. In D. Bjørner, M. Broy, and A. Zamulin, editors, Perspectives of System Informatics (PSI'99), volume 1755 of LNCS, pages 208–220. Springer-Verlag, 2000.
10. R. Milner, J. Parrow, and D. Walker. A calculus of mobile processes, parts 1-2. Information and Computation, 100(1):1–77, 1992.
11. W. Reisig. Petri Nets, volume 4 of EATCS Monographs on Theoretical Computer Science. Springer-Verlag, 1985.
12. R. Valk. On processes of object Petri nets. Technical Report FB-185, Fachbereich Informatik, Universität Hamburg, 1996.
13. R. Valk. Petri nets as token objects: An introduction to elementary object nets. In W. van der Aalst and E. Best, editors, Applications and Theory of Petri Nets 1998, Proceedings, volume 1420 of LNCS, pages 1–25. Springer-Verlag, 1998.
14. R. Valk. Concurrency in communicating object Petri nets. In G. Agha, F. De Cindio, and G. Rozenberg, editors, Concurrent Object-Oriented Programming and Petri Nets: Advances in Petri Nets, volume 2001 of LNCS, pages 164–195. Springer-Verlag, 2001.

Cryptomorphisms at Work[*]

Carlos Caleiro and Jaime Ramos

CLC - Departamento de Matemática, IST - UTL,
Av. Rovisco Pais, 1049-001 Lisboa, Portugal

Abstract. We show that the category proposed in [5] of logic system presentations equipped with *cryptomorphisms* gives rise to a category of parchments that is both complete[1] and translatable to the category of institutions, improving on previous work [15]. We argue that limits in this category of parchments constitute a very powerful mechanism for combining logics.

1   Introduction

The importance of studying combined logics and, specially, general mechanisms for combining logics is widely recognized [1]. This happens not only because of the theoretical interest and technical difficulties of the subject, but also for practical reasons. In many fields, the need for working with several logics at the same time is the rule rather than the exception. Among the various approaches to the combination of logics, two deserve our close attention. One has been developed within the general theory of institutions [12, 18], and focuses on the categorial combination of parchments [11, 13, 14, 15]. Another, very successful, approach is fibring [8, 7, 9, 16, 21, 2, 17]. The two approaches have also met each other in [4, 3], where some of the very general preservation results already identified for fibring have been brought to the level of parchments.

This work was motivated by a recent development in the theory of fibring which consisted on adopting a novel category of logic system presentations. Still, while the similarities between this category and the ones used in the parchment framework were evident, their properties seemed to be quite distinct. Our initial aim was to explore the precise relationship between them. Below, we shall briefly overview the problem at hand and the context in which the relevant concepts have appeared.

Ever since the first accounts of fibring, it could be noticed that fibring two logics could sometimes lead to the collapse of one logic into the other (for instance, fibring classical with intuitionistic logic would collapse into just classical logic

---

[*] This work was partially supported by FCT and FEDER, namely, via the Project FibLog POCTI/MAT/37239/2001, and the QuantLog initiative of CLC.

[1] In this paper, we shall only be interested in set-indexed categorial constructions. Therefore, all occurrences of categorial (co)completeness, (co)limit, etc. should be understood as small.

J.L. Fiadeiro, P. Mosses, and F. Orejas (Eds.): WADT 2004, LNCS 3423, pp. 45–60, 2005.
© Springer-Verlag Berlin Heidelberg 2005

[8, 6]). In [5], cryptofibring was proposed as an extension of fibring and shown to keep its general metatheoretical properties, like soundness and completeness preservation results, while also attacking this so-called "collapsing problem". In [17] another variant of fibring, modulated fibring, had been introduced and shown to avoid these collapses by means of a very careful use of adjunctions between lattice structured models. However, cryptofibring presents a structurally simpler independent solution to the problem, interesting in its own right, that encompasses the original definition of fibred model but also admits amalgamated models that can be used to show that the above mentioned collapses are no longer present. Cryptofibring is characterized categorially as a special kind of pushout in a suitable category of logic system presentations. Its objects are simple algebraic presentations of both the syntax and semantics of logic systems. The main novelty concerns its morphisms, that have been called *cryptomorphisms*, from where cryptofibring borrows its name. In the sequel, we shall recall their precise definitions, along with a short hint on the origin of the "collapsing problem" for fibring. Further details on the theory of fibring and the "collapsing problem" are however out of the scope of this paper. We encourage the interested reader to browse through the cited literature, namely for motivation and examples.

It turns out, as we show here, that the cryptomorphisms introduced in [5] set up a category of logic system presentations which is, modulo presentation details, half-way between the categories of *rooms* used in [15] to build the categories of *model-theoretic* parchments and its *logical* large subcategory. The main aim of its authors was to obtain a framework for combining logics using limits of parchments, and a smooth way of presenting them as institutions [12], following earlier work [11, 13, 14]. However, they could prove that model-theoretic parchments form a complete category but fail in general to present institutions, whereas logical model-theoretic parchments present institutions in a smooth way but do lack certain small limits. The counterexample used in [13, 14, 15], with minor variations, concerns the nonexistence of a certain limit combining total equational logic and partiality, whose intended result should help to grasp the meaning of equations involving undefined terms.

In this paper we further show that cryptomorphisms really work. Not only they extend fibring as was already known from [5], but, contrarily to what happened with model-theoretic parchments, logical or not, they also give rise to a category of parchments which is simultaneously complete and easily translatable to institutions. Indeed, both these properties follow directly from properties of cryptomorphisms: they always fulfill the necessary *satisfaction condition*, and they constitute a cocomplete category of logic system presentations. Arguably, parchments based on cryptomorphisms are therefore an extremely powerful tool for combining logics. As an application of cryptomorphisms, and with the main purpose of stressing the differences with respect to (logical) model-theoretic parchments, we shall revisit the partial equational logic example and show that the corresponding colimit encompasses models that are compatible with each possible interpretation of equality involving undefinedness, be it strong, weak, existential, three-valued, or even other.

We proceed as follows. In Section 2 we introduce the category of logic system presentations with cryptomorphisms and explore its relationship to the categories of rooms used to build model-theoretic parchments and logical model-theoretic parchments. Then, in Section 3, we show that cryptomorphisms indeed build up a cocomplete category, and highlight a few differences with respect to the other cases. The Section 4 is devoted to presenting the details of the example and discussing its result. We conclude, in Section 5, with a discussion of the results obtained and an outline of future work.

2 Cryptomorphisms

To combine logics and achieve a meaningful interplay between them we need to work with presentations that pinpoint the fine details of the logic's syntax and semantics. A usual approach, underlying the notion of parchment [11] as well as an essential dimension of fibring [16], is to consider some kind of algebraic presentation a logic. Another common feature of both approaches is to adopt a categorial setting where the combination mechanisms should be characterized as universal constructions [10]. Before we proceed to the definition of our working category of logic system presentations and *cryptomorphisms*, we start by recalling, or introducing, some notions and notation.

In the sequel, **AlgSig** is the category of many-sorted signatures, and **AlgSig**$_\phi$ its subcategory whose signatures have a distinguished sort $\phi \in S$ (for *formulas*) and whose morphisms preserve ϕ. We denote by **Alg** the flat category of many-sorted algebras and homomorphisms, and by **Alg**(Σ) the category of Σ-algebras and Σ-homomorphisms, for each signature Σ. We use $\mathcal{W}_\Sigma$ to denote the free Σ-algebra (the word algebra), and $[\![\_]\!]_\mathcal{A}$ (for word *interpretation*) to denote the unique Σ-homomorphism from $\mathcal{W}_\Sigma$ to a given Σ-algebra $\mathcal{A}$. Elements of $|\mathcal{W}_\Sigma|_s$ are referred to as *terms* and denoted by t. Every **AlgSig**-morphism $\sigma : \Sigma_1 \to \Sigma_2$ has an associated reduct functor $\_|_\sigma : \mathbf{Alg}(\Sigma_2) \to \mathbf{Alg}(\Sigma_1)$. Note that $[\![t]\!]_{\mathcal{A}|_\sigma} = [\![\sigma(t)]\!]_\mathcal{A}$ for each $t \in |\mathcal{W}_{\Sigma_1}|_s$ and Σ_2-algebra $\mathcal{A}$. As usual, we overload the notation and write σ for word *translation* instead of $[\![\_]\!]_{\mathcal{W}_{\Sigma_2}|_\sigma}$ to denote the unique Σ_1-homomorphism from $\mathcal{W}_{\Sigma_1}$ to $\mathcal{W}_{\Sigma_2}|_\sigma$. If Σ has a distinguished sort ϕ, Form$_\Sigma$ stands for the set $|\mathcal{W}_\Sigma|_\phi$ of *formulas*. We use φ to denote a formula.

Definition 1. A *logic system presentation* is a triple $\langle \Sigma, M, \mathbb{A} \rangle$ where $\Sigma \in |\mathbf{AlgSig}_\phi|$, M is a class (of *models*), and $\mathbb{A}$ associates to each $m \in M$ a Σ-*interpretation structure* $\mathbb{A}(m) = \langle \mathcal{A}_m, T_m \rangle$, where $\mathcal{A}_m$ is a Σ-algebra and $T_m \subseteq |\mathcal{A}_m|_\phi$ (the *designated* subset of the set of *truth-values*).

Given a model $m \in M$, we shall simply write $[\![\_]\!]_m$ instead of $[\![\_]\!]_{\mathcal{A}_m}$.

This kind of interpretation structure, featuring a set of designated truth-values, is commonly known as a *logical matrix* in the logic literature (see, for instance, [20]). We can define the usual *satisfaction* of a formula $\varphi \in$ Form$_\Sigma$ by $m \Vdash \varphi$ if $[\![\varphi]\!]_m \in T_m$.

The authors of [15] considered parchments built over a category **MPRoom** whose objects can be seen as logic system presentations, modulo presentation details that we shall ignore. They also considered a *logical* version of these parchments, which can similarly be built over a large subcategory **LogMPRoom** of **MPRoom**. We recall the precise definition of their morphisms.

Definition 2. A *morphism* $\langle \sigma, \mu, \eta \rangle$: $\langle \Sigma_1, M_1, \mathbb{A}_1 \rangle \rightarrow \langle \Sigma_2, M_2, \mathbb{A}_2 \rangle$ of logic system presentations consists of an **AlgSig**$_\phi$-morphism $\sigma : \Sigma_1 \rightarrow \Sigma_2$, a map $\mu : M_2 \rightarrow M_1$, and a family $\eta = \{\eta_m : \mathbb{A}_1(\mu(m)) \rightarrow \mathbb{A}_2(m)\}_{m \in M_2}$ where each η_m is a Σ_1-homomorphism from $\mathcal{A}_{\mu(m)}$ to $\mathcal{A}_m|_\sigma$ that preserves designated values, that is, $\eta_m(T_{\mu(m)}) \subseteq T_m$.

A morphism is said to be *closed* if each η_m also reflects designated values, that is, $\eta_m^{-1}(T_m) = T_{\mu(m)}$ for every $m \in M_2$.

A closed morphism is said to be *logical* if η_m is injective for every $m \in M_2$.

Let $\langle \sigma, \mu, \eta \rangle : \langle \Sigma_1, M_1, \mathbb{A}_1 \rangle \rightarrow \langle \Sigma_2, M_2, \mathbb{A}_2 \rangle$ be a morphism, $\varphi \in \text{Form}_{\Sigma_1}$ and $m \in M_2$. In general, it is clear that if $\mu(m) \Vdash_1 \varphi$ then $m \Vdash_2 \sigma(\varphi)$. However the converse does not hold, in general, since designated values are preserved but may not be reflected. For closed morphisms, however, we obtain the usual *satisfaction condition*: $\mu(m) \Vdash_1 \varphi$ if and only if $m \Vdash_2 \sigma(\varphi)$.

With the obvious definitions of identity and composition, logic system presentations and morphisms constitute the category **MPRoom**. **LogMPRoom** is the subcategory of **MPRoom** with only logical morphisms. But what is more, closed morphisms are precisely what have been called *cryptomorphisms* in [5]. Requiring the weakest possible condition that ensures the satisfaction condition was indeed the main reason for their precise formulation. In the remainder of the paper, we shall call **Crypt** to the corresponding category. It is worthwhile recalling that **Crypt** was proposed in order to characterize cryptofibring, a generalization of fibring aimed at solving an anomaly known as the "collapsing problem". Indeed, at this level of abstraction, cryptofibring can be seen to extend fibring in exactly the same proportion as the notion of cryptomorphism extends the notion of arrow between logic system presentations used to characterize fibring. For fibring, we just have $\langle \sigma, \mu \rangle : \langle \Sigma_1, M_1, \mathbb{A}_1 \rangle \rightarrow \langle \Sigma_2, M_2, \mathbb{A}_2 \rangle$ and require that $\mathcal{A}_{\mu(m)} = \mathcal{A}_m|_\sigma$ and $T_{\mu(m)} = T_m$. It is not difficult to understand that this strict condition is the main reason why collapses may occur in the first place [5]. In this context, it is also easy to understand the baptism of cryptomorphisms. Each η_m is precisely a "homomorphism" between $\mathcal{A}_{\mu(m)}$ and $\mathcal{A}_m$, algebras over distinct signatures, mediated by the signature morphism σ, as in **Alg**. This kind of "homomorphism" has been called a cryptomorphism before (see, for instance, [19]).

If one wishes to work with logics at the level of institutions it is essential that one considers parchments [11]. Fixing a base category **B** as a category of *rooms*, we can build up a corresponding category of **B**-*parchments*. We just define a **B**-parchment to be a functor $P : \textbf{Sig} \rightarrow \textbf{B}$, where **Sig** is some category of abstract signatures. A *morphism of* **B**-*parchments* from $P_1 : \textbf{Sig}_1 \rightarrow \textbf{B}$ to $P_2 : \textbf{Sig}_2 \rightarrow \textbf{B}$ is now just a pair $\langle \Phi, \alpha \rangle$ where $\Phi : \textbf{Sig}_1 \rightarrow \textbf{Sig}_2$ is a functor and $\alpha : P_2 \circ \Phi \rightarrow P_1$ is a natural transformation. Using **MPRoom** and **LogMPRoom** as a basis we obtain precisely the categories of model-theoretic parchments and logical model-theoretic parchments of [15]. These constructions mimic precisely the construction of the category of institutions and institution morphisms using (the dual of) the category of twisted relations [12] as a base. There are two very interesting features that these categories of parchments may enjoy: one is the possibility of setting up a functor to institutions, thus showing that the parchments at hand are indeed good ways of representing logics; the other is the possibility of combining logics using limits of parchments, when they exist. It is a straightforward property of the general construction of these categories of parchments that a translation to institutions can be obtained from a translation of the base category considered to (the dual of) the category of twisted relations. In the case of our logic system presentations, this amounts to choosing a notion of arrow that fulfills the satisfaction condition mentioned before. Moreover, well known results on indexed categories [19] show that this construction always yields a complete category of parchments if the base category considered is cocomplete.

Despite its problem with the satisfaction condition, **MPRoom** is a cocomplete category. However, the combinations obtained often feature combined models with a diversity of newly generated truth-values corresponding to the previously unknown result of applying an operation of one of the logics being combined to a value of another. These were called "junk" values and considered harmful in [13, 14, 15], which led its authors to restrict attention not to closed, but directly to logical morphisms. The category **LogMPRoom** is however not cocomplete (as usual, injectivity does not go along too well with coequalizers), as they have shown in a very interesting example. The category **Crypt** is already known to have some colimits, at least precisely those used for characterizing cryptofibring. Our job, in the next section, will be to show that **Crypt** is indeed cocomplete. While doing that we shall see that the previously mentioned "junk" values are an essential ingredient of the envisaged free interplay of concepts, if measured along the differences between colimits in **Crypt** and **MPRoom**, that we shall pinpoint.

3   Cocompleteness

Our task now is to prove the cocompleteness of **Crypt**, while highlighting the main differences between its colimits and those that can be obtained in **MPRoom**.

As argued in [19], many categories arising in computer science can be seen as indexed categories. Our categories of interest for now, **MPRoom** and **Crypt**, but also **LogMPRoom**, are good examples of that. We could explore this fact and the well known results about indexed categories to attack the cocompleteness of **Crypt** and compare its colimits with those of **MPRoom**. Indeed, at a first level, both can be seen as categories indexed by $\mathbf{AlgSig}_\phi$, and at a second level, the component categories of each of them can be seen as categories of interpretation structures indexed by the dual of the category of classes. However, there is a major difference between the two cases: whereas in the case of **MPRoom** each category of Σ-interpretation structures ($\mathbf{Str}(\Sigma)$ in the terminology of [15]) is also cocomplete, the relevant categories of structures in the case of **Crypt** (of course, the subcategories **CryptStr**(Σ) of each **Str**(Σ) with homomorphisms that do not only preserve but also reflect designated values) are not cocomplete. This fact implies that, at the second level, the sufficient conditions for the cocompleteness of indexed categories of [19] apply to **MPRoom** but not to **Crypt**. Still we do not start from scratch. In the sequel, we shall capitalize on two well known results (see, for instance, [19]): the cocompleteness of **Alg**, and the fact that the forgetful functor from **Alg** to **AlgSig** preserves colimits. Although we do not want to get into the fine details of the cocompleteness of **Alg** here, we shall at least have a brief look at their essential aspects by analyzing some very simple but critical examples. This exercise will not only provide further insight to the forthcoming construction of colimits in **Crypt**, but also help us in making the contrast between **Crypt** and **MPRoom**, by emphasizing the differences between **CryptStr**(Σ) and **Str**(Σ).

Colimits in **Alg** have two base pillars: colimits in each category **Alg**(Σ), and the existence of a left adjoint F_σ of the reduct functor $\_|_\sigma : \textbf{Alg}(\Sigma_2) \to \textbf{Alg}(\Sigma_1)$ associated to each signature morphism $\sigma : \Sigma_1 \to \Sigma_2$. Coproducts in **Alg**($\Sigma$), for a given signature Σ, are built by taking the free Σ-algebra over the disjoint union of all the carrier sets of the different algebras, and then making its quotient under the congruence generated by the interpretation of terms in each of the algebras. We next consider two contrasting situations.

Example 1. Let $\Sigma \in |\textbf{AlgSig}_\phi|$ be the signature with only the sort ϕ and the constant operation $c : \phi$, and consider the Σ-algebras $\mathcal{A}$ and $\mathcal{B}$, respectively with $|\mathcal{A}|_\phi = \{*\}$ and $c_\mathcal{A} = *$, and $|\mathcal{B}|_\phi = \{0,1\}$ and $c_\mathcal{B} = 1$. Their coproduct $\mathcal{A} \coprod \mathcal{B}$ in **Alg**(Σ) is (up to isomorphism) $\mathcal{B}$ itself, along with the homomorphisms $h : \mathcal{A} \to \mathcal{B}$ such that $h(*) = 1$ and $id_\mathcal{B} : \mathcal{B} \to \mathcal{B}$: just put together $*$, 0 and 1 with the only Σ-term c, and collapse $*$, 1 and c by noting that $[\![c]\!]_\mathcal{A} = *$ and $[\![c]\!]_\mathcal{B} = 1$.

Consider now the Σ-structures $\langle \mathcal{A}, \emptyset \rangle$ and $\langle \mathcal{B}, \{1\} \rangle$. In **Str**($\Sigma$) there is a canonical way of endowing the coproduct $\mathcal{B}$ of the algebras with a set T of designated values that makes both homomorphisms preserve designated values, that is, $h(\emptyset) \subseteq T$ and $id_\mathcal{B}(\{1\}) \subseteq T$. Since morphisms in **Str**(Σ) only need to preserve designated values, the minimal choice $T = \{1\}$ is canonical because any other designation preserving choice must be bigger. This means that $id_\mathcal{C}$ is a designation preserving homomorphism from $\langle \mathcal{C}, \{1\} \rangle$ to $\langle \mathcal{C}, T \rangle$ for every $T \supseteq \{1\}$. Indeed, the coproduct $\langle \mathcal{A}, \emptyset \rangle \coprod \langle \mathcal{B}, \{1\} \rangle$ in **Str**(Σ) is precisely $\langle \mathcal{B}, \{1\} \rangle$, along with h and $id_\mathcal{B}$, as depicted in Figure 1.

However, while $id_\mathcal{B} : \langle \mathcal{B}, \{1\} \rangle \to \langle \mathcal{B}, \{1\} \rangle$ also reflects designated values, $h : \langle \mathcal{A}, \emptyset \rangle \to \langle \mathcal{B}, \{1\} \rangle$ does not because $*$ is not designated but $h(*) = 1$ is. Actually, there is no possible choice of T that makes both homomorphisms preserve and reflect designated values. This is the reason why a coproduct $\langle \mathcal{A}, \emptyset \rangle \coprod \langle \mathcal{B}, \{1\} \rangle$ in **CryptStr**(Σ) does not exist. $\triangle$

Example 2. Now, let $\Sigma \in |\textbf{AlgSig}_\phi|$ be the signature with only the sort ϕ and a binary operation $f : \phi \times \phi \to \phi$, and consider the Σ-algebras $\mathcal{A}$ and $\mathcal{B}$ with

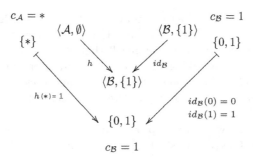

Fig. 1. Coproduct in $\mathbf{Str}(\Sigma)$

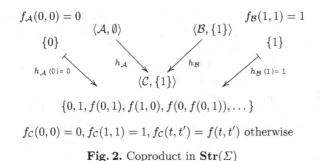

$$\{0, 1, f(0,1), f(1,0), f(0, f(0,1)), \dots\}$$

$$f_C(0,0) = 0, \ f_C(1,1) = 1, \ f_C(t,t') = f(t,t') \text{ otherwise}$$

Fig. 2. Coproduct in $\mathbf{Str}(\Sigma)$

$|\mathcal{A}|_\phi = \{0\}$ and $f_\mathcal{A}(0,0) = 0$, and $|\mathcal{B}|_\phi = \{1\}$ and $f_\mathcal{B}(1,1) = 1$. The coproduct $\mathcal{A} \coprod \mathcal{B}$ in $\mathbf{Alg}(\Sigma)$ is in this case (up to isomorphism) the free Σ-algebra over $\{0,1\}$ but with $f(0,0) \approx 0$ and $f(1,1) \approx 1$, let us call it $\mathcal{C}$, along with the injection homomorphisms $h_\mathcal{A} : \mathcal{A} \to \mathcal{C}$ and $h_\mathcal{B} : \mathcal{B} \to \mathcal{C}$. Note that $|\mathcal{C}|_\phi$ is infinite and contains 0, 1, $f(0,1)$, $f(1,0)$, $f(0, f(0,1))$, and so on, but not $f(0,0)$ nor $f(1,1)$.

Consider now the Σ-structures $\langle \mathcal{A}, \emptyset \rangle$ and $\langle \mathcal{B}, \{1\} \rangle$. In $\mathbf{Str}(\Sigma)$ there are now many ways of endowing the coproduct $\mathcal{C}$ of the algebras with a set T of designated values that makes both homomorphisms preserve designated values. Indeed any $T \subseteq |\mathcal{C}|_\phi$ such that $1 \in T$ will do. However, as in the previous example, the minimal choice $T = \{1\}$ is canonical. Indeed, the coproduct $\langle \mathcal{A}, \emptyset \rangle \coprod \langle \mathcal{B}, \{1\} \rangle$ in $\mathbf{Str}(\Sigma)$ is precisely $\langle \mathcal{C}, \{1\} \rangle$, along with $h_\mathcal{A}$ and $h_\mathcal{B}$.

Easily, now, both $h_\mathcal{A}$ and $h_\mathcal{B}$ also reflect designated values. However, the choice of $T = \{1\}$ is not canonical in $\mathbf{CryptStr}(\Sigma)$ because any (bigger) choice will preserve but not reflect it. This now means that $id_\mathcal{C}$ is not a designation reflecting homomorphism from $\langle \mathcal{C}, \{1\} \rangle$ to $\langle \mathcal{C}, T \rangle$ for every $T \supseteq \{1\}$ with $0 \notin T$. This is the reason why a coproduct $\langle \mathcal{A}, \emptyset \rangle \coprod \langle \mathcal{B}, \{1\} \rangle$ in $\mathbf{CryptStr}(\Sigma)$ does not exist. △

It turns out that coequalizers do not raise these problems. They always exist in $\mathbf{CryptStr}(\Sigma)$ and coincide with those obtained in $\mathbf{Str}(\Sigma)$, by capitalizing on coequalizers in $\mathbf{Alg}(\Sigma)$ which are simple quotients. Absolutely similar situations reappear, however, when we need to change signatures, namely by considering the left adjoints F_σ of the reduct functors $\_|_\sigma : \mathbf{Alg}(\Sigma_2) \to \mathbf{Alg}(\Sigma_1)$ associated to signature morphisms $\sigma : \Sigma_1 \to \Sigma_2$. We do not show any examples here, due to space limitations, but we can just recall that given a Σ_1-algebra $\mathcal{A}$, $F_\sigma(\mathcal{A})$ is the Σ_2-algebra built by taking the free Σ_2-algebra over the disjoint union of all the carrier sets of $\mathcal{A}$ that σ maps to each sort, and then making its quotient under the congruence generated by the interpretation of terms in $\mathcal{A}$ translated by σ.

Despite all of this, we can still prove that coequalizers and arbitrary small coproducts exist in **Crypt**. We start with coproducts.

Proposition 1. Crypt *has small coproducts.*

Proof. Let I be a set and $\{\langle \Sigma_i, M_i, \mathbb{A}_i \rangle\}_{i \in I}$ a family of logic system presentations. Since we need to share the distinguished sort ϕ, we can consider the canonical signature $\Sigma^\phi = \langle \{\phi\}, \emptyset \rangle$ and the corresponding injections $\sigma_i^\phi : \Sigma^\phi \to \Sigma_i$. Since **Sig** is cocomplete, we can build its pushout $\{\sigma_i : \Sigma_i \to \Sigma\}_{i \in I}$. Now, we must define the class of combined models. For ease of notation we assume that $\mathbb{A}_i(m_i) = \langle \mathcal{A}_{i,m_i}, T_{i,m_i} \rangle$. For each tuple of models $m = \langle m_i \rangle_{i \in I} \in \prod_{i \in I} M_i$, we first need to combine the family $\{\langle \Sigma_i, \mathcal{A}_{i,m_i} \rangle\}_{i \in I}$ in **Alg**. Taking into account that the word algebra over Σ^ϕ is empty, we consider the corresponding injections $\langle \sigma_i^\phi, \emptyset \rangle : \langle \Sigma^\phi, \emptyset \rangle \to \langle \Sigma_i, \mathcal{A}_{i,m_i} \rangle$, and use the cocompleteness of **Alg** to compute their pushout $\{\langle \sigma_i, \eta_{i,m} \rangle : \langle \Sigma_i, \mathcal{A}_{i,m_i} \rangle \to \langle \Sigma, \mathcal{A}_m \rangle\}_{i \in I}$. Now we need to consider all possible compatible choices of designated values in $\mathcal{A}_m$, and define $M = \{\langle m, T \rangle : m \in \prod_{i \in I} M_i, T \subseteq |\mathcal{A}_m|_\phi, \eta_{i,m}^{-1}(T) = T_{i,m_i}$ for every $i \in I\}$, $\mu_i : M \to M_i$ such that $\mu_i(m, T) = m_i$, and $\mathbb{A}(m, T) = \langle \mathcal{A}_m, T \rangle$. We claim that $\{\langle \sigma_i, \mu_i, \eta_i \rangle : \langle \Sigma_i, M_i, \mathbb{A}_i \rangle \to \langle \Sigma, M, \mathbb{A} \rangle\}_{i \in I}$ is a coproduct of $\{\langle \Sigma_i, M_i, \mathbb{A}_i \rangle\}_{i \in I}$ in **Crypt**. Let us prove the corresponding universal property.

Assume given a logic system presentation $\langle \widehat{\Sigma}, \widehat{M}, \widehat{\mathbb{A}} \rangle$ and a family of cryptomorphisms $\{\langle \widehat{\sigma}_i, \widehat{\mu}_i, \widehat{\eta}_i \rangle : \langle \Sigma_i, M_i, \mathbb{A}_i \rangle \to \langle \widehat{\Sigma}, \widehat{M}, \widehat{\mathbb{A}} \rangle\}_{i \in I}$. For each $\widehat{m} \in \widehat{M}$, let $m = \langle m_i \rangle_{i \in I}$ with each $m_i = \widehat{\mu}_i(\widehat{m})$. Clearly, each composition $\widehat{\sigma}_i \circ \sigma_i^\phi$ maps ϕ to the distinguished sort of $\widehat{\Sigma}$. Thus, the universal property of the pushout in **Alg** guarantees the existence of a unique morphism $\langle \widehat{\sigma}, \widehat{\eta}_{\widehat{m}} \rangle : \langle \Sigma, \mathcal{A}_m \rangle \to \langle \widehat{\Sigma}, \widehat{\mathcal{A}}_{\widehat{m}} \rangle$ such that $\widehat{\sigma} \circ \sigma_i = \widehat{\sigma}_i$ and $\widehat{\eta}_{\widehat{m}} \circ \eta_{i,m} = \widehat{\eta}_{i,\widehat{m}}$ for each $i \in I$. The universal property of the pushout in **Sig** guarantees that the signature morphism $\widehat{\sigma}$ is the same for every $\widehat{m}$. We can now define $\widehat{\mu} : \widehat{M} \to M$ by $\widehat{\mu}(\widehat{m}) = \langle m, T \rangle$ with $T = \widehat{\eta}_{\widehat{m}}^{-1}(\widehat{T}_{\widehat{m}})$. This is well defined because each $\eta_{i,m}^{-1}(\widehat{\eta}_{\widehat{m}}^{-1}(\widehat{T}_{\widehat{m}})) = (\widehat{\eta}_{\widehat{m}} \circ \eta_{i,m})^{-1}(\widehat{T}_{\widehat{m}}) = \widehat{\eta}_{i,\widehat{m}}^{-1}(\widehat{T}_{\widehat{m}}) = T_{i,m}$. So, it is immediate that $\langle \widehat{\sigma}, \widehat{\mu}, \widehat{\eta} \rangle$ constitutes a cryptomorphism and composes with each $\langle \sigma_i, \mu_i, h_i \rangle$ into $\langle \widehat{\sigma}_i, \widehat{\mu}_i, \widehat{\eta}_i \rangle$. Uniqueness follows from the fact that $T = \widehat{\eta}_{\widehat{m}}^{-1}(\widehat{T}_{\widehat{m}})$ is the unique possible choice that fulfills the closedness condition for each $\widehat{\eta}_{\widehat{m}}$. See Figure 3 for a graphical representation of the construction.   □

We can now turn to coequalizers. The construction is a little simpler because there is no need to share the canonical signature Σ^ϕ: we start with a pair of parallel arrows in $\mathbf{AlgSig}_\phi$ that already preserve ϕ.

Alg   **Crypt**

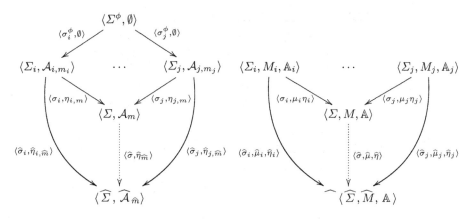

Fig. 3. Coproducts in **Crypt**

Proposition 2. Crypt *has coequalizers.*

Proof. Let $\langle \Sigma_1, M_1, \mathbb{A}_1 \rangle$ and $\langle \Sigma_2, M_2, \mathbb{A}_2 \rangle$ be logic system presentations, and consider a pair $\langle \sigma', \mu', \eta' \rangle, \langle \sigma'', \mu'', \eta'' \rangle : \langle \Sigma_1, M_1, \mathbb{A}_1 \rangle \to \langle \Sigma_2, M_2, \mathbb{A}_2 \rangle$ of cryptomorphisms. We first use the cocompleteness of **Sig** to build a coequalizer $\sigma : \Sigma_2 \to \Sigma$ of $\sigma', \sigma'' : \Sigma_1 \to \Sigma_2$. Then, for each model $m_2 \in M_2$ such that $\mu'(m_2) = \mu''(m_2) = m_1$, we can form in **Alg** the pair of arrows $\langle \sigma', \eta'_{m_2} \rangle, \langle \sigma'', \eta''_{m_2} \rangle : \langle \Sigma_1, \mathcal{A}_{1,m_1} \rangle \to \langle \Sigma_2, \mathcal{A}_{2,m_2} \rangle$ and take their coequalizer $\langle \sigma, \eta_{m_2} \rangle : \langle \Sigma_2, \mathcal{A}_{2,m_2} \rangle \to \langle \Sigma, \mathcal{A}_{m_2} \rangle$. Now we need to consider all possible compatible choices of designated values, and define $M = \{ \langle m_2, T \rangle : m_2 \in M_2, \mu'(m_2) = \mu''(m_2), T \subseteq |\mathcal{A}_{m_2}|_\phi, \eta^{-1}_{m_2}(T) = T_{2,m_2} \}$, $\mu : M \to M_2$ such that $\mu(m_2, T) = m_2$, and $\mathbb{A}(m_2, T) = \langle \mathcal{A}_{m_2}, T \rangle$. We claim that $\langle \sigma, \mu, \eta \rangle : \langle \Sigma_2, M_2, \mathbb{A}_2 \rangle \to \langle \Sigma, M, \mathbb{A} \rangle$ is a coequalizer of $\langle \sigma', \mu', \eta' \rangle$ and $\langle \sigma'', \mu'', \eta'' \rangle$ in **Crypt**. Checking that $\langle \sigma, \mu, \eta \rangle$ indeed coequalizes $\langle \sigma', \mu', \eta' \rangle$ and $\langle \sigma'', \mu'', \eta'' \rangle$ is routine. We are left with proving the corresponding universal property.

Assume that $\langle \widehat{\sigma}, \widehat{\mu}, \widehat{\eta} \rangle : \langle \Sigma_2, M_2, \mathbb{A}_2 \rangle \to \langle \widehat{\Sigma}, \widehat{M}, \widehat{\mathbb{A}} \rangle$ also coequalizes $\langle \sigma', \mu', \eta' \rangle$ and $\langle \sigma'', \mu'', \eta'' \rangle$. For each $\widehat{m} \in \widehat{M}$, let $m_2 = \widehat{\mu}(\widehat{m})$. It is clear that $\mu'(m_2) = \mu''(m_2)$. Since it must also be the case that $\langle \widehat{\sigma}, \widehat{\eta}_{\widehat{m}} \rangle : \langle \Sigma_2, \mathcal{A}_{2,m_2} \rangle \to \langle \widehat{\Sigma}, \widehat{\mathcal{A}}_{\widehat{m}} \rangle$ coequalizes $\langle \sigma', \eta'_{m_2} \rangle$ and $\langle \sigma'', \eta''_{m_2} \rangle$ in **Alg**, there exists a unique morphism $\langle \widehat{\widehat{\sigma}}, \widehat{\widehat{\eta}}_{\widehat{m}} \rangle : \langle \Sigma, \mathcal{A}_{m_2} \rangle \to \langle \widehat{\Sigma}, \widehat{\mathcal{A}}_{\widehat{m}} \rangle$ such that $\widehat{\widehat{\sigma}} \circ \sigma = \widehat{\sigma}$ and $\widehat{\widehat{\eta}}_{\widehat{m}} \circ \eta_{m_2} = \widehat{\eta}_{\widehat{m}}$. The universal property of the coequalizer in **Sig** guarantees that the signature morphism $\widehat{\widehat{\sigma}}$ is the same for every $\widehat{m}$. We can now define $\widehat{\widehat{\mu}} : \widehat{M} \to M$ by $\widehat{\widehat{\mu}}(\widehat{m}) = \langle m_2, T \rangle$ with $T = \widehat{\widehat{\eta}}_{\widehat{m}}^{-1}(\widehat{T}_{\widehat{m}})$. This is well defined because $\eta^{-1}_{m_2}(\widehat{\widehat{\eta}}_{\widehat{m}}^{-1}(\widehat{T}_{\widehat{m}})) = (\widehat{\widehat{\eta}}_{\widehat{m}} \circ \eta_{m_2})^{-1}(\widehat{T}_{\widehat{m}}) = \widehat{\eta}_{\widehat{m}}^{-1}(\widehat{T}_{\widehat{m}}) = T_{2,m_2}$. So, it is immediate that $\langle \widehat{\sigma}, \widehat{\mu}, \widehat{\eta} \rangle$ constitutes a cryptomorphism and composes with $\langle \sigma, \mu, \eta \rangle$ into $\langle \widehat{\sigma}, \widehat{\mu}, \widehat{\eta} \rangle$. Uniqueness

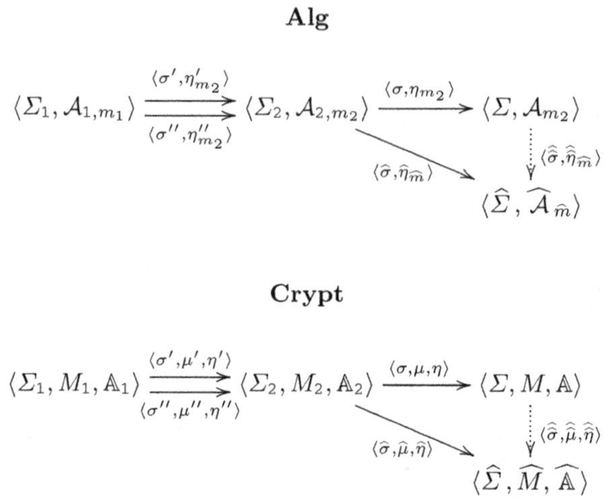

Alg

$$\langle \Sigma_1, \mathcal{A}_{1,m_1} \rangle \underset{\langle \sigma'', \eta''_{m_2} \rangle}{\overset{\langle \sigma', \eta'_{m_2} \rangle}{\rightrightarrows}} \langle \Sigma_2, \mathcal{A}_{2,m_2} \rangle \xrightarrow{\langle \sigma, \eta_{m_2} \rangle} \langle \Sigma, \mathcal{A}_{m_2} \rangle$$

$$\langle \widehat{\sigma}, \widehat{\eta}_{\widehat{m}} \rangle \searrow \qquad \Big\downarrow \langle \widehat{\sigma}, \widehat{\eta}_{\widehat{m}} \rangle$$

$$\langle \widehat{\Sigma}, \widehat{\mathcal{A}}_{\widehat{m}} \rangle$$

Crypt

$$\langle \Sigma_1, M_1, \mathbb{A}_1 \rangle \underset{\langle \sigma'', \mu'', \eta'' \rangle}{\overset{\langle \sigma', \mu', \eta' \rangle}{\rightrightarrows}} \langle \Sigma_2, M_2, \mathbb{A}_2 \rangle \xrightarrow{\langle \sigma, \mu, \eta \rangle} \langle \Sigma, M, \mathbb{A} \rangle$$

$$\langle \widehat{\sigma}, \widehat{\mu}, \widehat{\eta} \rangle \searrow \qquad \Big\downarrow \langle \widehat{\sigma}, \widehat{\mu}, \widehat{\eta} \rangle$$

$$\langle \widehat{\Sigma}, \widehat{M}, \widehat{\mathbb{A}} \rangle$$

Fig. 4. Coequalizers in **Crypt**

follows from the fact that $T = \widehat{\eta}_{\widehat{m}}^{-1}(\widehat{T}_{\widehat{m}})$ is the unique possible choice that fulfills the closedness condition for each $\widehat{\eta}_{\widehat{m}}$. See Figure 4 for a graphical representation of the construction.   □

Finally we can state the desired result.

Theorem 1. Crypt *is cocomplete.*

Of course the proofs above are not too informative with respect to the concrete result obtained in specific examples. To provide a better understanding of the power of the colimit construction in **Crypt** and work out a meaningful application of cryptomorphisms we shall revisit the example of [13, 14, 15] in the next section. However, we can already analyze the essential differences between colimits in **Crypt** and **MPRoom**, at the light of Examples 1 and 2.

In **MPRoom** it is always possible to obtain one canonical combined model: it features the minimal possible set of designated values, possibly at the expense of designating a value that was previously not designated. In **Crypt** it depends: no previously undesignated values can become designated by the construction, which means that in some cases the combined structure must simply be ignored; still, if that is not the case, any choice of newly generated values will provide a relevant set of designated values. We claim that this is precisely where the free interaction of the logics being combined emerges, as a result of the absolutely fundamental role played by the "junk" values. If we are combining logics that share a formula that is valid in one of them but has no model in the other then we are (and should be) in trouble, and the combination will trivialize. However, if no such inconsistencies exist, the resulting combined logic will encompass enough models to guarantee that any choice of new combined formulas is satisfiable, while still keeping intact the validities of each of the given logics.

4   Partial Equational Logic

We shall now borrow the example from [13, 14], used in [15] precisely to show that **LogMPar** was not complete. We choose this example not only because it was developed in this context, thus allowing us to sharpen the distinctions with respect to (logical) model theoretic parchments, but also because it concerns the relevant question of assigning a meaning to equations in the presence of undefined operations. Last but not least, although it is very simple, the end result provides a good illustration of the power of cryptomorphisms.

Of course, we recast the example at the level of rooms, and not of parchments as in the original formulation, and work it out in the category **Crypt**. Therefore, we consider fixed a many-sorted signature with partial operations, that is, a triple $\langle S, TO, PO \rangle$ such that both $\langle S, TO \rangle$ and $\langle S, PO \rangle$ are many-sorted signatures, respectively of total and partial operations, with $TO \cap PO = \emptyset$. The logic system presentation ALG of *total equational logic without equations* has sorts $S \cup \{\phi\}$ and operations TO, its models are precisely the $\langle S, TO \rangle$-algebras, and each $\langle S, TO \rangle$-algebra $\mathcal{A}$ is endowed with the structure $\langle \mathcal{A}^2, \{1\} \rangle$ where $|\mathcal{A}^2|_\phi = \{0, 1\}$ and $|\mathcal{A}^2|_s = |\mathcal{A}|_s$ for $s \in S$, with $f_{\mathcal{A}^2} = f_\mathcal{A}$ for each $f \in TO$. Although not very interesting *per se*, this logic system presentation is the common part of two other logic system presentations: ALG($\approx$) for *total equational logic*, and PALG for *partial equational logic without equations*. The idea is precisely to obtain a free combined semantics for *partial equational logic*. The example is particularly well set since the colimit of ALG($\approx$) and ALG while sharing ALG focuses precisely on the missing bit: the interpretation of equations involving undefined values.

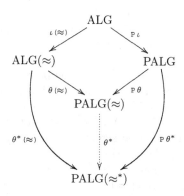

Fig. 5. The partial equational logic pushout

Now, ALG($\approx$) has sorts $S \cup \{\phi\}$ and operations TO plus $\approx: s \times s \to \phi$ for each $s \in S$, the models are also the $\langle S, TO \rangle$-algebras, and each $\langle S, TO \rangle$-algebra $\mathcal{A}$ is endowed with the structure $\langle \mathcal{A}^2_\approx, \{1\} \rangle$ where $\mathcal{A}^2_\approx$ extends $\mathcal{A}^2$ by

$$\approx_{\mathcal{A}^2_\approx} (x, y) = \begin{cases} 1 & \text{if} \quad x = y \\ 0 & \text{if} \quad x \neq y \end{cases}.$$

Clearly, $\mathcal{A} \Vdash_{\text{ALG}(\approx)} t \approx t'$ if and only if $[\![t]\!]_{\mathcal{A}} = [\![t']\!]_{\mathcal{A}}$. We denote by $\iota(\approx)$: ALG $\rightarrow$ ALG$(\approx)$ the obvious cryptomorphism that injects the signatures, is the identity on models, and also the identity on each structure.

On its turn, PALG has sorts $S \cup \{\phi\}$ and operations $TO \cup PO$, the models are also precisely the $\langle S, TO, PO \rangle$-partial algebras, and each $\langle S, TO, PO \rangle$-partial algebra $\mathcal{B}$ is endowed with the structure $\langle \mathcal{B}^2, \{1\} \rangle$ where $|\mathcal{B}^2|_{\phi} = \{0, 1\}$ and $|\mathcal{B}^2|_s = |\mathcal{B}|_s \uplus \{\perp_s\}$ for $s \in S$, with

$$f_{\mathcal{B}^2}(\overrightarrow{x}) = \begin{cases} f_{\mathcal{B}}(\overrightarrow{x}) & \text{if all } x_i \neq \perp_{s_i} \text{ and } f_{\mathcal{B}}(\overrightarrow{x}) \downarrow \\ \perp_s & \text{otherwise} \end{cases}$$

for each $f : s_1 \times \cdots \times s_n \rightarrow s$ in $TO \cup PO$. We denote by $\mathrm{P}\iota :$ ALG $\rightarrow$ PALG the obvious cryptomorphism that injects the signatures, forgets the partial operations, and injects the corresponding structures.

The desired result should therefore correspond to the pushout of $\iota(\approx)$ and $\mathrm{P}\iota$ in **Crypt**, which is actually not very difficult to compute. First of all, we have to combine the signatures of ALG$(\approx)$ and PALG while sharing their common subsignature ALG. We end up with sorts $S \cup \{\phi\}$ and operations $TO \cup PO$ plus $\approx: s \times s \rightarrow \phi$ for each $s \in S$. Let us consider a pair of models, $\mathcal{A}$ from ALG$(\approx)$ and $\mathcal{B}$ from PALG that coincide when mapped to ALG, that is, $\mathcal{A}$ is precisely the restriction of $\mathcal{B}$ to the total operations. The corresponding combined algebra $\mathcal{C}$ will include, besides $\{0, 1\}$, a whole new set of freely generated truth-values corresponding to the new denotations of $\approx$ involving undefined values, that is, $V = \bigcup_{s \in S} \{x \approx \perp_s : x \in |\mathcal{B}|_s\} \cup \{\perp_s \approx x : x \in |\mathcal{B}|_s\} \cup \{\perp_s \approx \perp_s\}$. In detail, $\mathcal{C}$ is such that $|\mathcal{C}|_s = |\mathcal{B}|_s \uplus \{\perp_s\}$ for $s \in S$, $|\mathcal{C}|_{\phi} = \{0, 1\} \cup V$, $f_{\mathcal{C}} = f_{\mathcal{B}^2}$ and

$$\approx_{\mathcal{C}} (x, y) = \begin{cases} 1 & \text{if } x = y \text{ and } x \neq \perp_s \text{ and } y \neq \perp_s \\ 0 & \text{if } x \neq y \text{ and } x \neq \perp_s \text{ and } y \neq \perp_s \\ x \approx y & \text{otherwise} \end{cases}$$

Therefore, the combined models can be seen as pairs $\langle \mathcal{B}, T \rangle$ with $T \subseteq V$ representing any possible choice of new designated values. The structure associated to each pair $\langle \mathcal{B}, T \rangle$ is precisely $\langle \mathcal{C}, \{1\} \cup T \rangle$. It is straightforward to set up the inclusion cryptomorphisms $\theta(\approx) :$ ALG$(\approx) \rightarrow$ PALG$(\approx)$ and $\mathrm{P}\theta :$ PALG $\rightarrow$ PALG$(\approx)$: $\theta(\approx)$ is the inclusion on signatures, maps each model $\langle \mathcal{B}, T \rangle$ to the restriction $\mathcal{A}$ of $\mathcal{B}$ to the total operations, and then injects $\mathcal{A}_{\approx}^2$ into $\mathcal{C}$; $\mathrm{P}\theta$ is also the inclusion on signatures, maps each model $\langle \mathcal{B}, T \rangle$ to $\mathcal{B}$, and then injects $\mathcal{B}^2$ into $\mathcal{C}$. It is clear that $\theta(\approx) \circ \iota(\approx) = \mathrm{P}\theta \circ \mathrm{P}\iota$.

Proposition 3. *The logic system presentation* PALG$(\approx)$ *together with the cryptomorphisms* $\theta(\approx)$ *and* $\mathrm{P}\theta$ *constitutes a pushout of* $\iota(\approx)$ *and* $\mathrm{P}\iota$ *in* **Crypt***.*

The universal property of the construction of PALG$(\approx)$ can be interpreted as follows. Choose your favorite interpretation of partial equations, and define with it a logic system presentation PALG$(\approx^*)$. One can of course imagine very strange situations, but one can impose as a minimal requirement that the choice is at least based on partial algebras, and that it extends the usual interpretation of total equations. In that case, it should be routine to define two cryptomorphisms $\theta^*(\approx) : \mathrm{ALG}(\approx) \to \mathrm{PALG}(\approx^*)$ and $\mathrm{P}\theta^* : \mathrm{PALG} \to \mathrm{PALG}(\approx^*)$ such that $\theta^*(\approx) \circ \iota(\approx) = \mathrm{P}\theta^* \circ \mathrm{P}\iota$. Therefore, the construction guarantees that there exists precisely one compatible cryptomorphism $\theta^* : \mathrm{PALG}(\approx) \to \mathrm{PALG}(\approx^*)$, which means that the chosen interpretation of partial equations corresponds to a particular choice of models in $\mathrm{PALG}(\approx)$. Let us see how this goes for some of the most common interpretations of partial equations.

Example 3. The logic system presentations $\mathrm{PALG}(\approx^*)$, with $* \in \{w, s, e, 3\}$, of *weak, strong, existential, and strict three-valued partial equational logic*, respectively, all have sorts $S \cup \{\phi\}$ and operations $TO \cup PO$ plus $\approx: s \times s \to \phi$ for each $s \in S$, and the $\langle S, TO, PO \rangle$-partial algebras as models. They differ from $\mathrm{PALG}(\approx)$, and between each other, on the interpretation structures associated to each model $\mathcal{B}$. In the sequel, $\mathcal{A}$ always stands for the restriction of $\mathcal{B}$ to the partial operations.

In the weak case, $\mathcal{B}$ is endowed with $\langle \mathcal{B}^{2,w}_{\approx}, \{1\} \rangle$ where $\mathcal{B}^{2,w}_{\approx}$ extends $\mathcal{B}^2$ by

$$\approx_{\mathcal{B}^{2,w}_{\approx}} (x, y) = \begin{cases} 1 & \text{if } x = y \text{ or } x = \bot_s \text{ or } y = \bot_s \\ 0 & \text{otherwise} \end{cases}.$$

Note that $\mathcal{B} \Vdash_{\mathrm{PALG}(\approx^w)} t \approx t'$ if and only if $[\![t]\!]_{\mathcal{B}} = [\![t']\!]_{\mathcal{B}}$ or at least one of them is undefined. The cryptomorphisms $\theta^w(\approx)$ and $\mathrm{P}\theta^w$ simply inject $\mathcal{A}^2_{\approx}$ and $\mathcal{B}^2$ into $\mathcal{B}^{2,w}_{\approx}$. The unique compatible cryptomorphism θ^w maps each $\mathcal{B}$ to $\langle \mathcal{C}, \{1\} \cup V \rangle$, and then all the values in V, from $\mathcal{C}$, to 1 in $\mathcal{B}^{2,w}_{\approx}$.

In the strong case, the structure is $\langle \mathcal{B}^{2,s}_{\approx}, \{1\} \rangle$ where $\mathcal{B}^{2,s}_{\approx}$ extends $\mathcal{B}^2$ by

$$\approx_{\mathcal{B}^{2,s}_{\approx}} (x, y) = \begin{cases} 1 & \text{if } x = y \\ 0 & \text{otherwise} \end{cases}.$$

Note that $\mathcal{B} \Vdash_{\mathrm{PALG}(\approx^s)} t \approx t'$ if and only if $[\![t]\!]_{\mathcal{B}} = [\![t']\!]_{\mathcal{B}}$ or both are undefined. The cryptomorphisms $\theta^s(\approx)$ and $\mathrm{P}\theta^s$ simply inject $\mathcal{A}^2_{\approx}$ and $\mathcal{B}^2$ into $\mathcal{B}^{2,s}_{\approx}$. The cryptomorphism θ^s maps each $\mathcal{B}$ to $\langle \mathcal{C}, \{1\} \cup \{\bot_s \approx \bot_s : s \in S\} \rangle$, and then all the values in $V \setminus \{\bot_s \approx \bot_s : s \in S\}$ to 0 in $\mathcal{B}^{2,w}_{\approx}$, and $\{\bot_s \approx \bot_s : s \in S\}$ to 1.

In the existential case, the structure is $\langle \mathcal{B}^{2,e}_{\approx}, \{1\} \rangle$ where $\mathcal{B}^{2,e}_{\approx}$ extends $\mathcal{B}^2$ by

$$\approx_{\mathcal{B}^{2,e}_{\approx}} (x, y) = \begin{cases} 1 & \text{if } x = y \text{ and } x \neq \bot_s \text{ and } y \neq \bot_s \\ 0 & \text{otherwise} \end{cases}.$$

Note that $\mathcal{B} \Vdash_{\mathrm{PALG}(\approx^e)} t \approx t'$ if and only if $[\![t]\!]_{\mathcal{B}} = [\![t']\!]_{\mathcal{B}}$ with both values defined. The cryptomorphisms $\theta^e(\approx)$ and $\mathrm{P}\theta^e$ again simply inject $\mathcal{A}^2_{\approx}$ and $\mathcal{B}^2$ into $\mathcal{B}^{2,e}_{\approx}$. The unique cryptomorphism θ^e now maps each $\mathcal{B}$ to $\langle \mathcal{C}, \{1\} \rangle$, and then all the values in V to 0 in $\mathcal{B}^{2,e}_{\approx}$.

The strict three-valued case is slightly different. Each $\mathcal{B}$ is now endowed with the structure $\langle \mathcal{B}^3_{\approx}, \{1\} \rangle$ where $|\mathcal{B}^3_{\approx}|_{\phi} = \{0, 1\} \uplus \{\bot\}$ and $|\mathcal{B}^3_{\approx}|_s = |\mathcal{B}|_s \uplus \{\bot_s\}$ for $s \in S$, with $f_{\mathcal{B}^3_{\approx}} = f_{\mathcal{B}^2}$ for each $f \in TO \cup PO$ and

$$\approx_{\mathcal{B}^3_{\approx}} (x, y) = \begin{cases} 1 & \text{if } x = y \text{ and } x \neq \bot_s \text{ and } y \neq \bot_s \\ 0 & \text{if } x \neq y \text{ and } x \neq \bot_s \text{ and } y \neq \bot_s \\ \bot & \text{otherwise} \end{cases}.$$

Note that $\mathcal{B} \Vdash_{\mathrm{PALG}(\approx^3)} t \approx t'$ if and only if $\mathcal{B} \Vdash_{\mathrm{PALG}(\approx^e)} t \approx t'$. The cryptomorphisms $\theta^3(\approx)$ and $\mathrm{P}\theta^3$ inject $\mathcal{A}_\approx^2$ and $\mathcal{B}^2$ into $\mathcal{B}_\approx^3$. The cryptomorphism θ^3 maps each $\mathcal{B}$ to $\langle \mathcal{C}, \{1\} \rangle$, and then all the values in V from $\mathcal{C}$ to $\bot$ in $\mathcal{B}_\approx^3$. $\qquad\triangle$

Note however that the combination obtained is so absolutely free that less orthodox choices are also possible, namely asymmetric ones, or choices that consider different solutions for each sort.

5   Conclusion

We have shown that the cryptomorphisms proposed in [5] really work, in the sense that they set up a category of logic system presentations that is cocomplete, together with the fact that they always fulfill the usual satisfaction condition. This implies that cryptomorphisms give rise to a complete category of parchments that easily translates to the category of institutions. Therefore, limits in this category of parchments always exist, and constitute a very powerful mechanism for combining logics that extends fibring along the lines of [5]. Not only the syntaxes of the given logics are freely combined, but also their semantics. Undesired collapses are also avoided, as long as shared formulas have a uniform semantics in the logics being combined. A solution to the paradigmatic collapsing situation in the combination of classical and intuitionistic logics, using cryptomorphisms, can be found in [5]. We have also put cryptomorphisms in context with the notions of morphism and logical morphism arising from the work on model-theoretic parchments, and explained the absolutely fundamental role played by "junk" values in the freeness of the colimits obtained using cryptomorphisms, in contrast with the logicality constraints advocated in [15]. The approach was illustrated using a simple but meaningful partial equational logic example, whose result encompasses models that are compatible with every possible interpretation of equality involving undefinedness, even if less standard.

Nevertheless, we agree that the proliferation of truth-values can be seen, at least, as annoying. Moreover, the freeness of the construction really takes advantage of this fact in allowing possibly less orthodox choices of newly designated values. But there are certainly other ways to prevent unorthodox choices, given any reasonable notion of orthodoxy. The subsequent use of congruence relations, as in [13], is one of them. As usual, each uniform congruence on the resulting combined structures can be seen as the outcome of a corresponding cryptomorphism. The cryptomorphisms θ^* do precisely that in the example of the previous section. Still, there are more interesting possibilities. One of them, certainly worth pursuing, is to consider representations of all the logics involved in a *universal* logic, as proposed in [14, 15]. In alternative, we can work along with calculi associated to each of the logics, and require their soundness as a minimal requirement, as done in [5] for cryptofibring. This last possibility also opens the way to incorporating and extending the soundness and completeness preservation results well-known for fibring to this wider context.

Acknowledgment

The authors are grateful to Till Mossakowski and Andrzej Tarlecki for useful discussions on this topic.

References

1. P. Blackburn and M. deRijke. Why combine logics? *Studia Logica*, 59(1):5–27, 1997.
2. C. Caleiro. *Combining Logics*. PhD thesis, IST, TU Lisbon, 2000.
3. C. Caleiro, P. Gouveia, and J. Ramos. Completeness results for fibred parchments: Beyond the propositional base. In *Recent Trends in Algebraic Development Techniques*, volume 2755 of *LNCS*, pages 185–200. Springer, 2003.
4. C. Caleiro, P. Mateus, J. Ramos, and A. Sernadas. Combining logics: Parchments revisited. In *Recent Trends in Algebraic Development Techniques*, volume 2267 of *LNCS*, pages 48–70. Springer, 2001.
5. C. Caleiro and J. Ramos. Cryptofibring. In *Proceedings of CombLog'04, Workshop on Combination of Logics: Theory and Applications*, pages 87–92, 2004. Extended abstract. Available at http://wslc.math.ist.utl.pt/ftp/pub/CaleiroC/04-CR-fiblog25s.pdf.
6. L. del Cerro and A. Herzig. Combining classical and intuitionistic logic. In *Frontiers of Combining Systems*, pages 93–102. Kluwer, 1996.
7. D. Gabbay. Fibred semantics and the weaving of logics: part 1. *Journal of Symbolic Logic*, 61(4):1057–1120, 1996.
8. D. Gabbay. An overview of fibred semantics and the combination of logics. In *Frontiers of Combining Systems*, pages 1–55. Kluwer, 1996.
9. D. Gabbay. *Fibring Logics*. Clarendon, 1999.
10. J. Goguen. A categorical manifesto. *Mathematical Structures in Computer Science*, 1:49–67, 1991.
11. J. Goguen and R. Burstall. A study in the foundations of programming methodology: specifications, institutions, charters and parchments. In *Category Theory and Computer Programming*, volume 240 of *LNCS*, pages 313–333. Springer, 1986.
12. J. Goguen and R. Burstall. Institutions: abstract model theory for specification and programming. *Journal of the ACM*, 39(1):95–146, 1992.
13. T. Mossakowski. Using limits of parchments to systematically construct institutions of partial algebras. In *Recent Trends in Data Type Specification*, volume 1130 of *LNCS*, pages 379–393. Springer, 1996.
14. T. Mossakowski, A. Tarlecki, and W. Pawłowski. Combining and representing logical systems. In *Category Theory and Computer Science 97*, volume 1290 of *LNCS*, pages 177–196. Springer, 1997.
15. T. Mossakowski, A. Tarlecki, and W. Pawłowski. Combining and representing logical systems using model-theoretic parchments. In *Recent Trends in Algebraic Development Techniques*, volume 1376 of *LNCS*, pages 349–364. Springer, 1998.
16. A. Sernadas, C. Sernadas, and C. Caleiro. Fibring of logics as a categorial construction. *Journal of Logic and Computation*, 9(2):149–179, 1999.
17. C. Sernadas, J. Rasga, and W. Carnielli. Modulated fibring and the collapsing problem. *Journal of Symbolic Logic*, 67(4):1541–1569, 2002.
18. A. Tarlecki. Moving between logical systems. In *Recent Trends in Data Type Specification*, volume 1130 of *Lecture Notes in Computer Science*, pages 478–502. Springer, 1996.

19. A. Tarlecki, R. Burstall, and J. Goguen. Some fundamental algebraic tools for the semantics of computation: Part 3. indexed categories. *Theoretical Computer Science*, 91:239–264, 1991.

20. R. Wójcicki. *Theory of Logical Calculi*. Kluwer, 1988.

21. A. Zanardo, A. Sernadas, and C. Sernadas. Fibring: Completeness preservation. *Journal of Symbolic Logic*, 66(1):414–439, 2001.

Towards a Formal Specification of an Electronic Payment System in CSP-CASL

Andy Gimblett[1], Markus Roggenbach[1], and Bernd-Holger Schlingloff[2]

[1] University of Wales Swansea,
Department of Computer Science, United Kingdom
[2] Fraunhofer Institute FIRST and Humboldt University at Berlin, Germany

Abstract. This paper describes the formal specification of a future banking system by abstract data types and process algebra. In contrast to previous exercises (e.g., [1]), the system's description is an actual industrial standard which is being used to develop the next generation of automatic banking machines. The specification language CSP-CASL is particularly well suited to this type of problem, since it combines both control and data aspects and allows loose specification of data types for later refinement. During the formalisation, several inconsistencies and ambiguities were exhibited. The obtained specification serves as a starting point for further validation.

1   Introduction

Electronic payment systems represent an important application area for both the theory and practice of system specification. In theory, they provide a suitable benchmark to demonstrate the abilities of a certain specification method (consider e.g. [1, 5, 7]). In practice, they are classified as safety critical systems and thus must be developed with due diligence. In this paper we consider such an application by studying in detail how to build a formal specification for the electronic payment system ep2 [2], a new international standard developed by a consortium of leading Swiss finance institutes.

ep2 is typical of a number of similar applications. The system consists of seven autonomous entities centred around the ep2 *Terminal*: Cardholder (i.e., customer), Point of Service (i.e., cash register), Attendant, POS Management System, Acquirer, Service Center, and Card, see Fig. 1. These entities communicate with the Terminal and, to a certain extent, with one another via *XML-messages* in a fixed format. These messages contain information about authorisation, financial transactions, as well as initialisation and status data. The state of each component heavily depends on the content of the exchanged data. Each component is a *reactive system* defined by a number of *use cases*. Thus, there are both reactive parts and data parts which need to be modelled, and these parts are heavily intertwined.

The ep2 system also represents a typical industrial case study. The specification consists of roughly 600 pages of text, which is a mixture of plain English and

J.L. Fiadeiro, P. Mosses, and F. Orejas (Eds.): WADT 2004, LNCS 3423, pp. 61–78, 2005.

other semi-formal notation. Some parts are specified up to a bit encoding level, while others are left open and referred to common understanding. It is, however, an actual international standard which is used to implement and validate banking machines from different manufacturers.

In the formalisation, we use the specification language Csp-Casl [22]. This language combines process algebraic specification of reactive behaviour and algebraic specification of data types at various levels of detail. Csp-Casl uses the process algebra Csp [10, 23] for the modelling of reactive behaviour, whereas the properties of the communications are specified in Casl [3, 16]. Csp-Casl is generic in the Csp semantics. Furthermore, Csp-Casl offers a notion of refinement with clear relations to both data refinement in Casl and process refinement in Csp.

Structuring our Csp-Casl specifications in nearly the same way as the original ep2 documents allows us to exhibit several ambiguities, omissions, and contradictions in the documents. Here, especially Csp-Casl's loose specification of data types plays an important role. Often, the top level ep2 documents provide only an overview of the data involved, while the presentation of further details for a specific type is delayed to separate low-level documents. Csp-Casl is able to match such a document structure by a library of specifications, where the informal design steps of the ep2 specification are mirrored in terms of a formal refinement relation.

The paper is structured as follows. First, we give an overview of the ep2 system, where we focus on the existing specification and the shortcomings thereof. Then, we quickly review the specification language Csp-Casl. In section 3, we describe our formalization, and in section 4 we report on our results with this formalization. Finally, we summarize our results, discuss related approaches, and conlude with hints on future work and perspectives.

2   The ep2 System

ep2 stands for 'EFT/POS 2000', short for 'Electronic Fund Transfer/Point Of Service 2000', and is a joint project established by a number of (mainly Swiss) financial institutes and companies in order to define EFT/POS infrastructure for credit, debit, and electronic purse terminals in Switzerland (www.eftpos2000.ch). ep2 builds on a number of other standards, most notably EMV 2000 (the Europay/Mastercard/Visa Integrated Circuit Card standard, see www.emvco.com) and various ISO standards. An overview of ep2 is shown in Fig 1.

2.1   ep2 Document Structure

The ep2 specification consists of twelve documents, each of which either considers some particular component of the system in detail, or considers some aspect common to many or all components. The Terminal, Acquirer, POS Management System, Point of Service (POS), and Service Center components all have specification documents setting out 'general', 'functional', and 'supplementary' requirements, where the functional requirements carry the most detail, and consist mainly of use cases discussing how that particular component behaves in various situations. As well as the specifications of particular components, there is a Security Specification, an Interface Specification, and a Data Dictionary.

One obvious characteristic of such a document structure is that, when considering some aspect of the system, the information required to understand that aspect is contained in several different documents, each of which has its own things to say about the situation in question. For example, in order to gather all information about the SI-Init interface between Terminal and Acquirer, see Fig. 1, one has to examine the Terminal Specification, the Acquirer Specification, the Interface Specification, and the Data Dictionary. As we will see, this approach easily leads to inconsistencies and ambiguities.

2.2 ep2 Specification Style

The original ep2 documents are comprised of a number of different specification notations: plain English; UML-like graphics (use cases, activity diagrams, mes-

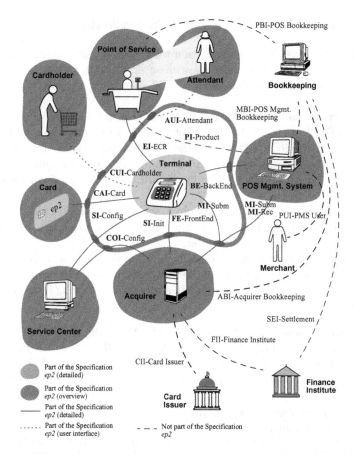

Fig. 1. Overview of the ep2 System, following closely [2]

sage sequence charts, class models, etc.); pictures; tables; lists; file descriptions; encoding rules.

Subsequently, we will focus on the *SI-Init* connection between *Terminal* and *Acquirer* (see Fig. 1).

The Acquirer is defined in a table of roles as a *"Card processor, which runs a system for processing of electronic payment transactions. The Acquirer is in contact with the merchant."* Later, in another table describing the main system features, the functionality of the Acquirer is classified into four subsystems:

- *Acquirer Initialisation System:* Supports remote SW-parameter initialisation. Exchanges Terminal configuration data with the Service Center.
- *Authorisation System:* Processes Terminal on-line authorisation requests, as well as transaction reversal requests. Forwards issuer scripts to the Terminal.
- *Submission System:* Processes transactions.
- *Reconciliation System:* Provides reconciliation[1] data to the merchant.

In the Acquirer general requirements document, a fifth subsystem is identified:

- *COI* [2] *server:* Used for data exchange with the Service Center.

Another table lists the communication interfaces; in particular, *"The SI-Init interface is used by the Acquirer to download application specific initialisation data which include Acquirer data necessary for Acquirer authentication and data submission."*

Later in the System specification, this communication is depicted in a use case, seen in Fig 2. It shows that the "Get Initialization Function" can be called by the service man either directly at the Terminal, or via an "Initiate Terminal Setup" at the Point of Service. Additionally, the function can be called in cyclic intervals by a timer process, or by an authentication server process at the Acquirer's site.

For both the Terminal and the Acquirer, activity diagrams are given describing the flow of control on the receipt of messages. For conciseness, in Fig. 3 we only show the diagram for the Acquirer.

For each state in this activity diagram, a verbal description is given of which message parameters are admissible in this state, and what the appropriate answer messages are composed of. For example, in state "Send ≪Config Data Request≫ Message":

> The Acquirer shall send the message ≪Config Data Request≫ to the Terminal. The Acquirer shall set <Config Data Object> to the configuration data object which the Acquirer is interested in. For CPTD, TACD and CAD the Acquirer shall specify with an AID resp. a RID which table

---

[1] Reconciliation: to compare the business undertaken at the terminal with that recorded by the acquirer and credited to the merchant's bank account.

[2] COI stands for *configuration and initialisation* (of the terminal) within the ep2 specification.

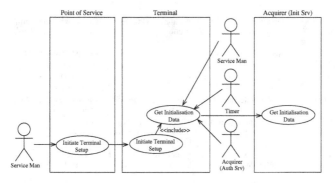

Fig. 2. Part of use case for Get Initialisation Function as shown in [2]

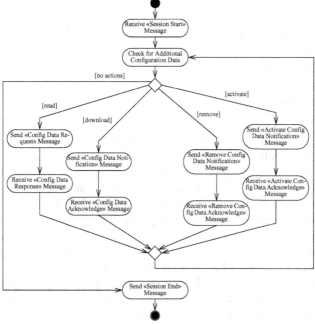

Fig. 3. Activity diagram for Acquirer getting initialisation data, as shown in [2]

*exactly it wants to receive. If the Acquirer sets <Config Data Object> to
LAID, he receives a list of all AID's supported by him from the Terminal.*
. . .

The appropriate parameter values are informally described in another table, the
beginning of which is given in Fig. 4.

On the concrete data encoding level, the SI-Init connection is constrained by
the following requirements in the system description:

<Config Data Object> value	Object Name	Additional Data Element	Returned by Terminal
ACD	Acquirer Config Data	-	One ACD object of the requesting acquirer
AISD	Acquirer Init Srv Data	-	One AISD object of the requesting acquirer
CPTD	Card Profile Table Data	<Application Identifier (AID)>	One CPTD object of the requested AID.
CAD	Certification Auth Data	<Registered Application Provider Identifier (RID)>	One CAD object of the requested RID.
TACD	Terminal Application Config Data	<Application Identifier (AID)>	One TACD object of the requested AID.
...	...	...	...

Fig. 4. Excerpt of message parameters and expected answers for initialisation [2]

- ep2 interface.
- Uses XML based on TCP/IP.
- Message based.
- Uses strong security mechanisms.

2.3   Shortcomings

The above specification style is typical for a number of today's industrial developments. As described above, it uses a number of up-to-date specification notations, and has additional verbal explanations and cross references throughout the books. However, for a team of developers which has to rely on this specification as a sole basis for an implementation it may be hard to produce a correct result. (A typical scenario would be a company which is not part of the consortium and wishes to produce a compliant device). Some of the reasons for this are:

First, there are several *ambiguities* within the documents which could lead to interoperability problems between different implementations. Ambiguities are inherent in all natural-language documents, since human language is subject to individual interpretation. As an example, consider the expected answer *"One ACD object of the requesting Acquirer"* in Fig. 4. This could mean

- One object, and it must be the one of the requesting Acquirer.
- One of all the objects belonging to the requesting Acquirer.

(In mathematical logic the difference is formalized by Russell's jota- and Hilbert's eta-operators.) Different opinions about the meaning of this requirement could lead to incompatible implementations.

Worse, there are some *inconsistencies* within the documents themselves. In fact, the data flow for the "Acquirer Init Srv Data" message is specified in the data dictionary as *from* the Acquirer via Service Center *into* the Terminal. In the above activity diagram, the Acquirer is allowed to *read* this data object *from* the Terminal. It contains the Acquirer's identifier, public key and communication address. The only plausible reason for the Acquirer to receive this data is to check its consistency. However, the Acquirer has no way to initiate a correction of these data, even if an inconsistency is detected. Since the specification is rather large and was written by several authors, such situations cannot be avoided.

Third, the ep2 documents are not suitable for *tool supported software development*. In particular, since the various requirements are intermingled, they cannot be easily input into an automated requirement management system such as Telelogic's DOORS or IBM/Rational's Requisite Pro. Thus, it is hard to assure that all required functionality is present in an implementation. Moreover, it is not possible to automatically check consistency of the requirements with one another, or to prove the conformance of a particular implementation with respect to the specification.

Last but not least, the given documents *interleave different levels of abstraction*. For example, the above mentioned architecture of the Acquirer is augmented by "logical component requirements" such as permanent accessibility, as well as use cases and a data model. Thus, it is not easy to use the specification in a structured development process. In fact, since implementation details are to be found throughout the specification, a programmer might be forced to reinvent parts which have already been developed by others. Moreover, implementation details are often subject to change; thus, the whole ep2 specification must be updated whenever some detail is modified. This can result in serious version compatibility problems.

3   CSP-CASL

CSP-CASL [22] is a comprehensive language which combines the specification of *data types* in CASL [3, 16] with *processes* written in CSP [10, 23]. The general idea of this language combination is to describe reactive systems in the form of processes based on CSP operators, but where the communications between these processes are the values of data types, which are loosely specified in CASL. All standard CASL features are available for the specification of these data types, namely many-sorted FOL with sort-generation constraints, partiality, and subsorting[3]. Furthermore, the various CASL structuring constructs can be used to describe data types within CSP-CASL. This includes the structured **free** construct, which adds the possibility to specify data types with initial semantics. For the description of processes, the typical CSP operators are included in CSP-CASL: there are for instance internal choice and external choice; the various parallel operators like the interleaving operator, the alphabetized parallel operator, and the general parallel operator; also communication over channels is included. Similarly to CASL, CSP-CASL specifications can be organized in libraries. Indeed, it is possible to mix CASL specifications and CSP-CASL specifications in

---

[3] For technical reasons, in CSP-CASL sub-sorting is restricted to subsort relations with so-called top elements. As it turns out e.g. in our current case study of specifying ep2, this restriction is of no practical relevance.

one library, separating the development of data types in CASL from their use within CSP-CASL. This allows the specification of a complex system like ep2 in a modular way.

Syntactically, a CSP-CASL specification with name N consists of a data part Sp, which is a structured CASL specification, an (optional) channel part Ch to declare channels, which are typed according to the data specification, and a process part P written in CSP, within which CASL terms are used as communications, CASL sorts denote sets of communications, relational renaming is described by a binary CASL predicate, and the CSP conditional construct uses CASL formulae as conditions:

ccspec N = data Sp channel Ch process P end

See Fig. 6 for a concrete instance of such a scheme. In the process part, the `let ... in ...` construct offers the possibility for recursive process definitions. Processes can also be parameterized with variables typed by CASL sorts. In general, this combination of recursion and parameterization leads to an infinite system of process equations. The theory of CSP offers syntactic characterizations for the existence and uniqueness of solutions for such systems of equations.

As a consequence of the loose semantics of CASL, *semantically* a CSP-CASL specification is a family of process denotations for a CSP process, where each model of the data part Sp gives rise to one process denotation. The definition of the language CSP-CASL is generic in the choice of a specific CSP semantics. For example, all denotational CSP models mentioned in [23] are possible parameters. For the purpose of specifying ep2 in CSP-CASL, we mainly use the CSP denotational stable-failures model. This model is able to distinguish between the different choice operators, and allows for infinite non-determinism as well as for infinite communication alphabets: features which naturally appear in abstract system descriptions involving loosely specified data types.

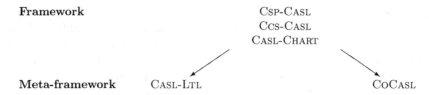

Fig. 5. Relationship between CSP-CASL and other reactive CASL extensions

Related Specification Languages. Within the context of CASL, various reactive extensions were proposed – see Figure 5 for a small selection and classification. Our definition of CSP-CASL, like CCS-CASL [24, 25] and CASL-CHART [20], combines CASL with a particular mechanism to describe reactive systems. This results in a *Framework* suitable to model real-world systems. CASL-LTL [19] and CoCASL [21, 15], on the other side, can be seen as a *Meta-framework* aiming

```
1   ccspec ep2 =
2   data sorts D_CAI_Card; D_SI_Config; D_SI_Init; D_FE_FrontEnd;
3               D_MI_Subm; D_BE_BackEnd; D_EI_ECR;  D_COI_Config; D_MI_Rec;
4       free type D_MI_Subm_or_Rec ::=
5           subm(select_subm:? D_MI_Subm) | rec (select_red:?  D_MI_Rec);
6   channels C_CAI_Card: D_CAI_Card;      C_SI_Config: D_SI_Config;
7            C_SI_Init: D_SI_init;        C_FE_FrontEnd;
8            C_MI_Subm: D_MI_Subm;        C_BE_BackEnd: D_BE_BackEnd;
9            C_EI_ECR: D_EI_ECR;          C_COI_Config: D_COI_Config;
10           C_MI_Subm_or_Rec: D_MI_Subm_or_Rec;
11  process
12    let Card          =    Run(C_CAI_Card)
13        ServiceCenter =    Run(C_SI_Config) ||| Run(C_COI_Config)
14        Acquirer      =    Run(C_COI_Config) ||| Run(C_SI_Init)
15                      ||| Run(C_FE_FrontEnd) ||| Run(C_MI_Subm)
16                      ||| Run(C_MI_Subm_or_Rec)
17        PosMgmtSystem =    Run(C_BE_BackEnd) ||| Run(C_MI_Subm_or_Rec)
18        PointOfService =   Run(C_EI_ECR)
19        Terminal      =    Run(C_CAI_Card)   ||| Run(C_SI_Config)
20                      ||| Run(C_SI_Init)     ||| Run(C_FE_FrontEnd)
21                      ||| Run(C_MI_Subm)     ||| Run(C_BE_BackEnd)
22                      ||| Run(C_EI_ECR)
23    in Terminal
24       [| C_CAI_Card, C_SI_Config, C_SI_Init, C_FE_FrontEnd,
25          C_MI_Subm, C_BE_BackEnd, C_EI_ECR |]
26       (Card
27       ||| ((ServiceCenter
28             [ C_COI_Config || C_COI_Config, C_MI_Subm_or_Rec ]
29           Acquirer)
30             [ C_COI_Config, C_MI_Subm_or_Rec || C_MI_Subm_or_Rec ]
31         PosMgmtSystem)
32       ||| PointOfService)
33  end
```

Fig. 6. Modelling ep2: The architectural level

more for the formalization of (the semantics of) different frameworks for reactive systems.

Outside the CASL context, e.g. μCRL [9], LOTOS [11], and E-LOTOS [13] provide other solutions for the integrated specification of data and processes within one language. Conceptually, μCRL and CSP-CASL are quite similar in their respective design. In the data part however, CSP-CASL is far more rich: among other features, it offers partiality and subsorting which are frequently used in the modelling of ep2. LOTOS [11] and its recently defined successor E-LOTOS [13] use for data description initial semantics and a functional programming language, respectively. Thus, these languages do not allow for the modelling of the abstract system layers of ep2 as presented here within CSP-CASL.

4 Formalizing ep2 in CSP-CASL

The present formalization of the ep2 system is the first major industrial case study in CSP-CASL. It was done with a number of different aims. Our main objective was to show the feasibility of the approach. This includes many aspects:

Scalability. We wanted to show that it is possible to completely specify a nontrivial system in this formalism. Previous approaches restricted themselves to academic toy examples or small fragments of actual systems.

Expressiveness. Another aim was to prove that CSP-CASL encompasses enough expressive power to deal with the given application. In particular, ep2 contains most aspects which can be found in typical present-day computational systems.

Usability. An important point was to demonstrate that CSP-CASL specifications are easy to write and easy to understand. Many specification formalisms are only targeted at experts and require intensive training and experience.

Adequacy. In order to investigate to what extent the informal and natural language descriptions can be formalized, we wanted to follow the original document structure as closely as possible.

A second objective relates to the actual ep2 system itself. We wanted to show how formal methods can help to improve the design.

Clarity. By structuring the formal specification appropriately, we wanted to untangle the different levels of abstraction in the documents. This could guide future implementors in building a modular implementation.

Precision. We wanted to exhibit ambiguities and inconsistencies within the informal descriptions, which facilitate implementations by third-party implementors.

Validation and Verification. In a second step, we want to use the resulting formal specification to validate actual implementations, prove their conformance with the standard and to generate test cases from the formal specification.

In this section, we give an overview on the structuring of our formalization. According to the general paradigm of CSP-CASL, there are two main aspects: the reactive behaviour of ep2 components and the data structures which are involved.

4.1   Reactive Behaviour

It is natural to model ep2 as a reactive system. In CSP-CASL, we describe its different components by CSP processes which interchange data over communication channels typed by CASL sorts.

On the *architectural level* in the center of the ep2 system there is a `Terminal` process – c.f. line 23 of Fig. 6. This `Terminal` communicates over channels with its environment, expressed here in terms of the CSP general parallel operator [| C_CAI_Card, ..., C_EI_ECR |] linking the `Terminal` with its environment. The environment consists of the processes `Card`, `ServiceCenter`, ..., `PointOfService`.

Note how this model directly corresponds to Fig. 1, which is the first and most abstract description of ep2 given in the ep2 System Specification. We express this correspondence by the choice of names: ep2 components become identically named *processes*, an ep2 interface is characterized by the possible data to be exchanged over it — prefix D_ for the corresponding *sort* providing the type of this data — and by the connection it represents — prefix C_ for the corresponding *channel*.

We do not model the Cardholder and the Attendant as processes as the ep2 specification covers their role only on the level of user interfaces. Most of the processes in the environment run independently of each other, expressed by the CSP interleaving operator ||| (lines 27 and 32). Some of them also interchange information which each other: the ServiceCenter, the Acquirer, and the PosMgmtSystem. Here, we use the CSP alphabetized parallel operator, e.g. [C_COI-Config || C_COI-Config, C_MI_Subm_or_Rec] (line 28), which synchronizes in the intersection of the two alphabets, i.e. in this example in C_COI-Config. On the architectural level, we leave the behaviour of the different processes completely unspecified, i.e. they are modelled by the CSP process RUN(A), which is the deadlock-free, non-terminating process able to engage in any event in a set of communications A. For any process of the ep2 system we choose this set A to consist of all messages, which it might send or receive over the channels which connect it to other processes. For instance, for the Terminal the set A consists of all messages which can be sent or received over any of the channels named in the general parallel [|C_CAI_Card, C_SI_Config, ..., C_EI_ECR|] which connects the Terminal with its environment. This is expressed here as the interleaving of several Run processes (lines 19 – 22).

On the *abstract component description level*, we refine the processes RUN(A) of the above architectural model without changing the overall communication structure. Our example stems from the Terminal specification, showing the Terminal's reactions to the Acquirer's requests on initialization data. In a first step, we specify only that the Terminal produces answers of the right kind, e.g. on a D_SI_ConfigDataRequest a D_SI_Init_ConfigDataResponse is sent:

```
TConfigManagement =  C_SI_Init ? x ->
 if x in D_SI_ConfigDataRequest
   then !y:C_SI_Init.D_SI_Init_ConfigDataResponse -> TConfigManagement
 else if x in D_SI_ConfigDataNotification
   then !y:C_SI_Init.D_SI_Init_ConfigDataAcknowledge -> TConfigManagement
 else ...
```

Here, !y:A -> P denotes the process which first communicates a value y out of the set A and then behaves like P; i.e. the ! operator is similar to the CSP prefix choice, but for the former the choice is internal, while for the latter the choice is external.

In the next step, the *concrete component description level*, we model which specific values the Terminal is going to send. It is at this level, that the process becomes *stateful*, i.e. it depends on a parameter p:Pair[TState][Trigger]. Here, TState represents the Terminal's memory, while Trigger says what kind of signal initiated the configuration management.

```
TConfigManagement (p:Pair[TState][Trigger]) = C_SI_Init ? x ->
 if x in D_SI_ConfigDataRequest
   then C_SI_Init ! configDataResponse(x,state(p))
     -> TConfigManagement(p)
 else if x in D_SI_ConfigDataNotification
   then C_SI_or_FE ! configDataAcknowledge
     -> TConfigManagement (pair(activateData(x,state(p)),trigger(p)))
 else ...
```

This example illustrates the interaction between the specification of reactive behaviour and the modelling of data types when studying the control flow within a component: A message x is received from the Acquirer over the channel C_SI_Init. Depending on the type of x, different answers are sent back to the Acquirer, e.g. information configDataResponse(x,state(p)) on the current configuration of the Terminal or a message configDataAcknowledge. Then the configuration management is continued, either without a state change or with a state change to pair(activateData(x,state(p)),trigger(p)).

It is at the component description levels that more information on data in terms of CASL elements come into play: for instance, there is the test if the value x belongs to a certain subsort D_SI_ConfigDataRequest. The response is computed by a function configDataResponse that takes the message x and the current state state(p) of the Terminal as parameters, or the new state is computed by a function activateData(x,state(p)).

4.2   Data on Different Levels of Abstraction

In direct correspondence to the development of ep2's reactive behaviour over different levels of abstraction, the data types involved are made more and more concrete.

On the *architectural level*, see Fig. 6, it is sufficient to speak merely about the existence of sets of values which are communicated over channels; e.g. the data sort D_CAI_Card is interchanged on a channel C_CAI_Card: D_CAI_Card between the Card and the Terminal. Or a channel shall be shared by different message types, as channel C_MI_Subm_or_Rec: D_MI_Subm_or_Rec. Here, the CASL free type construct ensures that the different kinds of data are kept separate.

If the *component specification level* is abstract, it is usually sufficient to introduce suitable subsorts. Consider for instance the communication between Acquirer and Terminal, see Fig. 3. To specify how the Acquirer interchanges initialisation data, it is enough to know the type of the data, i.e. whether it is a <<SessionStart>> Message or a <<ConfigDataRequest>> Message. In CASL, this can be specified by a free type construct

```
free type D_SI_Init ::= sort D_SI_Init_SessionStart
                      | sort D_SI_Init_ConfigDataRequest
                      | sort D_SI_Init_ConfigDataResponse
                      | ...
```

where each alternative corresponds to a message type occurring in the activity diagram.

But if on the component description level the concrete value of a message triggers a specific behaviour, it is necessary to specify the data types up to representation. Fig. 4 shows the different messages which the `Acquirer` might send to the `Terminal` in order to make requests on its configuration. These messages can be modelled by a CASL free type, and we can finally make concrete which data are involved in a **D_SI_Init_ConfigDataRequest**:

```
free type ConfDataObjRequest ::=
   ACD                     %% Acquirer Config Data
 | AISD                    %% Acquirer Init Srv Data
 | CPTD (ApplicationID)    %% Card Profile Table Data
 | CAD  (RegisteredApplicationProviderID) %% Certification Auth Data
 | TACD (ApplicationID)    %% Terminal Application Config Data
 ...
free type D_SI_Init_ConfigDataRequest ::=
 configDataRequest(ac:AcquirerID;term:TerminalID;conf:ConfDataObjRequest)
```

Up to now data modelling involved only sort declaration, sub-sorting and several forms of disjoint union via the free types construct. But on the component description level, also operations on data and axioms describing them come into play. We give a simple example, again from the context of the Terminal's initialisation. The ep2 documentation states here: *If the configuration download is started by the service man or the 'Use Case: Initiate Terminal Setup', the Terminal sets the* <Config Download Mode> *to '1' indicating 'Forced download' otherwise to '0' for 'Download check'.* We model this case distinction by a function `sessionStart` which is specified by the following axioms:

```
axioms sessionStart(serviceMan) = forcedDownload;
       sessionStart(initialTerminalSetup) = forcedDownload;
       sessionStart(others) = downloadCheck
```

Note that like in the modelling of the reactive behaviour, the different levels of data abstractions are clearly connected by refinement relations.

5   Results

Our overall experience of specifying ep2 in CSP-CASL is that while it's easy to formalize high level descriptions (e.g. the system architecture) from semi-formal descriptions (e.g. UML-like diagrams), writing specifications at the concrete level is more involved. At the more concrete levels, one has to deal with more unresolved and unclear descriptions (mostly presented as text), and decide which information must be formalized and what details should be ignored as they belong to other components or to another abstraction level. Having overcome these obstacles, the CSP-CASL formalization is again fairly straightforward. In this sense, our CSP-CASL specifications clearly mirror the ep2 document structure and specify at the different abstraction levels present therein.

As for CSP-CASL's expressivity, both the data types and the reactive behaviour present in ep2 can be adequately formalized. In modelling the data types, CASL's subsorting feature proved particularly helpful. In modeling the reactive behaviour, CSP's distinction between internal and external choice was similarly important.

5.1   Resolution of Shortcomings

Formalising ep2 in CSP-CASL leads to the partial or complete resolution of the problems outlined in section 2.3.

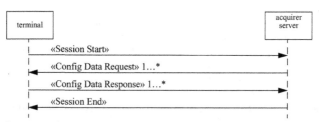

Fig. 7. Sequence diagram for requesting configuration data, as shown in [2]

One reason for this is that we are describing the system within *one* framework. In the *data modelling* for instance the possible values of the message type <<Config Data Request>> are described independently in various ep2 documents, where such different formats as text and tables are involved. Here, one of these texts mentions values LAID and LRID – see the excerpt in Section 2.2 – which do not appear anywhere else. CSP-CASL enables us to specify the corresponding data type only once and – via CSP-CASL's library mechanism – use it then in different contexts. If only data types are concerned, the CASL tool set CATS offers the possibility of static checks for inconsistencies. Looking on the *reactive side*, a comparison of the diagrams Fig. 3 and Fig. 7 shows that they specify the order of Requests and Responses differently: Fig. 3 requires that after one Request exactly one Response has to follow. In spite of this, Fig. 7 suggests that several Requests and several Responses can be 'bundled', and that a session might include different numbers of Requests and Responses. This inconsistency is clearly due to the change between the two formalisms involved (and maybe a weakness of the latter). In CSP-CASL, we can easily specify both variants; in our ep2 formalization we decided to follow Fig. 3.

Another aspect is that the specification language CSP-CASL itself guides us during the formalization process. CSP for instance is famous for its clarity concerning different forms of non-determinism. Thus, in modelling a diagram such as Fig. 3, it is natural to ask if the decision between the different branches is an internal or an external one. In this example, it is the Acquirer who takes the decision. Studying the documentation of the Acquirer further it turns out that for the purpose of the ep2 system it is unnecessary to model the database which is checked for 'Additional Configuration Data' in order to trigger the decision. This leads finally to a simple stateless process as a model for this part of the Acquirer. Interestingly enough, in the description of different parts of ep2 the 'decision points' depicted as diamond with outgoing arcs in this kind of diagram need different formalizations in CSP-CASL: as internal non-determinism, as external non-determinism, and sometimes it is even the case that there is no decision to make. Concerning data, the loose semantics of CASL allows us to postpone design decisions until they are actually required. As seen in Section 4.2 on the different levels of data abstraction, sub-sorting is a powerful mechanism in decomposing complex data type into subtypes of manageable size.

The formalization helps also to design a certain system aspect only once, with the consequence of avoiding possible source of inconsistencies. For data types, this has been illustrated above with the message type <<Config Data Request>>. Concerning reactive behaviour, writing a CSP-CASL specification often helped to avoid over-specification. For instance, in the ep2 documentation the Terminal's responses to a request from the Acquirer are described at least twice: in the Terminal documents and in the Acquirer documents. In the world of CSP processes this is unnecessary: after sending a request to the Terminal, the Acquirer process wants only to receive a message on the channel which is connected with the Terminal. Only in the formalization of the Terminal is it necessary to state which specific response has to be sent, and as we have seen in Section 4.1, this is only necessary at a quite concrete level of abstraction.

5.2   Access to Formal Proofs

One of the benefits of specifying ep2 formally is that it makes it possible to establish properties by formal proofs on the CSP-CASL specifications describing the system. First experiments in this direction include proofs of refinement relations, deadlock analysis and consistency checks of the data types.

For instance, with the newly developed CSP-Prover [12] we were able to prove that

- our CSP-CASL specification corresponding to the activity diagrams 'Get Initialisation Data' — see Fig. 3 for the Acquirer's side of the protocol — is deadlock-free, and
- that — concerning the reactive part of the CSP-CASL specification of the activity diagrams 'Get Initialisation Data' — the specification on the concrete component description level refines[4] the abstract component description level.

Concerning data types, we used the CASL consistency checker [14] in order to prove the consistency of data types on the component description level. Here, we concentrated on the simple case of data types corresponding to ep2 messages as e.g. the ConfigDataRequest. At first glance this seems to be trivial, as on the CASL side these data types involve just a free datatype construct. But as the components of such a free type refer to other specifications, the question of consistency becomes a more involved problem as checking for non-interference between several separate specifications is required.

---

[4] Here, we use CSP's notion of stable-failure refinement.

6   Discussion and Future Work

We have shown how to specify a non-trivial system in the formal specification language CSP-CASL. Since ep2 is a prototypical example, the obtained results also hold for other systems such as web services, communicating financial agents, etc. Reconsidering our original aims, the specification language turned out to be well-suited to "translate" informal and natural language constructs, and rich enough to cover most important aspects of this particular system. Furthermore, it turned out that it is mostly possible, but not always advisable to follow the original document structure in the formalization. Considering scalability, we found that it is neither much harder nor much simpler to write a formal than an informal specification. In fact, we think that both styles have their own benefits; ideally the formal text should accompany informal descriptions in a 'literate specification'.

Related Work. The specification and implementation of banking software belongs to the most widely used exercises in computer science education. For example, in [4] the implementation of automated teller machine (ATM) software from an object oriented analysis and design is described. This graduate-level tutorial comprises a nice example of current best practice in software engineering, from the informal requirements specification up to an executable applet which can be used by students for testing purposes.

Similarly, many efforts have been invested in the verification of basic principles of the communication protocols which are employed in banking software. For example, in [8], some aspects of the *Millicent micropayment protocol* are modelled in an abstract protocol notation which is close to CSP, and security aspects are verified from this. As another example, in [18] authentication issues in the *Secure Electronic Transaction* (SET) protocol of Visa/Mastercard are verified by model checking a multi-agent logic of belief and time.

Not much work, however, has been mentioned in the formal specification and verification of real banking software and standards such as EMV or ep2. As an early example, in [26], the UNITY-method is used to refine a high-level specification of an electronic funds transfer system into one that could in principle be turned into an executable program. A more recent example of a formal specification of an actual banking standard is reported in [27], where the *Mondex electronic purse system* was proven correct with respect to its CSP and Z specification and was certified according to UK ITSEC Level 6. In [17], the *Internet Open Trading Protocol* (IOTP) is specified with colored Petri nets from an Request for Comments (RFC) by the Internet Engineering Task Force (IETF). In [6], it is argued that an interdisciplinary approach is necessary in this field, where experts from business administration, computer science and electrical engineering specify different views of a system. As example, a real internet based CD retail store system is specified in an integrated system model.

Future Work. Our next steps on formalizing ep2 will be to complete the modelling as far as possible. In particular, up to now we have formalized only a significant part of the whole specification, where the main omissions are the low-level XML communication between actors and the security layer. In fact, the security part of ep2 heavily relies on common sense and external documents; in order to be able to prove security properties we will have to add certain assumptions about the underlying cryptographic methods. Other proofs on the formal model which we already started include refinement relations and deadlock analysis with CSP-Prover [12], as well as consistency of the data types [14]. Livelock analysis is to follow.

Finally, we want to use the model to automatically generate test cases for the different components of the ep2 system. It is an interesting research topic to define criteria which measure both data and control coverage of such test suites.

Acknowledgements. The authors would like to thank Yoshinao Isobe (AIST) for help with refinement proofs and deadlock analysis of our specifications with CSP-Prover, Erwin R. Catesbeiana Jr (USW) for his advice on the modelling approach, and Christoph Schmitz and Martin Osley (Zühlke Engineering AG) for the good cooperation on ep2.

References

1. FM'99 exhibition: Competition 'Cash-Point Service', 1999. Based on the "9th Problem" stated at the Workshop on "The Analysis of Concurrent Systems", Cambridge, September 1983, LNCS 207.
2. *eft/pos 2000 Specification, version 1.0.1.* EP2 Consortium, 2002.
3. M. Bidoit and P. D. Mosses. Casl *User Manual.* LNCS 2900. Springer, 2004.
4. R. C. Bjork. Course notes 'Object-Oriented Software Development'. Department of Mathematics and Computer Science, Gordon College, Fall 2004. http://www.math-cs.gordon.edu/local/courses/cs211/ATMExample.
5. B. T. Denvir, W. T. Harwood, M. I. Jackson, and M. J. Wray, editors. *The Analysis of Concurrent Systems,* LNCS 207. Springer, 1985.
6. A. Franz, P. Sties, and S. Vogel. Formal specification of e-commerce applications – an interdisciplinary approach. In K. Altinkemer and K. Chari, editors, *Proceedings of the Sixth INFORMS Conference on Information Systems and Technology.* ForSoft Publications, TU Munich, 2001.
7. M. Frappier and H. Habrias, editors. *Software Specification Methods.* Springer, 2001.
8. M. G. Gouda and A. X. Liu. Formal specification and verification of a micropayment protocol. In *Proceedings of the 13th IEEE International Conference on Computer Communications and networks, Chicago (Oct 2004).* IEEE Press. To appear.
9. J. F. Grote and A. Ponse. The syntax and semantics of μCRL. In A. Ponse, C. Verhoef, and S. F. M. van Vlijmen, editors, *Algebra of Communicating Processes '94,* Workshops in Computing. Springer, 1995.
10. C. A. R. Hoare. *Communicating Sequential Processes.* Prentice Hall, 1985.
11. ISO 8807. Lotos — a formal description technique based on the temporal ordering of observational behaviour, 1989.
12. Y. Isobe and M. Roggenbach. A generic theorem prover of CSP refinement. In *Proceedings of TACAS 2005,* LNCS. Springer, to appear.
13. JTCI/CS7/WG14. The E-LOTOS final draft international standard, 2001.

14. C. Lüth, M. Roggenbach, and L. Schröder. CCC —the CASL Consistency Checker. In J. L. Fiadeoiro, P. D. Mosses, and F. Orejas, editors, *Recent Trends in Algebraic Development Techniques, 17th International Workshop, WADT 2004, Barcelona, Spain, March 27-30, 2004, Revised Selected Papers*, LNCS. Springer, to appear.
15. T. Mossakowski, M. Roggenbach, and L. Schröder. CoCASL at work — Modelling Process Algebra. In *Coalgebraic Methods in Computer Science*, volume 82 of *Electronic Notes Theoretical Computer Science*, 2003.
16. P. D. Mosses, editor. CASL *Reference Manual*. LNCS 2960. Springer, 2004.
17. C. Ouyang, L. M. Kristensen, and J. Billington. A formal and executable specification of the internet open trading protocol. In K. Bauknecht, A. M. Tjoa, and G. Quirchmayr, editors, *Proceedings of 3rd International Conference on E-commerce and Web Technology*, LNCS 2455, pages 377–387. Springer, 2002.
18. M. Panti, L. Spalazzi, and S. Tacconi. Verification of security properties in electronic payment protocols. In *Workshop on Issues in the Theory of Security (WITS '02); Co-located with IEEE POPL, Portland (Jan. 2002)*, 2002.
19. G. Reggio, E. Astesiano, and C. Choppy. CASL-LTL — a CASL extension for dynamic Reactive Systems — Summary. Technical Report DISI-TR-99-34, Università di Genova, 2000.
20. G. Reggio and L. Repetto. CASL-CHART: a combination of statecharts and of the algebraic specification language CASL. In *Algebraic Methodology and Software Technology*, volume 1816 of *LNCS*, pages 243–257. Springer, 2000,.
21. H. Reichel, T. Mossakowski, M. Roggenbach, and L. Schröder. Algebraic-coalgebraic specification in CoCASL. In *Recent Developments in Algebraic Development Techniques, 16th International Workshop (WADT 02)*, LNCS 2755. Springer, 2003.
22. M. Roggenbach. CSP-CASL – A new integration of process algebra and algebraic specification. *Theoretical Computer Science*, to appear.
23. A. Roscoe. *The theory and practice of concurrency*. Prentice Hall, 1998.
24. G. Salaün, M. Allemand, and C. Attiogbé. A formalism combining CCS and CASL. Technical Report 00.14, University of Nantes, 2001.
25. G. Salaün, M. Allemand, and C. Attiogbé. Specification of an access control system with a formalism combining CCS and CASL. In *Parallel and Distributed Processing*, pages 211–219. IEEE, 2002.
26. M. G. Staskauskas. The formal specification and design of a distributed electronic funds transfer system. *IEEE Transactions on Computers*, 37, 1988.
27. S. Stepney, D. Cooper, and J. Woodcock. An Electronic Purse: Specification, Refinement, and Proof. Technical Monograph PRG-126, Oxford University Computing Laboratory, 2000.

Algebraic Semantics of Design Abstractions for Context-Awareness

Antónia Lopes[1] and José Luiz Fiadeiro[2]

[1] Department of Informatics, Faculty of Sciences, University of Lisbon,
Campo Grande, 1749-016 Lisboa, Portugal
`mal@di.fc.ul.pt`
[2] Department of Computer Science, University of Leicester,
University Road, Leicester LE1 7RH, UK
`jose@fiadeiro.org`

Abstract. We investigate essential features of contexts and proper abstractions for modelling context-awareness within CommUnity, a language that we have been developing to support architectural design of distributed and mobile system. Under the assumption that the context that a component perceives is determined by its current position, we explore the use of abstract data types for defining design primitives through which different notions of context can be modelled explicitly according to the application domain.

1   Introduction

One of the major challenges in modern distributed computing is to deal with highly dynamic operation contexts. As components are entitled to move across networks whose nodes can themselves be mobile and required to execute in different locations, the availability and responsiveness of resources and services are often difficult to predict and out of control [1]. Computational resources such as CPU and memory are no longer fixed as in conventional computing. When visiting a site, a piece of mobile code may fail to link with libraries it requires for its computation. Network connectivity and bandwidth are other factors that can affect, in a fundamental way, the behaviour of mobile computing systems. In this setting, it is no longer reasonable to treat attempts at using absent resources or accessing unavailable services as exceptions. Systems should be provided with the means to observe the context in which they operate. They should also be developed taking into account the different conditions in which they can be required to operate.

The possibility of observing the context also opens new and more sophisticated ways of designing systems. For instance, as pointed out in [2], in order to scan a database of images stored at a remote site for which there is a cheap algorithm that quickly identifies those that are potentially interesting, we may conceive a solution that takes advantage of the perception of the context as follows. First, the cheap algorithm is executed remotely. Then, the size of the selected images, the current network latency and the processing power that is available in both hosts are used to decide if

J.L. Fiadeiro, P. Mosses, and F. Orejas (Eds.): WADT 2004, LNCS 3423, pp. 79–93, 2005.

the remaining (intensive) computation should be performed remotely or if the images selected by the cheap algorithm should be sent back.

Context-awareness is the emergent computing paradigm that is addressing this kind of approach to mobile systems construction. By *context*, one refers to the part of the operation conditions of a running system that may affect its behaviour. Typically, different kinds of applications require different notions of context. Hence, it is important that formalisms for designing mobile systems consider contexts as first-class design entities. If a specific notion of context is fixed, for instance as in Ambients [4], the encoding of a different notion of context can be cumbersome and entangled with other aspects, if at all possible. On the contrary, if we support the explicit modelling of notions of context according to the application domain, we make it possible for such aspects to be progressively refined through the addition of detail, without interfering with the parts of the system already designed.

In this paper, we investigate essential features of contexts and proper abstractions for modelling context-awareness. Under the assumption that the context that a component perceives is determined by its current position, we explore the use of abstract data types for defining design primitives through which different notions of context can be modelled explicitly according to the application domain.

Our approach provides means for describing explicitly how contexts affect the behaviour of systems. Any notion of context defines a specific set of constructs that can be used in the specification of system actions. These constructs allow us to enrich architectural models with context-aware patterns of computation, coordination and mobility in a non-intrusive way. That is, context-awareness can be added as an orthogonal dimension without interfering with context-independent decisions made at the level of computation (e.g. properties of data returned by services), coordination (e.g. interactions managed through a shared control unit) and mobility (e.g. shared control unit at a fixed location).

We present this approach over CommUnity, a language that we have been developing to support architectural design [11]. CommUnity was recently extended in order to support the description of mobile systems [14]. This extension addresses a specific notion of context that is centred on the notions of connectivity and reachability of positions. In this paper, we take this extension of CommUnity a step further in order to support the definition of application-specific notions of context. In particular, we will take into account the availability of computational resources and services at the locations in which components are placed.

2   Designing Mobile Systems in CommUnity

CommUnity is a parallel program design language similar to Unity [6] and IP [10] in its computational model but relying on communication rather than shared memory for interaction. In CommUnity, the individual components of a system are designed in terms of *channels* and *actions*. The role of the channels is to exchange data between different components. Actions are associated with guarded commands that manipulate and compute data and provide points for rendez-vous synchronisation with other components. In order to support distribution and mobility, CommUnity provides mechanisms for assigning channels to locations and distributing the execution of actions among different locations.

To illustrate the way components of mobile systems can be designed in CommUnity and, later on, other aspects of our model, we use the image search problem mentioned in the introduction. We start with a high-level description of a server that can be used to control the access to a shared resource:

```
design server[N:nat] is
outloc l
prv     gr@l:[0..N], rq_i@l:bool i=1,…,N
do      i=1,…,N
        req_i[rq_i]@l: ¬rq_i → rq_i:=true
[]      grt_i[gr]@l: rq_i∧gr=0 → gr:=i
[]      rel_i[gr,rq_i]@l: gr=i → rq_i:=false||gr:=0
```

This design models the basic functionalities of a server that supports up to N connections and acts like a scheduler by allowing only one connection to be on at any time. Through each action req_i it accepts requests for using the resource, which it records in the private channel rq_i. Private channels are internal in the sense that the data that they store is not available to the environment but only for interaction inside the component, namely among the actions that the component can perform. Through each action grt_i the server signals that access to the resource has been granted to the particular client that has requested it through the action req_i. This is because the condition rq_i is part of the guard of action grt_i, which means that a request must be pending on the i-th connection. The other conjunct of the guard – $gr=0$ – ensures that no other request has been granted. The private channel gr is used, precisely, to communicate the status of the connections: it takes the value 0 if the resource is free and the value $i:1..N$ if a request has been granted along the i-th connection. The server acknowledges the release of the resource through the actions rel_i by resetting both gr and rq_i.

A location variable l is declared for handling distribution and mobility. In fact, the server is modelled as a centralised component because all its constituents are located at l. Furthermore, the server is not mobile because this location variable is declared to be output – which means that it is under the control of the component – but is not updated by any of its actions – meaning that it remains invariant.

In CommUnity, a component is designed in terms of a set of channels V (declared as input, output or private), a set of location variables L (input or output) and a set of actions Γ (shared or private) that, together, constitute what we call a signature. We use X to denote $V \cup L$. Input channels are used for reading data from the environment. Output and private channels are controlled locally by the component. Output channels allow the environment to read data processed by the component.

Locations variables, or locations for short, are used as "containers" that may transport data and code while moving. The association of a channel x with a location l is described by $x@l$. Intuitively, this means that the position of the space where the values of x are made available is given by the value of l. Every action g is associated with a set of locations $\Lambda(g)$ meaning that the execution of g is distributed over those locations. Input locations are read from the environment and cannot be modified by the component and, hence, the movement of any constituent located at an input location is under the control of the environment. Output locations can only be modified locally and, hence, the movement of any constituent located at an output location is under the control of the component.

Private actions represent internal computations in the sense that their execution is uniquely under the control of the component; shared actions represent possible interactions between the component and the environment. The computational aspects are described by associating with each action g:

- a set $D(g)$ with the local channels and locations into which executions of g can write;
- for each $l \in \Lambda(g)$, an expression of the form

$$\texttt{L(g@l), U(g@l)} \rightarrow \texttt{R(g@l)}$$

Two conditions $L(g@l)$ and $U(g@l)$ on X establish the interval in which the enabling condition e of any guarded command that implements $g@l$ must lie: the *lower bound* $L(g@l)$ is implied by e, and the *upper bound* $U(g@l)$ implies e. When the enabling condition of g is fully determined we write only one condition. The parameter $R(g@l)$ is a condition on X and $D(g)'$ where by $D(g)'$ we denote the set of primed symbols in $D(g)$. As usual, these primed symbols account for references to the values that they take after the execution of the action. When $R(g)$ is such that the primed channels and locations in $D(g)$ are fully determined, we obtain a conditional multiple assignment, in which case we use the notation that is normally found in programming languages ($\|_{x \in D(g)}\ x := F(g,x)$).

A CommUnity design is called a program when, for every $g \in \Gamma$, $\bigwedge_{l \in \Lambda(g)} L(g@l)$ and $\bigwedge_{l \in \Lambda(g)} U(g@l)$ are equivalent, and the relation $\bigwedge_{l \in \Lambda(g)} R(g@l))$ defines a conditional multiple assignment.

In order to illustrate how CommUnity can handle distribution and mobility, we now address the design of a client whose purpose is to search a remote image database for a particular type of images:

```
design client1 is
outloc lf,lc
in      lr:Loc, db:set(image)
out     res@lc, img@lf:set(image), size@lf:nat
prv     st_f@lf:[0..5], st_c@lc:[0..4], home@lc:Loc
do      gof@lf: st_f=0 → st_f:=1
           @lc: st_c=0 → st_c:=1||lf:=lr
[]      req@lf: st_f=1 → st_f:=2
[]      filter@lf: st_f=2 → st_f:=3||img:=filterop(db)
[]      rel@lf: st_f=3 → st_f:=4||size:=imgsize(img)
[]      backf@lf: st_f=4 → st_f:=5
           @lc: st_c=1∧small(size) → st_c:=2||lf:=lc
[]      goc@lf: st_f=4 → st_f:=5
           @lc: st_c=1∧¬small(size) → st_c:=2||lc:=lr||home:=lc
[]      check@lc: st_c=2 → st_c:=3||res:=checkop(img)
[]      backc@lc: st_c=3∧lc≠home → st_c:=4||lc:=home
```

The images are modelled through an abstract data type that involves the domain *image*. The database itself is modelled as a value of type *set(image)* that the environment makes available through the channel *db*.

The client is a distributed component: it involves two locations, *lf* and *lc*, both under its control. The location *lf* is where the client runs, through action *filter*, the "cheap algorithm" – captured by the user-defined operation *filterop* on *set(image)* – that quickly identifies the images that are potentially interesting, which it makes available in the channel *img*. The size of this set is computed by the action *rel* using the operation *imgsize* and made available through the channel *size*. Access to the database itself is requested through action *req*. Release of the database takes place through the action *rel*, i.e. when the size of the extracted set is evaluated. The private channel st_f located at *lf* ensures the correct sequencing of these actions.

As motivated in the introduction, the filtering activity needs to be performed wherever the database is located. This location is made available by the environment through the input channel *lr*. Hence, through action *gof*, the client moves the filtering activity to this location, i.e. assigns *lr* to *lf*.

The action *gof* is distributed between *lf*, where it initialises the filtering process by setting st_f to *1*, and the second location *lc*, where the actual migration is performed. The location *lc* is where the client determines how to proceed. If *size* is small, the client, through action *backf*, sends back the filter with the extracted images by assigning *lc* to *lf*. Otherwise, the *checker* moves itself to the remote host by assigning *lr* to *lc*. The search of the interesting images, performed by the operation *checkop* on *set(img)*, is executed by action *check* at *lc*. The final result is returned, at *lc*, through the channel *res*. Finally, the client returns home if it ever migrated to the remote host. The private channel st_c located at *lc* ensures the correct sequencing of these actions.

Designs are defined over a collection of data types that are used for structuring the data that the channels transmit and define the operations that perform the computations that are required. Hence, the choice of data types determines, essentially, the nature of the elementary computations that can be performed locally by the components, which are abstracted as operations on data elements. We consider that the collection of data types appropriate for the design of a specific system is explicitly specified through a first-order algebraic specification. That is to say, we assume a data signature $\Sigma = \langle S, \Omega \rangle$, where S is a set (of sorts) and Ω is a $S^* \times S$-indexed family of sets (of operations), to be given together with a collection Φ of first-order sentences specifying the functionality of the operations.

In CommUnity, the space within which movement takes place is explicitly represented with a distinguished sort *Loc*. Location variables are all implicitly typed with this sort. *Loc* models the positions of the space in a way that is considered to be adequate for the particular application domain in which the system is or will be embedded. Together with the definition of the operations on locations, this provides a description mechanism that is expressive enough to establish, for instance, location hierarchies or taxonomies. The only requirement that we make is for a special position $\perp$ to be distinguished that accounts for a special position of the space where context-transparency is supported. That is, the behaviour of any component located at $\perp$ is context-unaware.

In the sequel, we use Θ to refer to the extension of the data type specification with what concerns the space of mobility. In the image search example, the specification Θ includes the specification of Booleans, natural numbers and sets, which are standard data types. Moreover, Θ includes a specific sort *img* accounting for images, the specific operations *filterop* and *checkop* over sets of images for the searching of the interesting images, and operations $imgsize:set(img) \rightarrow nat$ and $small:nat \rightarrow nat$ that establish the size of a set of images and a threshold over which the set of images is considered too large, respectively. In this way, we have a design of the system that abstracts from specific representation of images and the specific types of images that have to be filtered. In this example, the space of mobility just has to include three different positions. In addition to $\perp$, it is only necessary to account for the host where the client runs and the location of the database server. We have not identified any need for special operations over positions at this stage of the design.

3   Moving Contexts into the Picture

So far we have focused on the features that support the description of mobile systems. The question that concerns us most in this paper is the extent to which the behaviour of these systems is affected by the context surrounding them.

Most approaches to the specification of distributed and mobile systems adopt context-transparency as an abstraction principle and define the behaviour of systems regardless of their context (e.g., [12], [16], [20]). This implies that network connectivity between two hosts is guaranteed whenever it is needed, and that it is possible to migrate code anywhere and anytime without restrictions. The few formalisms that adopt a context-aware approach do not address contexts explicitly and assume specific notions of context (e.g., [3], [4], [5], [17]). This is also the case of our previous work in CommUnity [15]. In this section, we analyse several issues that are central to the choice of an abstract notion of context and the development of design primitives for context-awareness. This a first step towards a more expressive model for context-aware computing that supports the definition of application specific notions of context and the design of components that deal with changes in the operating conditions as part of their intrinsic behaviour.

We should start by making clear that, by context, we refer to any collection of characteristics and properties that are relevant to the system and are not under its direct control. For instance, in network applications that are able to establish firewalls or security policies, neither the locations of the firewalls nor the structure of the space can be considered as part of the context. However, latency can be considered as part of the context of systems that use the network. Although these systems necessarily affect latency, these effects are achieved in an indirect way.

We also assume that a system is not involved in the monitoring of the contextual information. Any such activity has to rely on another system – the *context-provider* – that supports the gathering of context information from relevant sources (e.g. from the network layer or physical sensors) and its delivery to the system. Such a clear separation of context monitoring from the rest of the system is important because it contributes to the taming of the complexity of designing and building context-aware mobile systems: software designers only have to define the notion of context that best fits the system at hand and do not have to be concerned with the way context information needs to be sensed. This separation is important also because it promotes the development of general context sensing systems that can be used (and reused) as the basic building blocks in the construction of application-specific context-provider systems.

The fact that we are considering distributed systems implies that contexts are also distributed, which raises the question of whether this distribution should be abstracted or not. The design primitives that can be made available for modelling context-awareness clearly depend on this decision. A common choice in this respect is to consider that a system has transparent access to the distributed context information (e.g., [8]). This approach abstracts the fact that some properties considered relevant for the system, for instance network connectivity, affect necessarily the gathering of remote context information. In CommUnity, we adopt an alternative approach in which all parts of a context can be sensed locally. Notice that this does not mean that distant entities cannot affect the behaviour of a component but, rather, that the transmission of any remote context information that is relevant has to be explicitly designed as part of the behaviour of the system.

Any notion of context is constrained by the unit of mobility. In models with a fine-grained unit of mobility, such as CommUnity, the software designer should write a single context definition that applies to the entire system. This is because a component may have several constituents distributed over different locations that can be moved independently and, hence, the context perceived by a component results from what its constituents perceive in their current locations.

The notion of context in mobile computing encompasses two different aspects. The *active* aspects includes the properties that affect the behaviour of a system even if the design of the system does not explicitly use any context information. In CommUnity, the identification of the *active* characteristics, and the definition of how these characteristics influence the behaviour of a system, are part of the formal semantics. For instance, a specific command may be defined to have different results according to the context in which it is executed. The *passive* part of the context includes the characteristics that only affect the behaviour of a system if we use them explicitly in the design. That is, the way in which each *passive* characteristic affects the behaviour of the system is explicitly coded in the design of the system. In CommUnity, we propose that abstract data types be used for defining design primitives through which different notions of context can be modelled explicitly as part of the application domain. Through the specification of abstract data types, software developers can define the structure of the contextual information demanded by the system and the operations that are needed to access this data.

To be more precise, in CommUnity, the context definition is an explicit and central part of the design of any context-aware system. A single notion of context, that applies to the entire system, is defined by a data type specification χ and a subset O of its operations. We take χ in the form of a first-order algebraic specification. Each operation symbol *obs* in O represents an observable that can be used to describe the behaviour of the system. More concretely, in a CommUnity design defined over a context specification $<\chi,O>$, the conditions that establish the guards and effects of the actions are built with terms involving operations of Θ and O.

We require that every context description χ includes standard specifications of sets and natural numbers (extended with ∞) and we assume they are represented by *set(T)* and nat^∞, respectively. This is because we require four special observables – $rs: nat^\infty$, $sv:set(\Omega)$, $bt: set(Loc)$ and $reach: set(Loc)$ – be included in O. These observables constitute the active aspect of the context and capture the fact that, in location-aware systems, regardless of their particular application domains:

- *Computations*, as performed by individual components, are constrained by the *resources* and *services* available at the positions where the components are located. For instance, a piece of mobile code that relies on high-precision numerical operations may fail when placed in a location where memory is scarce, computations will not be able to proceed if the operations that the code requires are not available.
- *Communication* among components can only take place when they are located in positions that are *"in touch"* with each other. For instance, the physical links that support communication between the positions of the space of mobility (e.g., wired networks, or wireless communications through infrared or radio links) may be subject to failures or interruptions, making communication temporarily impossible.
- *Movement* of components from one position to another is constrained by *"reachability"*. Typically, the space of mobility has some structure that can be given by walls and doors, barriers erected in communication networks by system administrators, or the simple fact that not every position of the space has a host where code can be executed.

The purpose of rs and sv is to represent respectively, the *resources* and *services* that are available for computation. The observable rs quantifies the resources available. It may be defined as a function of other observables in χ (for instance, the remaining lifetime of a battery or the amount of memory available) through the inclusion in χ of whatever axioms are appropriate. The observable sv represents the services available and it is taken as a subset of the operations of the data type signature Σ. This is because, as we have seen in Section 2, the services that perform the computations are abstracted as operations on data elements. The intuition behind bt and $reach$ is even simpler: both represent the set of locations within reach. The former represents the locations that can be reached through communication while the latter concerns reachability through movement. The actual meaning of these four special observers is defined by the formal semantics of CommUnity designs.

Before embarking on the definition of the semantics of CommUnity, we consider again the image search problem. We address the design of the solution presented in the introduction, which relies on the observation of two properties of the context: network latency and processing power. In this case, the specification of the context is rather simple. We define two specific observers, both of them constants and returning natural values.

```
observer operations
   lat: nat  // latency
   ppw: nat  // processing power available
```

Additionally, we may specify a relation between the possibility of communication and latency. In the same way, we can relate the processing power with the special observable rs.

```
axioms  l:Loc
        belongs(l,bt)∧l≠⊥ ⊃ positive(lat)
        positive(rs) ⊃ positive(ppw)
```

In the envisaged system, the choice of where to execute the compute-intensive algorithm is based on the size of the selected images, on the processing power available in the remote and local machines and on the latency measured in the local host. The concrete criteria have to be available in the form of an operation in the data type specification Θ.

```
operations
        crit: nat*nat*nat*nat->bool      //Is it ok to compute locally?
```

The design *client1* of the client that we gave in Section 2 is now modified in order to accommodate the new requirement.

```
design cwt-client is
outloc lf,lc
in      lr: Loc, db:set(img)
out     res@lc, img@lf:set(img),
        size@lf:nat, rpw@lf:nat
prv     st_f@lf:[0..5], st_c@lc:[0..4], home@lc:Loc
do      gof@lf: st_f=0 → st_f:=1
              @lc: st_c=0 → st_c:=1||lf:=lr
[]      req@lf: st_f=1 → st_f:=2
[]      filter@lf: st_f=2 → st_f:=3||img:=filterop(db)
[]      rel@lf: st_f=3 → st_f:=4||size:=imgsize(img)||rpw:=ppw
[]      backf@lf: st_f=4 → st_f:=5
              @lc: st_c=1∧crit(size,pw,rpw,lat) → st_c:=2||lf:=lc
[]      goc@lf: st_f=4 → st_f:=5
              @lc: st_c=1∧¬crit(size,pw,rpw,lat) → st_c:=2||lc:=lr||home:=lc
[]      check@lc: st_c=2 → st_c:=3||res:=checkop(img)
[]      backc@lc: st_c=3∧lc≠home → st_c:=4||lc:=home
```

This design introduces a new channel *rpw* through which the *filter*, once in the remote location, sends to the *checker* the processing power that is measured there. The enabling condition of the actions that model the return of the *filter* and the *migration* of the *checker* were changed to reflect the new criteria for migration.

Consider now a design *client* only differing from *client1* in the choice of where to run the checker operation, which is made nondeterministic. That is to say, in *client*, actions *backf* and *goc* have exactly the same enabling condition:

```
        backf@lf: st_f=4 → st_f:=5
              @lc: st_c=1 → st_c:=2||lf:=lc
[]      goc@lf: st_f=4 → st_f:=5
              @lc: st_c=2 → st_c:=2||lc:=lr||home:=lc
```

Both *client1* and *cwt-client* can be obtained from *client* through the superposition of additional behaviour. From a methodological point of view, what is interesting is that it is possible to capture the superposed aspects as an architectural element (a connector) that is plugged to the *client* to control in which situations the *checker* have to migrate. Changing from one design decision to another is then just a matter of unplugging a connector and plugging a new one.

For instance, the aspects that need to be superposed to the *client* in order to obtain *cwt-client* are captured by what in CommUnity is called a connector *Cwt* with the following glue:

```
design cwt-glue is
inloc  lc,lf
in     size:nat,
out    rpw@lf:nat
do     rel@lf: true → rpw:=ppw
[]     backf@lc: crit(size,pw,rpw,lat)→ skip
[]     goc@lc: ¬crit(size,pw,rpw,lat)→ skip
```

An architecture of the image search system that makes use of the connector *Cxt* is presented below. The other connector – *Comm* – accounts for the communication protocol that is adopted for the communication between the *filter* component of the *client* and the database server. This architecture shows how the introduction of context-awareness in the architectural model of a system can be achieved in a non intrusive way.

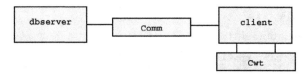

The advantage of this architecture is that in order to change the design decision we adopted for the migration of the checker, we just need to replace *Cwt* by an appropriate connector.

4   Semantic Aspects

Consider a CommUnity program *P* defined over a data specification $\Theta=<\Sigma,\Phi>$, where $\Sigma=<S,\Omega>$, and a context specification $Cxt=<\chi,O>$, where $\chi=<\Sigma^\chi,\Phi^\chi>$.

In order to define the behaviour of *P*, we have to fix, first of all, the carrier sets $\mathcal{U}_s$ that define the possible values of the each data sort *s*. In particular, the set $\mathcal{U}_{Loc}$ defines the positions of the space of mobility for the situation at hand. These sets of values are considered to be global and invariant over time. In contrast, as shown below, the interpretation of the operation symbols in Σ is considered to be local to each position of the space and may change over time. This accounts for the possible evolution of the actual *implementations* of the operations that perform the computations that are required in *P*.

We also consider that part of the data type specification that defines the context has a global and static interpretation. The exception is the interpretation of the observables that account for the actual contextual information and which is considered to be local and dynamic. More concretely, we fix an algebra $\mathcal{U}'$ for the sub-specification of χ that is obtained by forgetting the subset of the operation symbols of *O* and the axioms involving these symbols. In the sequel, we designate this specification by χ'. The algebra $\mathcal{U}'$ should provide the standard interpretation to *nat*$^\infty$ and *set(T)*.

Because *P* is a context-aware program, the surrounding context can affect its behaviour. Therefore, we also need to fix a model of the "world" where *P* is placed to run. In fact, we have to provide a model for the context defined by *Cxt*.

The model of the context should capture the fact that it may change continuously. In the trace-based semantics of CommUnity, we take context models in the form of infinite sequences of states; such states capture the contextual information at a particular instant of time. As motivated in Section 3, we consider that the state of contextual information is distributed and put together from what is sensed locally at each position of the space.

Local states have three dimensions. The first addresses the interpretation of the operations of Σ. It consists of an algebra $\mathcal{U}$ for the part of Θ that captures the operations that are available in the state. This algebra is based on the carriers sets $\mathcal{U}_s$ that were fixed before. This partiality captures that some locations may not have local implementations of some types of computations that are required in P.

The second dimension concerns the level of resources required for the computation of each operation in Σ. It consists of a partial function $\rho:\Omega\rightarrow N^\infty$ that must be defined for every operation symbol for which the algebra $\mathcal{U}$ establishes an interpretation.

The third dimension is about the contextual information that can be observed by the program P. It provides the current values of the observables in a particular location and consists of an O-indexed set $o=\{obs_{\mathcal{U}}\}_{obs\in O}$ of functions defining an interpretation for each observable obs in O based on the carrier sets defined by $\mathcal{U}'$. This set, together with the χ'-algebra $\mathcal{U}'$, should define a χ-algebra. Furthermore, the interpretation of the special observables rs, sv, bt and $reach$, provided by the constant functions $rs_{\mathcal{U}}:N^\infty$, $sv_{\mathcal{U}}:2^\Omega$, $bt_{\mathcal{U}}:2^{Loc_{\mathcal{U}}}$ and $reach_{\mathcal{U}}:2^{Loc_{\mathcal{U}}}$ is constrained as follows.

- The set of available services, $sv_{\mathcal{U}}$, must be the set of Σ-operations for which the algebra $\mathcal{U}$ establishes an interpretation.
- In any local state associated to position m, the sets of positions $bt_{\mathcal{U}}$ and $reach_{\mathcal{U}}$ must include m. Intuitively, this means that we require that *be in touch* and *reachability* are reflexive relations. Furthermore, $bt_{\mathcal{U}}$ must include the special position $\perp_{\mathcal{U}}$. This condition establishes part of the special role played by $\perp_{\mathcal{U}}$: at every position of the space, the position $\perp_{\mathcal{U}}$ is always "in touch". In addition, we require that in any local state associated to position $\perp_{\mathcal{U}}$, $bt_{\mathcal{U}}$ be the set $\mathcal{U}_{Loc}$. In this way, any entity located at $\perp_{\mathcal{U}}$ can communicate with any other entity in a location-transparent manner and vice-versa.
- The position $\perp_{\mathcal{U}}$ is also special because it supports context-transparent computation, i.e. a computation that takes place at $\perp_{\mathcal{U}}$ is not subject to any kind of restriction. This is achieved by requiring that the values of $rs_{\mathcal{U}}$ and $sv_{\mathcal{U}}$ in any state associated to the position $\perp_{\mathcal{U}}$ be $+\infty$ and Ω, respectively. In other words, the computational resources available at $\perp_{\mathcal{U}}$ are unlimited and all services are available.

In summary, a context model is an infinite sequence of functions $\mathcal{M}_0.\mathcal{M}_1....$ in which each $\mathcal{M}_i$ is a function over $\mathcal{U}_{Loc}$ that returns a three-dimensional state. For ease of presentation of the program behaviour, we use $\alpha_i(m)$ and $\rho_i(m)$ to denote the first and the second component of $\mathcal{M}_i(m)$, respectively. Moreover, we use $o_i(m)$ to denote the interpretation of the observable o provided by the third component of $\mathcal{M}_i(m)$.

The behaviour of the program P running in a world modelled by $\mathcal{M}_0.\mathcal{M}_1....$ is defined in terms of set of traces as follows.

We take traces in the form of $\mathcal{V}_0.\gamma_0.\mathcal{V}_1.\gamma_1....$ where each $\mathcal{V}_i$ is a valuation of the channels and locations of P (an S-indexed set of functions $\mathcal{V}_s:V_s\rightarrow \mathcal{U}_s$) and γ_i is a set of actions of P. Notice that transitions of the form $\mathcal{V}_i.\varnothing.\mathcal{V}_{i+1}$ capture state transitions that are performed by other components of the systems in which P is integrated as a component.

The terms and propositions used for defining the guards and effects of the actions of P are built over the signature Σ and the set of observables of Cxt. The local interpretation of these terms and propositions over a trace, i.e. from the point of view of a specific position of the space, can be defined in a straightforward way. It should just be noted that the interpretation $I_i(t)(m)$ of the term t at position m at time i over a trace depends not only on the valuation $\mathcal{V}_i$ of the channels and location variables of the program but also on the local state of the context at time i. This is because, on the one hand, the interpretation of the data operations of Σ is defined by $\alpha_i(m)$ and, on the other hand, the actual values of the observables are defined by $o_i(m)$. In particular, if t involves data operations for which $\alpha_i(m)$ does not establish an interpretation, then $I_i^m(t)$ is undefined.

The same applies to propositions. A proposition that involves one of these terms cannot be evaluated at position m and time i. We use $\mathcal{S},i,m \vdash \phi$ to denote that proposition ϕ is evaluated to true at position m and at time i of $\mathcal{S}$.

Formally,

Given a trace $\mathcal{V}_0.\gamma_0.\mathcal{V}_1.\gamma_1...$, an action g of P is enabled at time i iff
- *for every $l_1,l_2 \in \Lambda(g)$, $\mathcal{V}_i(l_2) \in bt_i(\mathcal{V}_i(l_1))$ and $\mathcal{V}_i(l_1) \in bt_i(\mathcal{V}_i(l_2))$)*
- *for every $l \in \Lambda(g)$, $g@l$ is enabled at time i, i.e.,*
    - *(a) $\mathcal{S},i,\mathcal{V}_i(l) \vdash L(g@l)$;*
    - *(b)   for every $x \in D(g)$, $I_i(F(g@l,x),\mathcal{V}_i(l))$ is defined;*
    - *(c)   for every $f \in \Omega$ used in $L(g@l)$ or $F(g@l,x)$ and every $x \in D(g)$, $\rho_i(m)(f) \leq rs_i(\mathcal{V}_i(l))$;*
    - *(d)   for every $x \in local(V)$ used in used in $L(g@l)$ or $F(g@l,x)$ and $x \in D(g)$, if $l' \in \Lambda(x)$ then $\mathcal{V}_i(l') \in bt_i(\mathcal{V}_i(l))$;*
    - *(e)   for every location $l' \in D(g)$, $I_i(F(g@l,l')) \in reach_i(\mathcal{V}_i(l'))$.*

The intuition behind these conditions, under which a distributed action g can be executed at time i, are the following:
- the execution of g involves the synchronisation of its local actions and, hence, their positions have to be mutually in touch;
- the local guards evaluate to true (in particular, they can be evaluated);
- the operations necessary to perform the computations that are required by $g@l$ are available as well as the resources they demand;
- the execution of the guarded command associated with $g@l$ requires that every channel in its frame can be accessed from its current position and, hence, l has to be in touch with the locations of each of these channels;
- if a location l' can be effected by the execution of $g@l$, then the new value of l' must be a position reachable from the current one.

Formally,

A trace $\mathcal{V}_0.\gamma_0.\mathcal{V}_1.\gamma_1...$ is a behaviour of P iff, for every
- *$i \in \omega$, $\mathcal{V}_{i+1}(x) = \mathcal{V}_i(x)$ for every $x \in local(L \cup X) \setminus \bigcup_{g \in \gamma_i} D(g)$ and for every $g \in \gamma_i$: (a) g is enabled at time i; (b) $\mathcal{S},i,\mathcal{V}_i(l) \vdash R(g@\lambda)$.*
- *$g \in prv(\Gamma)$, if g is infinitely often enabled then there are infinitely many i s.t. $g \in \gamma_i$.*

This defines that the execution of an action consists of the transactional execution of its guarded commands at their locations, which requires the atomic execution of the multiple assignments. Moreover, private actions are subject to a fairness requirement: if infinitely often enabled, they are guaranteed to be selected infinitely often.

5   Conclusions

In this paper, we addressed the design of context-aware systems by proposing design primitives that support the explicit description of contexts as part of the application domain. The idea of having individualized context definitions as part of system designs and the corresponding contextual information transparently provided by the context has many advantages. The maintenance of contextual information tends to be a complex task. For instance, it may require the interaction with heterogeneous physical sensors or the network layer. The separation of the design and construction of the context-provider system from the rest of the application helps to cope with this complexity and allows that context-provider systems be reused in different applications.

Having contexts defined through data type specifications, we showed that the mechanisms available in CommUnity to specify how a system should behave in different situations are applicable also when these situations are characterised by different context states. Moreover, we illustrated around an example, how these mechanisms support the introduction of context-awareness in architectural models in a nonintrusive way.

The importance of a clear separation of the context-aware aspects of system behaviour from the other aspects has been widely recognised. In infrastructure-centred approaches, e.g. [13] and [18], this separation is achieved through the adoption of special mechanisms for the specification of how context influences the behaviour of an application, different from the mechanisms available for the design of the application. However, in general, these approaches do not provide an abstract semantics of these mechanisms and, often, not even address their "physiological structure". For instance, in [18], context-awareness is specified through rules consisting of a context expression and a set of actions that must be performed when the context expression becomes true. However, the notion of action is left undefined and it is not explained to which extent the execution of these actions can interfere with the application behaviour.

In fact, much of the work that has been done in the area of context-aware computing has been devoted to the development of middleware infrastructures that facilitate the implementation of context-aware software by taking the responsibility for the gathering and dissemination of contextual information (e.g. [8], [9]). This work is generally based on rigid and narrow notions of context.

In what concerns the development of design frameworks that support the design of context-aware systems, we are only aware of Context Unity [19]. Context Unity also considers that context-aware systems should be designed assuming that context maintenance is provided by underlying support systems. The way the context can affect the behaviour of a component is, as in CommUnity, part of the component definition but with a completely different perspective on the notion of context. In Context Unity, the operational context with which a component may interact is defined by a set of observables whose values exclusively depend on the values of variables of other components in the system. However, in CommUnity, we consider that the operational context of a design is not under control of any part of the system in which the design is integrated as a component. The advantage of Context Unity is that it is possible to make precise for the context-provider system what the different observables have to be. The disadvantage is that it is only suitable for situations in which the context of a system is, to some extent, expressible in terms of the application domain. This is the case of the running example of [19], an application in which each component has to send messages to a group of components. This group is considered to be a part of the context of the component and is defined in terms of the messages that the component receives. For instance, any component from which a message is received is added to the group; any component that leaves a certain region around the component is removed from the group.

So far we have only addressed the use of context information at the level of the description of components and connectors, i.e., the building blocks of system architectures. It is also important to be able to take advantage of contextual information at the (re)configuration level, namely to use context information to program the dynamic reconfiguration of the system architecture. This includes, for instance, the possibility to react to context changes by removing deployed components or adding new ones, or yet by replacing the connectors in place. For instance, we may wish to specify that a GUI should be replaced by a TextualUI when the battery is low. Some of our future work will progress in this direction.

Acknowledgements

This work was partially supported through the IST-2001-32747 Project *AGILE – Architectures for Mobility*. We wish to thank our partners for much useful feedback.

References

[1]   IST Global Computing Initiative, http://www. cordis.lu/ist/fet/gc.html, 2001.

[2]   M.Acharya, M. Ranganathan and J. Saltz, "Sumatra: A Language for Resource-aware Mobile programs", *Mobile Object Systems: Towards the Programmable Internet*, 1997.

[3]   G.Boudol, "ULM A Core Programming Model for Global Computing", available at http://www-sop.inria.fr/mimosa/personnel/Gerard.Boudol.html.

[4]   L.Cardelli and A.Gordon, "Mobile Ambients", in Nivat (ed), *FoSSACs'98*, LNCS 1378, 140-155, Springer-Verlag, 1998.

[5]   L.Cardelli and R.Dawies, "Service combinators for web computing", *IEEE Transactions on Software Engineering* **25**(3), 303-316, 1999.

[6]   K.Chandy and J.Misra, *Parallel Program Design - A Foundation*, Addison-Wesley, 1988.

[7]   G.Chen and D.Kotz, "A Survey of Context-Aware Mobile Computing Survey", Dartmouth CS-TR 2000-381, 2000.

[8]   G.Chen and D.Kotz. "Context-sensitive resource discovery", *Proc. 1st IEEE International Conference on Pervasive Computing and Communications*, 243-252, 2003.

[9]   A.Dey, D.Salber and G.D.Abowd, "A conceptual framework and a toolkit for supporting the rapid prototyping of context-aware applications", *Human-Computer Interaction* **16**(2-4), 97-166, 2001.

[10] N.Francez and I.Forman, *Interacting Processes*, Addison-Wesley, 1996.

[11] J.L.Fiadeiro, A.Lopes and M.Wermelinger, "A Mathematical Semantics for Architectural Connectors", in R.Backhouse and J.Gibbons (eds), *Generic Programming, LNCS* 2793, 190-234, Springer-Verlag, 2003.

[12] C.Fournet, G.Gonthier, J.-J.Levy, L.Maragent and D.Remy, "A calculus of mobile agents", *CONCUR'96*, LNCS 1119, 315-330, Springer-Verlag, 1996.

[13] K.Henricksen, J.Indulska and A.Rakotonirainy, "Modeling Context Information in Pervasive Computing Systems", *Proc. of Pervasive 2002*, 167-180, 2002.

[14] A.Lopes, J.L.Fiadeiro and M.Wermelinger, "Architectural Primitives for Distribution and Mobility", *Proc. SIGSOFT 2002/FSE-10*, 41-50, ACM Press, 2002.

[15] A.Lopes and J. L. Fiadeiro, "Adding Mobility to Software Architectures", *ENCTS* **97**, 241-258, 2004.

[16] R.De Nicola, G.L.Ferrari and R.Pugliesi, "KLAIM: A Kernel Language for Agents Interaction and Mobility", *IEEE Transactions on Software Engineering* **24**(5), 315-330, 1998.

[17] G.P.Picco, A.L.Murphy and G.-C.Roman, "Lime: Linda meets Mobility", *Proc. ICSE 1999*, 368-377, 1999.

[18] A.Ranganathan and R.Campbell, "An infrastructure for context-awareness based on first order logic", *Pers Ubiquit Computing* **7**, 353-364, 2003.

[19] G.-C.Roman, C.Julien and J.Payton, "A Formal Treatment of Context-Awareness", *Proc. FASE 2004*, LNCS 2984, 12-36, Springer-Verlag, 2004.

[20] G.-C.Roman, P.J.McCann and J.Y.Plun, "Mobile UNITY: reasoning and specification in mobile computing", *ACM TOSEM*, **6**(3), 250-282, 1997.

CCC – The CASL Consistency Checker

Christoph Lüth[1], Markus Roggenbach[2], and Lutz Schröder[1]

[1] Department of Mathematics and Computer Science, Universität Bremen, Germany
[2] Department of Computer Science, University of Wales Swansea, United Kingdom

Abstract. We introduce the CASL Consistency Checker (CCC), a tool that supports consistency proofs in the algebraic specification language CASL. CCC is a faithful implementation of a previously described consistency calculus. Its system architecture combines flexibility with correctness ensured by encapsulation in a type system. CCC offers tactics, tactical combinators, forward and backward proof, and a number of specialised static checkers, as well as a connection to the CASL proof tool HOL-CASL to discharge proof obligations. We demonstrate the viability of CCC by an extended example taken from the CASL standard library of basic datatypes.

1 Introduction

Consistency of specifications is an important issue: validating a specification by proving intended consequences (sanity or conformance checking) is meaningless without a consistency proof – *ex falso quodlibet* –, and implementing a specification is impossible in the presence of inconsistencies. Some formal development paradigms and specification languages handle this problem by excluding inconsistent specifications. In contrast, algebraic specification languages such as CASL [4, 13] allow inconsistent specifications. This allows the developer to concentrate on the desired properties of the system during the requirements engineering phase, then validate their consistency in a separate, later step, before finally proceeding to implement the specification.

This paper describes a prototype of the CASL Consistency Checker (CCC), a tool that supports consistency proofs for CASL specifications. It is a faithful implementation of the previously introduced calculus for consistency proofs of CASL specifications [15]. CCC is part of wider effort to provide tool support for CASL, comprising the CASL tool set CATS [12] which includes a parser, static analysis, and an encoding into higher-order logic, which is used to embed CASL into Isabelle/HOL [14], thus providing proof support for CASL specifications (HOL-CASL).

The material is structured as follows: in Sect. 2, we review the basic concepts of the consistency calculus [15]. We then describe the system architecture in Sect. 3 and show the CCC at work with an extended example in Sect. 4.

J.L. Fiadeiro, P. Mosses, and F. Orejas (Eds.): WADT 2004, LNCS 3423, pp. 94–105, 2005.

2   The Consistency Calculus

The specification language CASL [4, 13] constitutes a standard in algebraic specification. Its features include total and partial functions, predicates, subsorted overloading, sort generation constraints, and structured and architectural specifications. A method for proving consistency of CASL specifications has been introduced in [15]; we briefly recall the main features of the calculus.

The consistency calculus comprises three parts, concerned with specification equivalence, conservativity of extensions, and definitionality of extensions, respectively. The core of the method is the conservativity calculus; consistency of a specification is encoded as conservativity over the empty specification. The implementation extends the calculus of [15] by well-formedness assertions, so that well-formedness also of unparsable specifications (namely, specifications that contain specification variables) can be guaranteed.

The Extension Calculus. This calculus handles extension judgements of the form $Sp_1 \preceq Sp_2$ which state that one has a signature inclusion which is a specification morphism $Sp_1 \to Sp_2$. Equivalence of specifications $Sp_1 \simeq Sp_2$ is defined as mutual extension. These notions of extension and equivalence are meant to be used only for minor syntactical adjustments; in particular, the extension calculus is not intended as a means to establish so-called views, which serve to describe general specification morphisms in CASL. Typical rules of the calculus state that $(Sp_1 \textbf{ then } Sp_2)$ extends Sp_1, that the union of specifications is idempotent, commutative, and associative, and that $(Sp_1 \textbf{ then } Sp_2)$ is equivalent to $(Sp_1 \textbf{ and } Sp_2)$, provided the latter is well-formed (this is an example where a well-formedness assertion is needed).

The Definitionality Calculus. An extension $Sp_1 \preceq Sp_2$ is called *definitional* if each model of Sp_1 extends *uniquely* to a model of Sp_2; in CASL, this is expressed by the semantic annotation %**def**. In particular, definitionality implies conservativity (see below). A definitionality assertion is written

$$def\,(Sp_1)(Sp_2).$$

The definitionality calculus plays an auxiliary role, since the main concern of the method is conservativity. It presently covers definition by abbreviation and primitive recursion; further extensions such as well-founded recursion are obvious, but require more elaborate tool support.

The Conservativity Calculus. The notion of conservativity denoted by the CASL annotation %**cons** is that of model extensivity: an extension $Sp_1 \preceq Sp_2$ is *conservative* if each model M of Sp_1 extends to a model of Sp_2; this is written

$$cons(Sp_1)(Sp_2).$$

The consistency assertion $c(Sp)$ abbreviates $cons(\{\})(Sp)$, where $\{\}$ denotes the empty specification. Since the empty specification has a unique model, $c(Sp)$ indeed states that Sp is consistent. The conservativity rules divide into three major groups:

- Basic language-independent rules, typical examples being a rule that states that conservative extensions compose and a rule which allows deducing conservativity from definitionality.
- Logic-independent rules that propagate conservativity along the various CASL structuring constructs. E.g., unions of specifications are treated by the rule

$$\textbf{(union)} \quad \frac{\begin{array}{cc} Sp_i \text{ defines the signature } \Sigma_i, i = 1, 2 & Sp \preceq Sp_1, Sp \preceq Sp_2 \\ \Sigma_1 \cup \Sigma_2 \text{ is amalgamable} & Sp \text{ defines } \Sigma_1 \cap \Sigma_2 \\ & cons(Sp)(Sp_1) \end{array}}{cons(Sp_2)(Sp_1 \textbf{ and } Sp_2)}$$

Approximative algorithms for checking amalgability are already implemented in CATS. This is a typical case where static side conditions are relegated to further tools integrated into CCC.

- Logic-specific rules that guarantee conservativity for certain syntactic patterns such as data types or positive Horn extensions. A simple example is

$$\textbf{(free)} \quad \frac{\begin{array}{c} newSort(DD_1 \ldots DD_n)(Sp) \\ Sp \textbf{ then types } DD_1; \ldots; DD_n \text{ has a closed term for each new sort} \end{array}}{cons(Sp)(Sp \textbf{ then free types } DD_1; \ldots; DD_n)}$$

where the DD_i are datatype declarations and the assertion $newSort(DD_1 \ldots DD_n)(Sp)$ states that the sorts declared in $DD_1 \ldots DD_n$ are not already in Sp – another example of a proof obligation that is discharged by a static checker.

The strategy for conservativity proofs is roughly as follows: the goal is split into parts using the logic-independent parts of the conservativity calculus, occasionally using the extension calculus for certain sideward steps; at the level of basic specifications, conservativity is then established by the definitionality calculus and the logic-specific rules of the conservativity calculus. This may involve the use of built-in static checkers, and, at eventually pinpointed hot spots, actual theorem proving.

3   System Architecture

For a tool such as a consistency checker or theorem prover, *correctness* is critical: if the tool asserts that a specification is consistent, we need to be sure that this follows from the consistency calculus, not from a bug in the implementation. On the other hand, *flexibility* is important as well: users should be as unconstrained as possible in the way which they conduct their consistency proofs.

CCC's design follows the so-called LCF design [8], where a rich logic (such as higher-order logic) is implemented by a small *logical core* of basic axioms and inference rules. In this design, the logical core implements an abstract datatype of theorems, with logically correct inference rules as operations. Other theorems can only be derived by applying these operations, i.e. by correct inferences; thus the correctness of the whole system is reduced to the correctness of the logical core. The logic encoded within the logical core is called the *meta-logic*, whereas the logic being modelled by the rules is the *object logic*.

Figure 1 shows the system architecture in three layers: innermost, we have the logical core, surrounded by the extended object logic which supplements the meta-logic with specialised proof procedures. The outermost layer is given by auxiliary proof infrastructure.

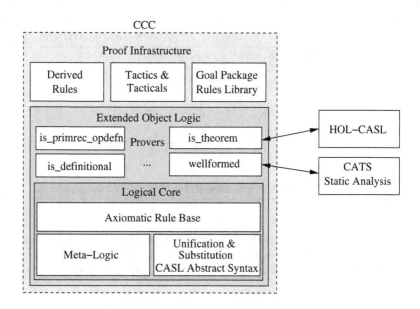

Fig. 1. CCC System Architecture

3.1   The Logical Core

The logical core of CCC implements the meta-logic, the axioms of the object logic, and the axiomatic rule base.

The *meta-logic* is a weak fragment of conjunctive logic. It formalises *rules* as we have seen in Sect. 2 above, and ways in which to manipulate them. A rule allows us to deduce a proposition, the *conclusion*, from a list of propositions, the *premises*. All deductions live in the context of a particular (global) environment which maps *names* to specifications; hence, all rules are parameterised by an environment. We write such a rule as $\Gamma \vdash P_1, \ldots, P_n \to Q$, where Γ is an environment, P_1 to P_n are the premises, and Q is the conclusion. Figure 2 shows the rules of the meta-logic (where $P\sigma$ is the application of a substitution σ to a proposition P, and mgu is the most general unifier of two propositions). This formulation of the meta-logic allows proofs by both forward and backward resolution. Forward resolution is application of the meta-rule $\texttt{compose}_i$ and allows us to derive a new rule from two old ones. Backward resolution allows goal-directed proof (see Sect. 3.3 below).

$$\frac{}{\Gamma \vdash P \to P}\ \texttt{assume}$$

$$\frac{\Gamma \vdash P_1, \ldots, P_n \to Q \quad \Delta \vdash R_1, \ldots, R_m \to S \quad \sigma = mgu(Q, R_i)}{\Delta \vdash R_1\sigma, \ldots, R_{i-1}\sigma, P_1\sigma, \ldots, P_n\sigma, R_{i+1}\sigma, \ldots, R_m\sigma \to S\sigma}\ \texttt{compose}_i \quad \begin{array}{c} 1 \leq i \leq m \\ \Gamma \subseteq \Delta \end{array}$$

$$\frac{\Gamma \vdash P_1, \ldots, P_n \to Q \qquad \begin{array}{c} i \neq j, 1 \leq i, j \leq n, \\ P_i = P_j \end{array}}{\Gamma \vdash P_1, \ldots, P_{j-1}, P_{j+1}, \ldots, P_n \to Q}\ \texttt{contract}_{i,j}$$

$$\frac{\Gamma \vdash P_1, \ldots, P_n \to Q}{\Gamma \vdash P_1\sigma, \ldots, P_n\sigma \to Q\sigma}\ \texttt{specialise}$$

Fig. 2. Inference Rules of the CCC meta-logic

The *object logic* implements the judgements of the conservativity calculus. They are modelled by a datatype **prop**, with CATS used to model the abstract syntax of CASL (in particular, AS.L_SPEC is the type of specifications):

```
datatype prop = consistent_SPEC of AS.L_SPEC
              | conservative of AS.L_SPEC * AS.L_SPEC
              | definitional of AS.L_SPEC * AS.L_SPEC
              | implicational of AS.L_SPEC *AS. L_SPEC
              | ext of AS.L_SPEC * AS.L_SPEC
              | equiv of AS.L_SPEC * AS.L_SPEC
              | provable of pprop
```

The datatype **pprop** is explained in detail below.

The third component of the logical core is the *rule base*. This is a collection of rules the correctness of which has to be proved outside of the system by appealing to the CASL semantics, as opposed to all other rules, which are derived from these rules using the meta-rules; in other words, the rule base contains all rules of the consistency calculus of [15] except the ones explicitly stated as derived.

3.2 Provable Propositions and Provers

The *extended object logic* adds decision procedures, called *provers*, to the logical core. Provers apply to specific proof obligations called *provable propositions* represented by the datatype `pprop`. There are about thirty kinds of provable propositions; an excerpt of the datatype `pprop` showing three typical cases is this:

```
datatype pprop = well_formed of AS.L_SPEC
               | is_just_signature of AS.L_SPEC
               | is_theorem of AS.L_SPEC * AS.FORMULA list | ...
```

The first type of proof obligations states that a particular specification is well-formed; this can be discharged by calling CATS' static analysis. The second states that a specification is merely a signature, and can be discharged by a straightforward recursive function which descends the syntax tree of the specification and returns false as soon as it finds something which does not belong into a signature (such as axioms or free datatypes). The third says that a list of formulae is provable from the given specification, and requires interactive theorem proving using HOL-CASL. A typical rule that has a provable proposition as a premise is the conservativity rule for subtype definitions,

$$\textbf{(sub)} \quad \frac{newSort(s)(Sp)}{cons(Sp)(Sp \textbf{ then sort } s = \{v : t \bullet F\})} \quad .$$

Here, the provable proposition $newSort(s)(Sp)$ states that the sort s is not already declared in Sp, a fact that is easily checked statically.

By distinguishing propositions (`prop`) and proof obligations (`pprop`), we restrict the potential harmful effects of wrongly implemented provers. For example, it is impossible to write a prover which returns `consistent_SPEC(`Sp`)` for every specification Sp. Note that provers are supplied when the system is built, never at run-time by the user.

3.3 Proof Infrastructure

The proof infrastructure contains further modules which facilitate interactive or semi-automatic proof. *Derived rules* are those of the rules from the calculus [15] which can be derived from the rule base. The *tactics package* allows us to write advanced proof procedures. A *tactic* is a function on rules. The rules of the meta-logic give us elementary tactics, which together with combinators such as case distinction or sequential composition can be composed to more sophisticated tactics such as one which handles all definitional extensions. The *rule library* stores and retrieves previously shown results, and the *goal package* allows backwards proof, starting from a stated goal and reducing it to the list of empty premises by tactics application.

Users interact with the system using the SML command line interface, or more comfortably using an instance of the Proof General interface [1]. The latter combines SML's flexibility and expressional power with script management and a comfortable interactive environment.

CCC consists of about 7500 lines of SML code (excluding CATS and HOL-CASL). It runs under SML of New Jersey, but should be easily portable to other SML implementations. Source code and binary builds can be downloaded from the CCC web site [5].

4   Extended Example

To demonstrate the CCC's capabilities, we will show the consistency of the specifications NAT of natural numbers (see Fig. 3) and CHAR of the datatype of ASCII characters (see Fig. 4), taken from the CASL standard library of Basic Datatypes [13][1]. The simple structure of these specifications allows a detailed discussion of their respective CCC proof scripts. However, the proofs involve non-trivial consistency arguments. Furthermore, in the case of the specification CHAR a complete consistency proof is not feasible without tool support due to the length of the specification which involves more than 1000 axioms.

```
spec Nat = free type Bool ::= TT | FF
      then free type Nat   ::= 0  | suc(pre:? Nat)
      then op __<=__: Nat * Nat -> Bool
            ...
      then op   __ * __ : Nat * Nat ->  Nat;
            forall m,n : Nat
            . 0 * m = 0
            . suc(n) * m = (n * m) + m
      then  op   1: Nat = suc (0); ...; op   9: Nat = suc (8);
            op   __ @@ __ (m:Nat;n:Nat): Nat = (m * suc(9)) + n
end
```

Fig. 3. The specification NAT

```
spec Char = Nat
  then sort Byte = { n: Nat . (n <= 255) = TT }
  then free type Char ::= chr(ord: Byte)
  then op  '\000' : Char = chr(0 as Byte);
    ...
  then op  NL:Char = LF;
  then op  '\n' : Char = NL;
end
```

Fig. 4. The specification CHAR

4.1   Consistency Proof of NAT

Figure 5 shows the CCC proof script. We start by loading the library containing NAT (load_lib "Numbers"), stating our goal, and unfolding the specification

---

[1] For the purposes of this paper, the specification text has been slightly modified to make the consistency proof more readable.

```
1    load_lib "Numbers"; ccc "Nat";                    (* start the proof *)
2    ap (compose' Struct.name1);                       (* unfold the spec *)
3    ap (Repeat(OpDefns));                         (* deal with Op defns *)
4    ap (Repeat prim_rec_defns);            (* deal with prim rec defs *)
5    ap (prove_free_type "0" 1);              (* deal with free type Nat,
6                        "0" as witness for non-empty carrier set *)
7    ap (prove 2 Prover.p_holcasl_auto);
8    ap (compose Struct.add_empty 1); (* add empty spec as start point *)
9    ap (prove 1 Prover.p_well_formed);
10   ap (prove_free_type "TT" 1); (* deal with free type Bool:
11                       "TT" as witness for non-empty carrier set *)
12   ap (prove 2 Prover.p_holcasl_auto);
13   ap (compose' Basic.triv_consistency); (* empty spec is consistent *)
14   ap (prove 1 Prover.p_is_just_signature);
15   qeccc "Nat";                                    (* store the result *)
```

Fig. 5. The CCC proof script for the specification NAT

(lines 1–2). The general idea of consistency proofs in CCC is to reduce the overall goal to simpler goals by working backwards through the specification text, reducing it to structures simple enough to show their consistency directly.

In our example, the first step is to show that the operation definition

op __ @@ __ (m:Nat;n:Nat): Nat = (m * suc(9)) + n

is definitional. If this is the case, the whole specification NAT is consistent if its specification text without the last line is consistent. This type of argument (the tactic OpDefns, line 3) can be repeated for all the digit definitions from op 1: Nat = suc (0) to op 9: Nat = suc (8). Here, we can use the tactical combinator Repeat, which applies its argument until it fails. Applying the composed tactic reduces our goal to consistency of this smaller specification:

```
      free type Bool ::= TT | FF
then ...
then op   __ * __ : Nat * Nat ->  Nat;
      forall m,n : Nat
     . 0 * m = 0
     . suc(n) * m = (n * m) + m
```

Here, multiplication on natural numbers is a new operation whose axioms are primitive recursive. This is verified by the tactic prim_rec_defns (line 4). Again, this type of argument can be repeated, as also __+__ and __<=__ are defined by primitive recursion. Hence, we have reduced the specification to be shown consistent to

```
            free type Bool ::= TT | FF
        then free type Nat  ::= 0  | suc(pre:? Nat)
```

```
1    load_lib "SimpleDatatypes"; ccc "Char";          (* start the proof *)
2    ap (compose' Struct.name1);                      (* unfold the spec *)
3    ap (Repeat(OpDefns));                         (* deal with the Op Defns *)
4    ...
5    ap (prove 3 Prover.p_new_sorts_closed_terms_dd);
6                                                 (* deal with free type *)
7    ap (specialize_with (("t", 0), "chr(0 as Byte)") 3);
8    ap (prove 3 Prover.p_closed_term_for_sort);
9    holcasl 3; ...; caslqed();
10
11   ap (compose SpecialExt.sub 2);        (*deal with subsort definition*)
12   ap (prove 2 Prover.p_new_sort);
13   ap (compose Imp.theorem_prover_basic 2 );
14   holcasl 2; by (rtac exI 1);              (* proof in HOL-CASL *)
15   by (rtac leq_def1_Nat 1); caslqed ();
16
17   ap (compose' Struct.named);              (* use the result c(Nat) *)
18   ap (compose' (get "Nat"));
19   qeccc "c_Char";                          (* store the proof *)
```

Fig. 6. The CCC proof script for the specification CHAR

Next, we deal with the definition of the natural numbers as a free type. The tactic **prove_free_type** (line 5) takes 0 as a witness that there exists a defined term of type Nat. In the next step, the definedness of 0 is verified by simple theorem proving in HOL-CASL. These arguments reduce the specification text relevant for consistency to

```
free type Bool ::= TT | FF
```

Now we add the empty specification (lines 8–9), as justified by our equivalence rules:

```
{} then free type Bool ::= TT | FF
```

This allows us to apply again the tactic **prove_free_type** (line 10), this time with TT as a witness. After discharging the proof obligation that TT is defined (line 12), it remains to prove that the empty specification is consistent (line 13). This is verified by the prover **Prover.p_is_just_signature**, which checks that the empty specification does not contain any axioms. Finally, we store our consistency result under the name Nat for later re-use (line 15).

4.2   Consistency Proof of CHAR

For this example, see the proof script in Figure 6, we need to load the library **SimpleDatatypes** (line 1), which imports the specification NAT the consistency of which we have shown in the previous section. After stating our proof goal (line

1) and unfolding the specification (line 2), the first actual proof steps consist of showing that all the operation definitions op `'\ n'` : Char = NL till `'\000'` : Char = chr(0 as Byte) are definitional (remember that we are working backwards). For this purpose we repeat again the tactic OpDefns (line 3). This reduces our goal to show the consistency of this smaller specification:

```
Nat then sort Byte = { n: Nat . (n <= 255) = TT }
    then free type Char ::= chr(ord: Byte)
```

To deal with free type Char ..., we have to show that the sort Char is new and non-empty. The prover Prover.p_new_sorts_closed_terms_dd checks the first condition (line 5) and generates a proof obligation, where the user has to provide a closed term as a witness that the carrier of the sort Char is non-empty; here, we choose "chr(0 as Byte)" (line 6) and can then discharge the proof obligation (line 7)[2]. This leaves us with the proof obligation that chr(0 as Byte) is actually defined, which we discharge with a small HOL-CASL proof (line 8; the details of the proof are elided here).

Similarly, to deal with the subsort definition Byte, we need to check that sort Byte is new (Prover.p_new_sort, line 11) and its carrier is non-empty. To this end, we need to show that there exists an element in the sort Nat which is less or equal to 255 (rule Imp.theorem_prover_basic, line 12), which requires more theorem proving in HOL-CASL (lines 13–14).

We finish the proof by recalling the consistency of Nat using the above established result (line 16–17). This is possible, because the specification Nat has been imported and hence is part of the global environment in which we prove the consistency of Char. Finally, the established theorem "c_Char" is stored with the command qeccc (line 18).

5   Conclusions and Future Work

CCC is a tool to support consistency proofs for specifications written in the standard algebraic specification language CASL. The calculus implemented by CCC supports a proof method where large specifications are split into parts along their explicit specification structure; trivial consistency issues are discharged along the way, leading to the real hot spots of the specification that possibly require actual theorem proving. As presented here, the tool should be seen as a prototype and research vehicle, which we can use to study how to conduct consistency proofs for large, realistic CASL specifications using our calculus; it is certainly not a ready-to-use industrial strength tool yet.

The design of CCC focuses on two main issues: firstly, by basing the design on a small and encapsulated logical core, *correctness* of the tool reduces to the correctness of this core, i.e. essentially correctness of the calculus in [15].

---

[2] These are the steps which we combined to the tactic prove_free_type used in the previous section in the consistency proof of NAT.

Secondly, tactics, tactical combinators, forward and backward proof give users the *flexibility* to conduct consistency proofs in a comfortable and extensible way, and to design powerful proof strategies. This allows us to gradually develop effective and efficient proof tactics for realistic specifications.

The use of CCC has been illustrated by means of an extended example. Further experiments include specifications from the libraries NUMBERS, RELATIONSANDORDERS, ALGEBRA_I from the CASL standard library of Basic Datatypes [13] as well as consistency checks of datatypes evolving in an industrial case study of specifying an electronic payment system [7]. While logically simple, these examples provide enough material both in terms of structure and size to show not only that the tool is able to deal with substantial specifications, but that its use indeed represents added value.

Related work: We can distinguish between approaches which avoid inconsistency by construction, and approaches which show consistency by showing satisfiability. The first approach comprises model-based specification formalisms such as Z [16] and VDM [10], where a model of the system is constructed rather than an axiomatic description; systems based on conservative extension such Isabelle [14], where specifications are built by conservatively extending consistent ones; and systems based on constructive type theory such as Coq [3] or Alfa/Agda [9, 6]. Following the second approach, there is a huge body of work on the satisfiability (and hence consistency) of first-order formulae, which is complimentary to our work; in our terminology, such automatic tools are provers which can be used to prove the consistency of a set of axioms, and we aim to integrate these tools into our system in the future. However, the contribution of our work is to provide a framework in which to conduct consistency proofs for large, structured specifications.

Future work: We will focus on designing more powerful tactics by testing the tool with more examples selected from a wide range of case studies including specifications found in [4, 13, 2, 7]. These examples will be more involved at the level of basic specifications (i.e. in terms of logic rather than in terms of structuring). On the other hand, more decision procedures will be provided in order to increase the degree of automation. Obvious candidates include decision procedures already implemented in the CASL tool set (for example concerning the search for witnesses of non-emptiness of types), as well as existing automatic consistency checkers or SAT solvers such as Chaff [11].

Acknowledgements

The authors would like to thank Janosch Neuweiler and Tobias Thiel for their help in implementing CCC, Erwin R. Catesbeiana for asking the right questions, Till Mossakowski for consultations on HOL-CASL and CATS, and David Aspinall for helping to set up the Proof General interface.

References

1. D. Aspinall, *Proof General: A generic tool for proof development*, Tools and Algorithms for the Construction and Analysis of Systems (TACAS), LNCS, vol. 1785, Springer, 2000, pp. 38–42.
2. H. Baumeister and D. Bert, *Algebraic specification in* CASL, Software Specification Methods: An Overview Using a Case Study (M. Frappier and H. Habrias, eds.), Springer, 2000.
3. Y. Bertot and P. Castéran, *Interactive Theorem Proving and Program Development. Coq'Art: The Calculus of Inductive Constructions*, Springer, 2004.
4. M. Bidoit and P. D. Mosses, CASL *User Manual*, LNCS, vol. 2900, Springer, 2004.
5. *The CCC homepage*, http://www.informatik.uni-bremen.de/cofi/ccc.
6. C. Coquand, *Agda homepage*, http://www.cs.chalmers.se/~catarina/agda.
7. A. Gimblett, M. Roggenbach, and H. Schlingloff, *Towards a formal specification of an electronic payment system in* CSP-CASL, Recent Trends in Algebraic Development Techniques (WADT 204) (José Luiz Fiadeiro, Peter Mosses, and Fernando Orejas, eds.), LNCS, Springer, To appear.
8. M. Gordon, R. Milner, and C. Wadsworth, *Edinburgh LCF: a Mechanised Logic of Computation*, LNCS, vol. 78, Springer, 1979.
9. T. Hallgren, *Alfa homepage*, http://www.cs.chalmers.se/~hallgren/Alfa.
10. C. B. Jones, *Systematic Software Development using VDM*, Prentice Hall, 1990.
11. M. Moskewicz, C. Madigan, Y. Zhao, L. Zhang, and S. Malik, *Chaff: Engineering an efficient SAT solver*, Design Automation, ACM, 2001, pp. 530– 535.
12. T. Mossakowski, CASL - *from semantics to tools*, Tools and Algorithms for the Construction and Analysis of Systems (TACAS), LNCS, vol. 1785, Springer, 2000, pp. 93–108.
13. P. D. Mosses (ed.), CASL *Reference Manual*, LNCS, vol. 2960, Springer, 2004.
14. T. Nipkow, L. C. Paulson, and M. Wenzel, *Isabelle/HOL — A Proof Assistant for Higher-Order Logic*, LNCS, vol. 2283, Springer, 2002.
15. M. Roggenbach and L. Schröder, *Towards trustworthy specifications I: Consistency checks*, Recent Trends in Algebraic Development Techniques (WADT 201), LNCS, vol. 2267, Springer, 2002, pp. 305–327.
16. M. Spivey, *The Z Notation: A Reference Manual*, Prentice Hall, 1992, 2nd edition.

Ontologies for the Semantic Web in CASL

Klaus Lüttich, Till Mossakowski, and Bernd Krieg-Brückner

BISS, FB3 – Dept. of Computer Science, Universität Bremen
{luettich, till, bkb}@tzi.de

Abstract. This paper describes a sublanguage of CASL, called CASL-DL, that corresponds to the Web Ontology Language (OWL) being used for the semantic web. OWL can thus benefit from CASL's strong typing discipline and powerful structuring concepts. Vice versa, the automatic decision procedures available for OWL DL (or more precisely, the underlying description logic $\mathcal{SHOIN}(\mathbf{D})$) become available for a sublanguage of CASL. This is achieved via translations between CASL-DL and $\mathcal{SHOIN}(\mathbf{D})$, formalized as so-called institution comorphisms.

1   Introduction

The internationally standardized Web Ontology Language (OWL) [10] is a major contribution to the upcoming Semantic Web [11, 5] that proposes a new form of web content meaningful to computers. One problem of the documents on the web is the restricted ability to search for certain topics without any knowledge how an author or organisation names the concept. Another problem results form multimedia files like audio or movie files which cannot be indexed by techniques available today; the meaning must be given by meta data. However, just giving a text describing a piece of multimedia yields only a very limited aid for searching.

Therefore, the W3C (World Wide Web Consortium) and Tim Berners Lee proposed the Semantic Web, where the meaning is given by shared and extended ontologies that provide organised knowledge about certain domains; thus the contents of the web is accessible by computers. Hence, it becomes possible e.g. to search for the least cost of a phone call from Singapore to Germany. The visitor from Europe does not need any knowledge of the foreign language, because the query is given in a semantic-based language that is also provided by the Singapore telephone company. Indeed, with Swoogle [1], a first search engine for OWL and RDF documents is available.

In this work, we interface OWL with the specification language CASL [6, 9]. CASL provides a strong typing discipline, which allows to find conceptual errors at an early phase. Moreover, powerful structuring constructs allow the modularization of large theories into manageable pieces. Both features are present in OWL in a very limited form only. We hence propose a sublanguage of CASL, called CASL-DL, which corresponds to OWL DL in expressive power, but which retains the above mentioned advantages. CASL-DL can also be used to interface CASL with efficient decision procedures that are available for description logics.

J.L. Fiadeiro, P. Mosses, and F. Orejas (Eds.): WADT 2004, LNCS 3423, pp. 106–125, 2005.

The paper is organised as follows: Section 2 recalls the underlying description logic $\mathcal{SHOIN}(\mathbf{D})$ of OWL DL. Section 3 describes the Web Ontology Language OWL DL. Section 4 introduces CASL and the sublanguage CASL-DL. Section 5 continues with translations between OWL DL and CASL-DL. Section 6 concludes the paper. Last but not least an appendix collects some tables showing the concrete translations between $\mathcal{SHOIN}(\mathbf{D})$ and CASL-DL constructs including semantics for the $\mathcal{SHOIN}(\mathbf{D})$ constructs.

2 $\mathcal{SHOIN}(\mathbf{D})$

$\mathcal{SHOIN}(\mathbf{D})$ is an expressive description logic [3, 19, 18]. Its main purpose is the definition of hierarchies of concepts and roles. In terms of logic, concepts are unary and roles are binary predicates. The general properties of concepts and roles are collected in a so-called TBox. By contrast, the ABox represents a particular database, i.e. defines individuals to belong to concepts and roles. It also defines concepts and roles involving predefined datatypes. See Fig. 1 for an example of a TBox describing the class definitions of a family.

$$Woman \equiv Person \sqcap Female$$
$$Man \equiv Person \sqcap \neg Woman$$
$$Mother \equiv Woman \sqcap \exists hasChild.Person$$
$$Father \equiv Man \sqcap \exists hasChild.Person$$
$$Parent \equiv Mother \sqcup Father$$
$$Grandmother \equiv Mother \sqcap \exists hasChild.Parent$$
$$MotherWithManyChildren \equiv Mother \sqcap \geqslant 3 \ hasChild$$
$$Wife \equiv Woman \sqcap \exists hasHusband.Man$$

Fig. 1. Example TBox: Family

The standard description logic that is the base of all description logics is called $\mathcal{ALC}$. $\mathcal{ALC}$ has a notation for the universal concept $\top$ and the bottom concept $\bot$ (representing the always true and the empty predicate). Moreover, new concepts can be built with unions, intersections and complements of concepts. Finally, concepts can be universally or existentially projected along roles (e.g. $\exists hasChild.Person$ means the concept that consists of all individuals having some person as their child).

The logic $\mathcal{ALC}_{R^+}$ adds the possibility to specify roles to be transitive. This logic is also abbreviated by $\mathcal{S}$, which is the first letter of the name $\mathcal{SHOIN}(\mathbf{D})$. Likewise, the other letters are used for various features of description logics [3, pp.494-495]. The letter $\mathcal{H}$ adds role hierarchies (i.e., the possibility to specify inclusions between roles) and the letter $\mathcal{I}$ adds inverse roles (i.e. the possibility to generate a new role by just swapping the arguments of a given role). Unqualified number restrictions and nominals are added by $\mathcal{N}$ and $\mathcal{O}$. The former allow stating that any individual is related to at most (or at least) n individuals by a given role. This way, the functionality of relations can be specified. Nominals allow for explicitly enumerating the members of a concept. Datatypes are added by the suffix $(\mathbf{D})$. The order of the letters does not really matter, so $\mathcal{SHOIN}(\mathbf{D})$ is sometimes called $\mathcal{SHION}(\mathbf{D})$. A $\mathcal{SHOIN}(\mathbf{D})$ axiom is either an equality or inclusion between concepts, or an inclusion between roles. A $\mathcal{SHOIN}(\mathbf{D})$ TBox consists of a set of $\mathcal{SHOIN}(\mathbf{D})$ axioms. The semantics of such a TBox is the set of all single-sorted first-order models interpreting concepts as unary and roles as binary relations, and satisfying the TBox axioms (see Figs. 12 and 13 for the $\mathcal{SHOIN}(\mathbf{D})$ syntax and Tables 1 to 7 for its semantics).

2.1   Tools for $\mathcal{SHOIN}(\mathbf{D})$

A crucial motivation for the use of description logics is that they usually correspond to decidable fragments of first-order logic. Indeed, there is a decision procedure for satisfiability and subsumption of $\mathcal{SHIN}(\mathbf{D})$ concepts ($\mathcal{SHIN}(\mathbf{D})$ is $\mathcal{SHOIN}(\mathbf{D})$ without nominals) that runs in non-deterministic exponential time, but performs much better for practical examples [25]. This has been the basis of efficient tools for OWL DL such as FaCT [17]. The tool Pellet [24] uses a combination of known algorithms for $\mathcal{SHIN}(\mathbf{D})$ and $\mathcal{SHON}(\mathbf{D})$ ($\mathcal{SHOIN}(\mathbf{D})$ without inverse properties). It is provably sound but incomplete with respect to the whole of $\mathcal{SHOIN}(\mathbf{D})$. The design of a decision procedure for the whole of $\mathcal{SHOIN}(\mathbf{D})$ is an open problem.

The Guarded Fragment of first-order logic has attracted much attention since it has shown to be decidable [2]. Note that many description logics (as well as propositional modal logic) can be translated into the guarded fragment. However, this does not hold for $\mathcal{SHOIN}(\mathbf{D})$ due to the presence of transitive roles and counting. Indeed, adding either of these features to the guarded fragment results in an undecidable logic.

3   OWL DL

OWL is not a single language, it has three sublanguages that can be ordered according to their expressiveness. The underlying idea of the W3C was to provide languages that were very expressive on the one hand but also useful in automated reasoning processes useful for the Semantic Web. The following species of OWL are available, starting with the most expressive language:

OWL Full provides unrestricted access to all OWL constructs. As RDF Schema, it does not enforce a strict separation of classes, properties, individuals and data values. Hence, there are e.g. no constraints on using a concept, called class in OWL, as an individual at the same time [10, Sect.8.1]. This corresponds to some untyped higher-order logic, leading to non-well-founded sets as semantics.

OWL DL restricts the constructs of OWL in such a way that they correspond to some fragment of first-order logic (actually, roughly to $\mathcal{SHOIN}(\mathbf{D})$). In particular, all axioms must form a tree-like structure. This means e.g. that every reference to a name in a "subclass of" axiom implies the presence of a declaration that this name refers to a class.

OWL Lite adds further constraints to OWL DL that lead to a straight-forward and easy implementation (whereas OWL DL tries to reach the very limits of description logics). For example, it disallows nominals (oneOf, hasValue), union and negation of class descriptions.

Note that neither OWL DL nor OWL Full provide full quantifier logic, but only some restricted forms of quantification corresponding to description logics. We will work with OWL DL here, since it comes closest to the typed first-order fragment of OWL, and we aim at a translation to the typed first-order language CASL. Moreover, apart from the need to involve non-well-founded sets, OWL Full has the additional drawback that no worked-out formal semantics exists, to our knowledge.

3.1 Classes (Concepts) in OWL DL

OWL DL can be understood as syntactic sugar on top of $\mathcal{SHOIN}(\mathbf{D})$, with a slightly new terminology. Concepts are called classes in OWL DL. The universal concept $\top$ is named class *Thing*, the empty concept $\bot$ is named class *Nothing*. New classes are introduced with either complete or partial descriptions. Complete descriptions are introduced by axioms stating equivalence of classes (=equality of concepts), while partial descriptions only specify a subclass (=subconcept) relation to a given class, which means that they are quite loose. Furthermore, the "one of" class axiom gives a complete definition by enumerating all individuals belonging to this class. Additionally, classes can be specified to be disjoint with other classes.

Classes can also be specified via restrictions. Cardinality restrictions specify a lower or upper bound or an exact number of properties that must be present. More precisely, this means that an individual belongs to such a class if and only if the number of individuals that it is related to (via some property, which is a binary relation) meets the specified bounds. "All values from" restrictions allow restricting the values that belong to a class or a role to come from another given class. It is also possible to demand a property to be present in relation to another class. A further restriction defines a class by having a property with (=relation to) a certain value. These restrictions are possible with object properties and datatype properties (see Sect. 3.2 and 3.3).

3.2 Properties (Roles) in OWL DL

Binary relations between individuals, called roles in $\mathcal{SHOIN}(\mathbf{D})$, are called properties in OWL DL. They are divided into two types: Datatype properties relate classes to datatypes (see Sect. 3.3 for the allowed datatypes), while object properties relate classes with classes. The first type of properties can only have other datatype properties as super-properties. Both types of properties can be restricted to a certain domain and range. Without the definition of a domain and range, every class can be related with every other class or datatype by a given property. By giving a domain and range, the property looses this overloadingpossibility. Another way of restricting a property to a certain datatype or class is possible by giving an "all values from" class axiom.

Datatype properties can be defined as the sub-property of and as equivalent to another datatype property. Furthermore, it is possible to specify a datatype property to be functional.

Object properties can have the same axioms regarding subsumption, equivalence and functionality as datatype properties. Additionally, properties can be defined to be symmetric, transitive, functional, inverse functional, or the inverse of another property. OWL also has annotation properties, which have no semantic meaning given by the OWL recommendation.

3.3   XML Schema Datatypes in OWL DL

OWL DL treats the datatypes allowed in $\mathcal{SHOIN}(\mathbf{D})$ somewhat differently than $\mathcal{SHOIN}(\mathbf{D})$. Datatypes are restricted to some XML Schema Datatypes. These are numeric datatypes for integers and various subsets of the integers and for decimal, float and double numbers. Strings and various specialized versions of strings are allowed as well as base64 and hexadecimal encoded binary data. Finally, various time and date specific datatypes are allowed. There is no way of defining further datatypes than those listed in [23, Sect.2.1].

3.4   Facts

Axioms describing the membership of individuals in classes and describing relations between individuals are called facts in OWL DL. In description logics, a set of such axioms is often called an ABox (Fig. 2). It is possible to state that an individual belongs to several classes. Implicitly, every individual is of class *Thing*, because this is the implicit maximal superclass. For properties, it is possible to provide either datatype or object values that are related. Finally, it is possible to state that several names denote different individuals, or the same individual.

Mother(MARY)   Father(PETER)
hasChild(MARY,PETER)   hasChild(PETER,HARRY)
hasChild(MARY,PAUL)

Fig. 2. Example ABox: Family individuals

4   CASL-DL

CASL, the *Common Algebraic Specification Language* [9], has been designed by CoFI, the *Common Framework Initiative* for algebraic specification and development. It has been designed by a large number of experts from different groups, and serves as a de-facto standard. The design of CASL has been approved by the IFIP WG 1.3 "Foundations of System Specification".

CASL consists of several major *levels*[1], which are quite independent and may be understood (and used) separately:

Basic specifications are written in many-sorted first-order logic. Subsorts (interpreted as injective embeddings) increase flexibility, while retaining a strong type system. Partial functions with possibly undefined values are distinguished from total functions. Finally, CASL basic specifications provide powerful and concise constructs for specifying datatypes, which, in the presence of subsorts, may also be used to specify disjoint, non-disjoint and exhaustive unions of sorts.

Structured specifications allow translation, reduction, union, and extension of specifications. A simple form of generic (parameterized) specifications is provided, allowing specifications to be re-used in different contexts.

Libraries allow the distributed storage and retrieval of (particular versions of) named specifications.

Major libraries of validated CASL specifications are freely available on the Internet, and the specifications can be reused simply by referring to their names. Tools are provided to support the practical use of CASL: checking the correctness of specifications, proving facts about them, etc.

While CASL has originally been designed for specifying requirements and design of software, we show here that CASL is also perfectly suited as a language for formalizing ontologies. In particular, we propose to use CASL sorts and subsorts for the development of a class hierarchy, and CASL binary predicates with axioms for the development of properties. This leads to a cleaner methodology for developing ontologies. With the aid of "strongly typed" ontologies it will be easier to avoid inconsistencies.

The sublanguage of CASL corresponding to $\mathcal{SHOIN}(\mathbf{D})$, called CASL-DL, is described as a syntactic restriction w.r.t. CASL. It contains sorts, subsorts, free and generated data types. The sorts *Thing* and *DATA* are assumed to be present in all signatures, and each sort must be a subsort of either of these. Figure 3 presents all declarations of sorts, predicates and functions possible in CASL-DL. Disjointness of concepts is defined easily with free type definitions with subsorts in CASL. Predicates are restricted to unary and binary predicates, where binary predicates relate *Thing* either with *Thing* or with *DATA* or relate subsorts of *Thing* with subsorts of *Thing* or *DATA*. Overloading is allowed, but the first position of the predicate must be a subsort of *Thing* and the second position must be filled either with a subsort of *Thing* or with one of the predefined

---

[1] We omit CASL architectural specifications, since these seem not so relevant here.

sorts $S_1, ..., S_n < S$	(subsort declaration)
sorts $S_1 = ... = S_n$	(sort equivalence)
sort $S_1 = \{\, x : S \bullet \phi(x) \,\}$	(subsort definition)
sort $D_{dr} = \{\, x : D \bullet x = v_1 \vee ... \vee x = v_n \,\}$	(data range subsort definition)
generated type $S ::= o_1|...|o_n$	(generated sort)
free type $S ::= o_1|...|o_n$	(enumerated free sort)
generated type $S ::= sorts\ S_1, ..., S_n$	(generated sort)
free type $S ::= sorts\ S_1, ..., S_n$	(free sort)
ops $\quad o_1, ..., o_n : Thing$	(function declaration)
ops $\quad o_1, ..., o_n : S$	(function declaration)
ops $\quad f_1, ..., f_n : S \rightarrow?\ SD$	(partial function declaration)
preds $P_1, ...P_n : Thing$	(predicate declaration)
preds $R_1, ..., R_n : S \times SD$	(predicate declaration)

where S_i are subsorts of *Thing* and S is *Thing* or a subsort of it, and
where $o_1, ..., o_n$ are constant operations, and
where SD stands for either *Thing* or *DATA* or a subsort of these
where D stands for a subsort of *DATA*

Fig. 3. Declarations, sort and type definitions of CASL-DL

datatypes that are subsorts of *DATA*. Figure 5 shows some of the predefined
data types in CASL-DL. Only partial functions with one argument of sort *Thing*
(or a subsort of it) and with result of either type *Thing* or *DATA* (or a subsort
of these) and 0-ary (constant) functions of type *Thing* (or typed with a subsort
of it) are allowed. Formulas for binary predicates are restricted to those given by
Fig. 4, with the further restriction that axioms for functional predicates cannot
be combined with the transitivity axiom. For predicates and functions which
relate to *DATA* only equivalence and implication axioms are allowed.

Descriptions are formulas that restrict the extension of a description, set-like
combinations of descriptions or assertions of membership in a named concept
as shown in Fig. 6. The formulas regarding the cardinality of predicates are
obtained through the instantiation of the predefined parameterized specification GENCARDINALITY (s. Fig. 10). Special axioms restricting the sort *DATA*
and subsorts of it are presented in Fig. 7. Implicit embedding to supersorts and
explicit downcast to subsorts in terms are allowed. Also, terms with nested function symbols are supported. Figure 8 shows all axioms of CASL-DL built upon
descriptions. Finally axioms relating constant operations are described in Fig. 9.

To illustrate the way in which we would like to write ontologies in CASL-DL,
we give a short example in Fig. 11. Actually, it has been obtained as a result
of a translation described in the next section; a direct specification in CASL-DL
may be more concise at some places, e.g.

$$\textbf{sort}\ Mother = \{x : Woman \bullet \exists y : Person \bullet hasChild(x, y)\}.$$

The example shows the definition of the concept Child that has exactly two
parents. In description logics, so-called cardinality constraints are often used to
denote this. In CASL-DL, a (predefined) generic specification is used to introduce

$$\forall x : S_1; y : S_3 \bullet R(x,y) \Rightarrow Q(x,y) \qquad \text{(implication)}$$
$$\forall x : S_1; y : S_3 \bullet R(x,y) \Leftrightarrow Q(x,y) \qquad \text{(equivalence)}$$
$$\forall x : S_1; y : S_2 \bullet R(x,y) \Leftrightarrow R(y,x) \qquad \text{(symmetry)}$$
$$\forall x : S_1; y : S_2 \bullet R(x,y) \Leftrightarrow Q(y,x) \qquad \text{(inverse)}$$
$$\forall x,y : S_1; z : S_2 \bullet R(x,z) \wedge R(y,z) \Rightarrow x = y \qquad \text{(inverse functional)}$$
$$\forall x : S_1; y : S_2 \bullet R(x,y) \Rightarrow \phi(x) \wedge \psi(y)^a \qquad \text{(argument restriction)}$$
$$\forall x : S_1; y : S_4 \bullet R(x,y) \Rightarrow \phi(x) \wedge \delta(y)^a \qquad \text{(argument restriction)}$$
$$\forall x : S_1; y : S_3 \bullet f(x) = y \Rightarrow Q(x,y) \qquad \text{(implication)}$$
$$\forall x : S_1; y : S_3 \bullet f(x) = y \Rightarrow g(x) = y \qquad \text{(implication)}$$
$$\forall x : S_1; y : S_3 \bullet f(x) = y \Leftrightarrow Q(x,y) \qquad \text{(equivalence)}$$
$$\forall x : S_1; y : S_3 \bullet f(x) = y \Leftrightarrow g(x) = y \qquad \text{(equivalence)}$$
$$\forall x : S_1; y : S_2 \bullet f(x) = y \Leftrightarrow f(y) = x \qquad \text{(symmetry)}$$
$$\forall x : S_1; y : S_2 \bullet f(x) = y \Leftrightarrow Q(y,x) \qquad \text{(inverse)}$$
$$\forall x : S_1; y : S_2 \bullet f(x) = y \Leftrightarrow g(y) = x \qquad \text{(inverse)}$$
$$\forall x,y : S_1; z : S_2 \bullet f(x) = z \wedge f(y) = z \Rightarrow x = y \qquad \text{(inverse functional)}$$
$$\forall x,y : S_1; z : S_2 \bullet R(x,y) \wedge R(y,z) \Rightarrow R(x,z) \qquad \text{(transitivity)}$$
$$\forall x : S_1; y : S_2 \bullet f(x) = y \Rightarrow \phi(x) \wedge \psi(y)^a \qquad \text{(argument restriction)}$$
$$\forall x : S_1; y : S_4 \bullet f(x) = y \Rightarrow \phi(x) \wedge \delta(y)^a \qquad \text{(argument restriction)}$$

where S_1, S_2 is either *Thing* or a subsort of it, and
where S_3 is either *DATA* or *Thing* or a subsort of these
where S_4 is either *DATA* or a subsort of it

---

^{a} one of the conjuncts may be omitted

Fig. 4. Predicate axioms in CASL-DL

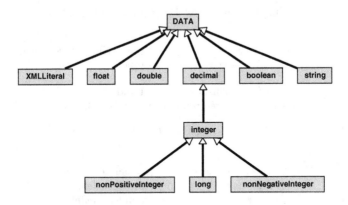

Fig. 5. Some of the predefined datatypes in CASL-DL and their subsort relations

cardinality predicates for a given predicate. So, the instantiation of GENCARDI-NALITY with **pred** *hasChild* yields a new predicate *cardinality*[*hasChild*](p, n) that holds if and only if the Person p is exactly related to n Persons with the predicate *hasChild*.

$$\phi(x), \psi(x) ::= true \mid false \mid$$

$x \in S \mid$	(sort membership)
$P(x) \mid$	(concept membership)
$\neg \phi(x) \mid$	(description negation)
$\phi(x) \wedge \psi(x) \mid$	(description union)
$\phi(x) \vee \psi(x) \mid$	(description intersection)
$\exists y : S \bullet R(x,y)^a \mid$	(existential quantification)
$\exists y : S \bullet R(x,y) \wedge \phi(y)^a \mid$	(existential quantification)
$\forall y : S \bullet R(x,y) \Rightarrow \phi(y)^a \mid$	(value restriction)
$R(x,o)^b \mid$	(has value restriction)
$R(x,v)^c \mid$	(has value restriction)
$x = o_1 \vee ... \vee x = o_n{}^b \mid$	(one of restriction)
$minCardinality[P](x,n)^d \mid$	(cardinality restriction)
$maxCardinality[P](x,n)^d \mid$	(cardinality restriction)
$cardinality[P](x,n)^d \mid$	(cardinality restriction)
$\exists y : D \bullet U(x,y)^e \mid$	(existential quantification)
$\exists y : DATA \bullet U(x,y) \wedge \delta(y)^e \mid$	(existential quantification)
$\forall y : DATA \bullet U(x,y) \Rightarrow \delta(y)^e \mid$	(value restriction)
$\phi(f(x)) \mid$	(existential quantification)
$def\, f(x) \Rightarrow \phi(f(x))$	(value restriction)

[a] where S is either *Thing* or a subsort of it
[b] where o, o_i is an individual, aka ground term, of sort *Thing*
[c] where v is a ground term of sort *DATA*
[d] where n is a ground term of sort *Nat*
[e] where D is either *DATA* or a subsort of it

Fig. 6. Formulas for Descriptions in CASL-DL

$$\delta(x) ::= x \in D \mid \qquad \text{(datatype membership)}$$
$$x = v_1 \vee ... \vee x = v_2 \quad \text{(one of)}$$

where D is a subsort of *DATA* and
where $v_1...v_n$ are ground terms of *DATA*

Fig. 7. Formulas for *DATA* in CASL-DL

$\forall x : Thing \bullet \phi(x) \Rightarrow \psi(x)$	(partial definition)
$\forall x : Thing \bullet \phi(x) \Leftrightarrow \psi(x)$	(complete definition)
$\forall x : S \bullet \psi(x)$	(complete definition)

where S is a subsort of *Thing*

Fig. 8. Description axioms allowed in CASL-DL

5 Translations Between CASL-DL and OWL DL

This paper provides a translation of OWL DL documents to CASL-DL and vice versa, in a way that an OWL DL ontology is optimised in the sense of

- $R(o_1, o_2)$ (predicate application)
- $R(o, v)$ (predicate application)
- $f(o_1) = o_2$ (function application)
- $f(o) = v$ (function application)
- $o \in S$ (membership in sort)
- $\phi(o)$ (description of o)
- $o_1 = o_2$ (same object)
- $\neg\, o_1 = o_2$ (different object)

where o_i and o denote 0-ary constant operations of sort *Thing* or a subsort of it, and

where v is a ground term of sort *DATA*, and

where S is a subsort of *Thing*

Fig. 9. Axioms Relating Constants

spec GENCARDINALITY [**sorts** *Subject, Object*
                        pred *predicate : Subject × Object*] =
{   SET [**sort** *Object*]
     reveal *Set[Object]*, ♯__, __ϵ__, *Nat, 0, 1, 2, 3, 4, 5, 6, 7, 8, 9,* __@@__,
                __≥__, __≤__
then op   *toSet : Subject → Set[Object]*
     $\forall$ *x: Subject; y: Object* • *predicate(x, y)* ⇔ *y* ϵ *toSet(x)*
     preds *minCardinality[predicate](s: Subject; n: Nat)* ⇔ ♯ *toSet(s)* ≥ *n;*
              maxCardinality[predicate](s: Subject; n: Nat) ⇔ ♯ *toSet(s)* ≤ *n;*
              cardinality[predicate](s: Subject; n: Nat) ⇔ ♯ *toSet(s) = n*
} **hide** *Pos, toSet, Set[Object]*, ♯__, __ϵ__, __≤__, __≥__

spec PREDEFINEDCONCEPTS =
     sort   *Thing*
     pred   *Nothing : Thing*
     $\forall$ *x: Thing* • $\neg$ *Nothing(x)*   %(empty_concept_Nothing)%

Fig. 10. CASL-DL Prelude

CASL's strong typing. By relying on the translations between OWL DL and $\mathcal{SHOIN}(\mathbf{D})$ from [18, 19], we just need to define the translation along the syntax of $\mathcal{SHOIN}(\mathbf{D})$. The remaining OWL DL specialties are dealt with as follows. An ontology imported via an import annotation is translated to a specification in CASL-DL and named with a part of the URI [4] of the OWL document. Then the "importing ontology" is an extension of the imported ontology. When defining the translation, we encountered the difficulty that in every OWL document other ontologies could be referenced directly as an URI pointing to a name defined in another OWL document. This problem is solved by localisation of the names as described in the semantics document of OWL [23, Sect.5.1]. The URIs are maintained via custom namespace annotations analogous to XML namespaces [7].

spec FAMILY =
    PREDEFINEDCONCEPTS
then sorts *Person, Female < Thing*;
              Woman < Person;
              Woman < Female;
              Woman = {x: Thing • x ∈ Person ∧ x ∈ Female};
              Man < Person;
              Man = {x: Thing • x ∈ Person ∧ ¬ x ∈ Woman}
      pred *hasChild : Person × Person*
      sorts *Mother < Woman*;
              Mother = {x: Thing • x ∈ Woman ∧
                      ∃ y: Person • hasChild(x as Woman, y)};
              Father < Man;
              Father = {x: Thing • x ∈ Man ∧
                      ∃ y: Person • hasChild(x as Man, y)};
              Parent < Person %% smallest common supersort
      generated type *Parent ::= sorts Mother, Father*
      sorts *Grandmother < Mother*;
              Grandmother = {x: Thing • x ∈ Mother ∧
                          ∃ y: Parent • hasChild(x as Mother, y)}
then GENCARDINALITY [**sort** *Person < Thing*
                      pred *hasChild : Person × Thing*]
then sorts *MotherWithManyChildren < Mother*;
              MotherWithManyChildren =
                  {x: Thing • (x ∈ Mother) ∧
                  minCardinality[hasChild](x as Mother, 3)}
      preds *hasHusband : Woman × Man*;
              Wife(w: Person) ⇔
              w ∈ Woman ∧ ∃ m: Man • hasHusband(w as Woman, m)
end

Fig. 11. Translation of the OWL-DL TBox from Fig.1 into CASL-DL

$$
\begin{array}{llll}
C, D ::= & A \mid & & \text{(atomic concept)} \\
& \top \mid & & \text{(universal concept)} \\
& \bot \mid & & \text{(bottom concept)} \\
& \neg A \mid & & \text{(atomic negation)} \\
& C \sqcap D \mid & & \text{(intersection)} \\
& C \sqcup D \mid & & \text{(union)} \\
& \{o_1, ...\} \mid & & \text{(nominals)} \\
& \forall R.C \mid & & \text{(value restriction)} \\
& \exists R.C \mid & & \text{(existential quantification)} \\
& R : o \mid & & \text{(has value restriction)} \\
& \geqslant n\, R \mid & & \text{(Unqualified} \\
& \leqslant n\, R \mid & & \text{number} \\
& = n\, R & & \text{restriction)}
\end{array}
$$

Fig. 12. Syntax for the Description of Concepts in $\mathcal{SHOIN}(\mathbf{D})$

$$R, S ::= S \mid \text{ (atomic role)}$$
$$R^- \text{ (inverse role)}$$

Fig. 13. Syntax for the Description of Roles in $\mathcal{SHOIN}(\mathbf{D})$

Figures 12 and 13 show the syntax production rules for the constructs of $\mathcal{SHOIN}(\mathbf{D})$. Tables 1 to 7 (see the appendix) describe in detail the $\mathcal{SHOIN}(\mathbf{D})$ constructs, their semantics (according to standard description logic semantics as given by [19]), and their translation to CASL-DL. Here, $\Delta^{\mathcal{I}}$ is the domain of individuals disjoint from the domain of data values $\Delta_D^{\mathcal{I}}$. All the sorts are subsorts either of *DATA* (for datatypes), or of *Thing*. Class descriptions denoted as C or $C_1...C_n$ are translated to formulas with one free variable in CASL-DL. $[\![C]\!](x)$ is the translation of description C in CASL-DL with the free variable x.

In principle, there are two variants of axioms: (1) partial concept axioms and (2) complete concept axioms, presented in Table 1. Table 2 shows axioms for some special cases that can be translated more succinctly (but note that these translations are semantically equivalent to those that could be derived from Tables 1 and 3). For example, a class axiom that defines a named concept can be translated either to a formula of the form $x \in A \Leftrightarrow x \in C_1 \wedge ... \wedge x \in C_n$, or to a subsort definition (where A stands for the name that is given to the class in definition). All variables in the tables without an explicit typing are of type *Thing*.

The translation of OWL DL to CASL-DL is not entirely adequate in the sense that OWL DL admits empty classes, while subsorts in CASL-DL must be non-empty. However, note that *unnamed* classes are translated to formulas in CASL-DL, and the latter may well be unsatisfiable, i.e. denoting the empty class. Only *named* classes are translated to subsorts. Generally, we feel that named classes should be non-empty from a conceptual point of view. However, for the case that the user really wants to have a possibly empty class, we provide an annotation in OWL DL that leads to a translation of the class to a unary predicate in CASL-DL.

The translation from CASL-DL back to $\mathcal{SHOIN}(\mathbf{D})$ (OWL DL) results from the same tables given for the translation from $\mathcal{SHOIN}(\mathbf{D})$ to CASL-DL by reading them from right to left. Of course, this is not well-defined in cases where a certain CASL-DL construct is reached (via the left-to-right translation) either from several $\mathcal{SHOIN}(\mathbf{D})$ constructs, or from none at all. In the first case, we just define the back translation to be the first $\mathcal{SHOIN}(\mathbf{D})$ construct in the table that fits. The second case can occur only for a limited number of constructs, for which we employ a special treatment: (1) Overloaded binary predicates are translated a bit differently from non-overloaded ones. The argument sorts yield two constructs of the form $S_1 \sqcup ... \sqcup S_n$ built from the sorts for the first and second argument. These constructs are then used in $\mathcal{SHOIN}(\mathbf{D})$ formulas for domain and range formulas (s. Tables 5 and 6). (2) The definitions for named classes resulting from unary predicates are marked with an OWL DL annotation. (3) Complete definitions of the form $\forall x : S \bullet \phi(x)$ are transformed to $\forall x : Thing \bullet x \in S \Leftrightarrow \phi(x)$ and this formula is then translated. (4) Subsort definitions of the form **sort** $S_1 = \{ x : S \bullet \phi(x)\}$ are transformed to **sort** $S_1 = \{ x : Thing \bullet \neg x \in S \vee \phi(x)\}$ before the translation.

5.1 Translations as Comorphisms

The translations can be shown to be institution comorphisms in the sense of [13]. First, we present briefly the involved signatures, sentences, models and satisfaction of $\mathcal{SHOIN}(\mathbf{D})$ and CASL-DL, giving rise to institutions in the sense of [14]. Then we show the signature, sentence and model mappings, giving rise to institution comorphisms.

The signatures in $\mathcal{SHOIN}(\mathbf{D})$ consist of several sets: (1) a set of concepts, (2) a subset of primary concepts, (3) a set of individual-valued roles, (4) a set of data-valued roles, (5) a set of individuals and (6) a set of axioms for subconcept relations, domain and range of roles, functional roles and concept membership. Productions of $\mathcal{SHOIN}(\mathbf{D})$ concepts are shown in Fig. 12 and 13. They are used to form sentences like $A \equiv C$ or $A \sqsubseteq C$, where A is a concept name and C is a concept. $\mathcal{SHOIN}(\mathbf{D})$ models consist of: (1) a set Δ^I of individuals and a set Δ^I_D of data values, (2) a subset $A^I \subseteq \Delta^I$ for each concept A in the signature, (3) an element of A^I for every individual, (4) a binary relation for every role: either (a) $R^I \subseteq A^I_1 \times A^I_2$ (individual valued roles) or (b) $R^I \subseteq A^I \times D^I$ with $D^I \subset \Delta^I_D$ (data-valued roles). The satisfaction relation between models and sentences is defined inductively in the obvious way.

Signatures of CASL-DL consist of: (1) a set of sorts, (2) a subsort hierarchy (which is just a pre-order) subsuming all sorts under the sort *Thing*, (3) a set of unary and binary typed relations, (4) a set of unary typed functions and (5) a set of typed constants. Productions of CASL-DL sentences are given in Fig. 4 to 9. CASL-DL models consist of non-empty carrier sets for each sort and an injective function for each subsort relation. Each predicate symbol is associated with a predicate declared on the appropriate carrier set. Each function symbol corresponds to a partial function between the appropriate carrier sets. Satisfaction is defined as standard in partial first-order logic.

Comorphism from $\mathcal{SHOIN}(\mathbf{D})$ to CASL-DL. Signature mappings: (1) Primary concepts are mapped to sorts and subconcept axioms of primary concepts yield a subsort declaration. (2) All non-primary concepts are mapped to unary predicates on either *Thing* or their supersort according to the subconcept axioms. (3) Individual-valued roles with functionality axiom yield a unary operation and the others a binary predicate. According to the type of a role the operation or predicate is typed either with *Thing* or a subsort of it. If a formula is given as type instead of a primary concept (mapped to a sort) argument restriction axioms are generated. (4) Data-valued roles with functionality axiom yield a unary operation and the others a binary predicate. According to the type of a role the operation or predicate is typed either with *Thing* or a subsort of it in the (first) argument position and with *DATA* or or a subsort of it as result / in second argument position. (5) Individuals are mapped to operations either typed with *Thing* or with a subsort of it. (6) From subconcept axioms, which involve concepts mapped to unary predicates, implications are generated. The mappings of sentences are shown in Tab. 1 to 7. A subsorted CASL-DL model with injections can be translated to a subsorted model with inclusions by taking the colimit of the inclusion diagram, see [15] for details. From this, it is straightforward to construct a $\mathcal{SHOIN}(\mathbf{D})$ model.

Comorphism from CASL-DL to $\mathcal{SHOIN}(\mathbf{D})$. Signature mapping: (1) All subsorts of *Thing* and sort *Thing* itself are mapped to primary concepts. The subsort hierarchy yields subsumption axioms. (2) Unary relations are mapped concepts as well. Each type of a predicate gives a subsumption axiom. (3) Binary relations are mapped to either individual-valued properties or data-valued properties according to their type. (4) Unary functions are mapped to either individual-valued properties or data-valued properties according to their type and yield a functionality axiom. (5) Constant operations are mapped to individuals of those primary concepts where their sorts have been mapped to. Tables. 1 to 7 show the mapping of sentences by reading them from right to left. Further details are given above how to map the sentences. A $\mathcal{SHOIN}(\mathbf{D})$-model is mapped to CASL-DL by interpreting the subsorts of *Thing* with interpretations of the corresponding concepts in $\mathcal{SHOIN}(\mathbf{D})$, and similarly for the predicates and individuals. Partial functions are interpreted by taking the interpretation of the corresponding binary role in $\mathcal{SHOIN}(\mathbf{D})$; by the axiomatization, it is the graph of a partial function.

The model translations of both comorphisms are inverse to each other. As a consequence, the institution comorphisms can be used to borrow, in a sound and complete way [8], any proof system that works for either of the two logics also for use with the other one.

6   Conclusion and Future Work

We have defined a sublanguage CASL-DL of CASL that corresponds to the web ontology language OWL DL (where DL refers to the sublanguage corresponding to a description logic), and we have described a translation between OWL DL and CASL-DL in detail. We believe that the main benefit of CASL-DL is the strong typing discipline, which may lead to detection of conceptual errors at a very early stage. By mapping OWL DL's classes and subclasses to CASL's sorts and subsorts, we inherit the flexibility of subsorting while retaining the strong type system. Moreover, CASL-DL offers the possibility to distinguish between a subsort and a unary predicate. Although this distinction is not relevant semantically, it has some conceptual importance [16]. We therefore propose CASL-DL as a language that can be used directly to specify ontologies. Moreover, via the correspondence to OWL DL, efficient tools (like FaCT [17] and Pellet [24]) providing decision procedures for (a large fragment of) OWL DL can also be used for CASL-DL.

A further advantage of CASL-DL is that it naturally comes with CASL's powerful structuring constructs. By contrast, OWL DL and other description logic languages only have quite limited ways to structure and combine ontologies. In fact, the only way that is recommended is a transitive import of other ontologies. To stay within OWL DL requires not to import the OWL Ontology, but to reference all names via URIs. Here CASL's structuring capabilities provide a much more elaborate and cleaner approach, by using hiding and translation of symbols. An interesting question is then the interaction of CASL's structuring concepts with the use of tools like FaCT and Pellet. We expect that the heterogeneous tool set HETS [22] will be a good starting point for answering this question. We also plan to implement the translations described in this paper within HETS.

From an ontological point of view, it would be better not to use CASL-DL for the development of ontologies but a more expressive language. Often KIF [12] (a first order language with Lisp-like syntax) is used for this purpose. But KIF lacks support for structuring ontologies in a nice way. Hence, it would be a good idea to use (full first-order) CASL instead of KIF to develop ontologies: as stated above, CASL provides strong typing and good structuring facilities. In this case, efficient tools like FaCT and Pellet are no longer applicable. Some ontology designers therefore provide their ontology both in a first-order version (e.g. in KIF) and in a description logic version (e.g. in OWL DL) [21]. However, the process of restricting first-order logic to description logic can hardly be automated, and keeping two different version of a document consistent manually is tedious and error-prone. We believe that it is more promising to use tools like the SPASS prover [26] that both support full first-order logic and, when applied to formulas from a suitable description logic fragment, such as CASL-DL, reach the efficiency of specialized tools like FaCT.

In general, we think that for mediating between different languages and tools, translations (formally realized as institution comorphisms) are extremely important. We have sketched two such comorphisms (between CASL-DL and $\mathcal{SHOIN}(\mathbf{D})$); future work will extend this to a graph of formalisms and translations.

References

1. Swoogle search engine. http://swoogle.umbc.edu.
2. H. Andréka, I. Németi, and J. van Benthem. Modal logic and bounded fragments of predicate logic. *Journal of Philosophical Logic*, 27(3):217–274, 1998.
3. F. Baader, D. Calvanese, D. McGuinness, D. Nardi, and P. F. Patel-Schneider, editors. *The Description Logic Handbook*. Cambridge University Press, 2003.
4. T. Berners-Lee, R. Fielding, and L. Masinter. Uniform Resource Identifiers (URI): Generic syntax. IETF Draft Standard (RFC 2396) http://www.ietf.org/rfc/rfc2396.txt, August 1998.

5. T. Berners-Lee, J. Hendler, and O. Lassila. The semantic web. *Scientific American*, May 2001.
6. Michel Bidoit and Peter D. Mosses. CASL *User Manual*. LNCS 2900 (IFIP Series). Springer, 2004. With chapters by Till Mossakowski, Donald Sannella, and Andrzej Tarlecki.
7. T. Bray, D. Hollander, and A. Layman, editors. Namespaces in XML. W3C Recommendation http://www.w3.org/TR/REC-xml-names/, 14 January 1999.
8. M. Cerioli and J. Meseguer. May I borrow your logic? (transporting logical structures along maps). *Theoretical Computer Science*, 173:311–347, 1997.
9. CoFI (The Common Framework Initiative). CASL *Reference Manual*. LNCS 2960 (IFIP Series). Springer, 2004.
10. M. Dean and G. Schreiber, editors. OWL Web Ontology Language – Reference. W3C Recommendation http://www.w3.org/TR/owl-ref/, 10 February 2004.
11. D. Fensel, J. Hendler, H. Liebermann, and W. Wahlster, editors. *Spinning the Semantic Web: Bringing the World Wide Web to Its Full Potential*. The MIT Press, Cambridge, MA, 2003.
12. M. R. Genesereth and R. E. Fikes. Knowlegde interchange format version 3.0 reference manual. Stanford Logic Group, Report Logic-92-1 http://logic.stanford.edu/sharing/papers/kif.ps, June 1992.
13. J. Goguen and G. Rosu. Institution morphisms. *Formal aspects of computing*, 13:274–307, 2002.
14. J. A. Goguen and R. M. Burstall. Institutions: Abstract model theory for specification and programming. *Journal of the Association for Computing Machinery*, 39:95–146, 1992. Predecessor in: LNCS 164, 221–256, 1984.
15. J. A. Goguen and J. Meseguer. Order-sorted algebra I: equational deduction for multiple inheritance, overloading, exceptions and partial operations. *Theoretical Computer Science*, 105:217–273, 1992.
16. N. Guarino. Personal communication.
17. I. Horrocks. FaCT and iFaCT. In Lambrix et al. [20], pages 133–135.
18. I. Horrocks and P. F. Patel-Schneider. Reducing OWL entailment to description logic satisfiability. In D. Fensel, K. Sycara, and J. Mylopoulos, editors, *Proc. of the 2003 International Semantic Web Conference (ISWC 2003)*, number 2870 in Lecture Notes in Computer Science, pages 17–29. Springer, 2003.
19. I. Horrocks, P. F. Patel-Schneider, and F. van Harmelen. From SHIQ and RDF to OWL: The making of a web ontology language. *Journal of Web Semantics*, 1(1):7–26, 2003.
20. P. Lambrix, A. Borgida, M. Lenzerini, R. Möller, and P. F. Patel-Schneider, editors. *Proceedings of the International Workshop on Description Logics (DL'99)*, 1999.
21. C. Masolo, S. Borgo, A. Gangemi, N. Guarino, A. Oltramari, and L. Schneider. Wonderweb deliverable D17. The wonderweb library of foundational ontologies and the DOLCE ontology. November 29 2002. Preliminary Report (ver. 2.0, 15-08-2002).
22. T. Mossakowski. The heterogeneous tool set. Available at www.tzi.de/cofi/hets, University of Bremen.
23. P. F. Patel-Schneider, P. Hayes, and I. Horrocks, editors. OWL Web Ontology Language – Semantics and Abstract Syntax. W3C Recommendation http://www.w3.org/TR/owl-semantics/, 10 February 2004.
24. E. Sirin, M. Grove, B. Parsia, and R. Alford. Pellet OWL reasoner. http://www.mindswap.org/2003/pellet/index.shtml, May 2004.

25. S. Tobies. *Complexity Results and Practical Algorithms for Logics in Knowledge Representation*. PhD thesis, RWTH Aachen, 2001.
26. C. Weidenbach, U. Brahm, T. Hillenbrand, E. Keen, C. Theobalt, and D. Topic. SPASS version 2.0. In Andrei Voronkov, editor, *Automated Deduction – CADE-18*, volume 2392 of *Lecture Notes in Computer Science*, pages 275–279. Springer-Verlag, July 27-30 2002.

Appendix: Translation Tables

Table 1. Base Concept Axioms for the TBox

$\mathcal{SHOIN}(\mathbf{D})$ Semantics		CASL-DL
$C_1 \sqsubseteq C_2$	$C_1^{\mathcal{I}} \subseteq C_2^{\mathcal{I}}$	$\forall x : Thing \bullet [\![C_1]\!](x) \Rightarrow [\![C_2]\!](x)$
$C_1 \equiv C_2$	$C_1^{\mathcal{I}} = C_2^{\mathcal{I}}$	$\forall x : Thing \bullet [\![C_1]\!](x) \Leftrightarrow [\![C_2]\!](x)$

Table 2. Derived Concept Axioms for the TBox

$\mathcal{SHOIN}(\mathbf{D})$	Semantics	CASL-DL
$\forall R.\neg A \sqcup C$	$(\forall R.\neg A \sqcup C)^{\mathcal{I}} =$ $\{x \vert \forall y.\langle x, y\rangle \in R^{\mathcal{I}} \to y \in (\neg A^{\mathcal{I}} \cup C)^{\mathcal{I}}\}$	$\forall y : A \bullet R(x,y) \Rightarrow [\![C]\!](y)$
$\exists R.A \sqcap C$	$(\exists R.A \sqcap C)^{\mathcal{I}} =$ $\{x \vert \exists y.\langle x, y\rangle \in R^{\mathcal{I}} \to y \in A^{\mathcal{I}} \cap C^{\mathcal{I}}\}$	$\exists y : A \bullet R(x,y) \wedge [\![C]\!](y)$

$\mathcal{SHOIN}(\mathbf{D})$	Semantics	CASL-DL
$A \sqsubseteq C_1 \sqcap ... \sqcap C_n$	$A^{\mathcal{I}} \subseteq C_1^{\mathcal{I}} \sqcap ... \sqcap C_n^{\mathcal{I}}$	$\forall x : Thing \bullet x \in A \Rightarrow [\![C_1]\!](x) \wedge ... \wedge [\![C_n]\!](x)$
$A \equiv C_1 \sqcap ... \sqcap C_n$	$A^{\mathcal{I}} = C_1^{\mathcal{I}} \sqcap ... \sqcap C_n^{\mathcal{I}}$	**sort** $A = \{x : Thing$ $\bullet [\![C_1]\!](x) \wedge ... \wedge [\![C_n]\!](x)\}^{a}$
$A_1 \sqsubseteq A_2$	$A_1^{\mathcal{I}} \subseteq A_2^{\mathcal{I}}$	**sorts** $A_1 < A_2$
$A \equiv A_1 \sqcup ... \sqcup A_n$	$A^{\mathcal{I}} = A_1^{\mathcal{I}} \cup ... \cup A_n^{\mathcal{I}}$	**generated type** $A ::= $ **sorts** $A_1, ..., A_n{}^{b}$
$A_1 \equiv A_2$	$A_1^{\mathcal{I}} = A_2^{\mathcal{I}}$	**sorts** $A_1 = A_2$
$A \equiv \{o_1, ..., o_n\}$	$A^{\mathcal{I}} = \{o_1^{\mathcal{I}}, ..., o_n^{\mathcal{I}}\}$	**generated type** $A ::= o_1 \vert ... \vert o_n{}^{c}$
$C_1 \sqcap C_2 \equiv \bot$	$C_1^{\mathcal{I}} \cap C_2^{\mathcal{I}} = \varnothing$	$\forall x : Thing \bullet [\![C_1]\!](x) \Rightarrow \neg[\![C_2]\!](x)$

[a] if C_i is an atomic concept, a subsort declaration for A is generated.

[b] if pairwise disjointness axioms for $A_1, ..., A_n$ are present a **free type** definition is used; an axiom is generated that A is a subsort of the smallest common supersort of $A_1, ..., A_n$.

[c] if axioms are present that $o_1, ..., o_n$ are all different individuals, a **free type** definition is used.

Table 3. Descriptions of Concepts

$\mathcal{SHOIN}(\mathbf{D})$	Semantics	CASL-DL
Descriptions (C)		$[\![C]\!]$
A	$A^{\mathcal{I}} \subseteq \Delta^{\mathcal{I}}$	$x \in A$ for sorts $A(x)$ for predicates
$\top$	$\top^{\mathcal{I}} = \Delta^{\mathcal{I}}$	true
$\bot$	$\bot^{\mathcal{I}} = \varnothing$	false
$C_1 \sqcap ... \sqcap C_n$	$C_1^{\mathcal{I}} \cap ... \cap C_n^{\mathcal{I}}$	$[\![C_1]\!](x) \wedge ... \wedge [\![C_n]\!](x)$
$C_1 \sqcup ... \sqcup C_n$	$C_1^{\mathcal{I}} \cup ... \cup C_n^{\mathcal{I}}$	$[\![C_1]\!](x) \vee ... \vee [\![C_n]\!](x)$
$\neg C$	$(\neg C)^{\mathcal{I}} = \Delta^{\mathcal{I}} \setminus C^{\mathcal{I}}$	$\neg [\![C]\!](x)$
$\{o_1, ..., o_n\}$	$(\{o_1, ..., o_n\})^{\mathcal{I}} = \{o_1^{\mathcal{I}}, ..., o_n^{\mathcal{I}}\}$	$x = o_1 \vee ... \vee x = o_n$
$\forall R.C$	$(\forall R.C)^{\mathcal{I}} = \{x \mid \forall y. \langle x, y \rangle \in R^{\mathcal{I}} \to y \in C^{\mathcal{I}}\}$	$\forall y : Thing \bullet R(x,y) \Rightarrow [\![C]\!](y)^a$
$\exists R.C$	$(\exists R.C)^{\mathcal{I}} = \{x \mid \exists y. \langle x, y \rangle \in R^{\mathcal{I}} \wedge y \in C^{\mathcal{I}}\}$	$\exists y : Thing \bullet R(x,y) \wedge [\![C]\!](y)^b$
$\exists R.A$	$(\exists R.A)^{\mathcal{I}} = \{x \mid \exists y. \langle x, y \rangle \in R^{\mathcal{I}} \wedge y \in A^{\mathcal{I}}\}$	$\exists y : A \bullet R(x,y)$
$R : o$	$(R : o)^{\mathcal{I}} = \{x \mid \langle x, o^{\mathcal{I}} \rangle \in R^{\mathcal{I}}\}$	$R(x,o)$
$\geqslant n\ R$ $\leqslant n\ R$ $= n\ R$	$(\geqslant n\ R)^{\mathcal{I}} = \{x \mid \sharp(\{y. \langle x, y \rangle \in R^{\mathcal{I}}\}) \geq n\}$ $(\leqslant n\ R)^{\mathcal{I}} = \{x \mid \sharp(\{y. \langle x, y \rangle \in R^{\mathcal{I}}\}) \leq n\}$ $(= n\ R)^{\mathcal{I}} = \{x \mid \sharp(\{y. \langle x, y \rangle \in R^{\mathcal{I}}\}) = n\}$	see discussion of GenCardinality in Fig. 10 for the definition of number restrictions in CASL-DL
$\forall U.D$	$(\forall U.D)^{\mathcal{I}} = \{x \mid \forall y. \langle x, y \rangle \in U^{\mathcal{I}} \to y \in D^{\mathcal{I}}\}$	$\forall y : DATA \bullet U(x,y) \Rightarrow [\![D]\!](y)^c$
$\exists U.D$	$(\exists U.D)^{\mathcal{I}} = \{x \mid \exists y. \langle x, y \rangle \in U^{\mathcal{I}} \wedge y \in D^{\mathcal{I}}\}$	$\exists y : DATA \bullet U(x,y) \wedge [\![D]\!](y)^d$
$\exists U.D_0$	$(\exists U.D_0)^{\mathcal{I}} = \{x \mid \exists y. \langle x, y \rangle \in U^{\mathcal{I}} \wedge y \in D_0^{\mathcal{I}}\}$	$\exists y : D_0 \bullet U(x,y)^e$
$U : v$	$(U : v)^{\mathcal{I}} = \{x \mid \langle x, v^{\mathcal{I}} \rangle \in U^{\mathcal{I}}\}$	$U(x,v)$
$\geqslant n\ U$ $\leqslant n\ U$ $= n\ U$	$(\geqslant n\ U)^{\mathcal{I}} = \{x \mid \sharp(\{y. \langle x, y \rangle \in U^{\mathcal{I}}\}) \geq n\}$ $(\leqslant n\ U)^{\mathcal{I}} = \{x \mid \sharp(\{y. \langle x, y \rangle \in U^{\mathcal{I}}\}) \leq n\}$ $(= n\ U)^{\mathcal{I}} = \{x \mid \sharp(\{y. \langle x, y \rangle \in U^{\mathcal{I}}\}) = n\}$	see discussion of GenCardinality and Fig. 10 for the definition of number restrictions in CASL-DL

---

[a] if R is declared as functional the formula is $def\ R(x) \Rightarrow [\![C]\!](R(x))$.

[b] if R is declared as functional the formula is $[\![C]\!](R(x))$.

[c] if U is declared as functional the formula is $def\ U(x) \Rightarrow [\![D]\!](U(x))$.

[d] if U is declared as functional the formula is $[\![D]\!](U(x))$.

[e] where D_0 is a data type name.

Table 4. Declaration of Data Ranges, Roles, Individuals and Data Values

$\mathcal{SHOIN}(\mathbf{D})$	Semantics	CASL-DL
Data Ranges (D)		$[\![D]\!]$
D	$D^{\mathcal{I}} \subseteq \Delta_D^{\mathcal{I}}$	$x \in D$
$\{v_1, ..., v_n\}$	$\{v_1, ..., v_n\}^{\mathcal{I}} = \{v_1^{\mathcal{I}}, ..., v_n^{\mathcal{I}}\}$	$y = v_1 \vee ... \vee y = v_n{}^a$
Object Properties (R)		
R	$R^{\mathcal{I}} \subseteq \Delta^{\mathcal{I}} \times \Delta^{\mathcal{I}}$	**pred** $R : Thing \times Thing$
Datatype Properties (U)		
U	$U^{\mathcal{I}} \subseteq \Delta^{\mathcal{I}} \times \Delta_D^{\mathcal{I}}$	**pred** $U : Thing \times DATA$
Individuals (o)		
o	$o^{\mathcal{I}} \in \Delta^{\mathcal{I}}$	**op** $o : Thing$
Data Values (v)		
v	$v^{\mathcal{I}} \in \Delta_D^{\mathcal{I}}$	some constant term of a pre-defined datatype

[a] for typing of binary predicates they are named subsorts of $DATA$ (s. Fig. 3 (data range subsort definition)).

Table 5. TBox Axioms for General Properties

$\mathcal{SHOIN}(\mathbf{D})$	Semantics	CASL-DL
$U \sqsubseteq U_i$	$U^{\mathcal{I}} \subseteq U_i^{\mathcal{I}}$	$\forall y : DATA \bullet U(x,y) \Rightarrow U_i(x,y)$
$\geqslant 1\, U \sqsubseteq C_i$	$U^{\mathcal{I}} \subseteq C_i^{\mathcal{I}} \times \Delta_D^{\mathcal{I}}$	$\forall x : Thing$ $\bullet\, minCardinality[U](x,1) \Rightarrow [\![C_i]\!](x)^a$
$\geqslant 1\, U \sqsubseteq A_i$	$U^{\mathcal{I}} \subseteq A_i^{\mathcal{I}} \times \Delta_D^{\mathcal{I}}$	**pred** $U : A_i \times DATA^b$
$\top \sqsubseteq \forall U.D_i$	$U^{\mathcal{I}} \subseteq \Delta^{\mathcal{I}} \times D_i^{\mathcal{I}}$	**pred** $U : Thing \times D^{bc}$
$U_1 = U_2$	$U_1^{\mathcal{I}} = U_2^{\mathcal{I}}$	$U_1(x,y) \Leftrightarrow U_2(x,y)$
$R \sqsubseteq R_i$	$R^{\mathcal{I}} \subseteq R_i^{\mathcal{I}}$	$R(x,y) \Rightarrow R_i(x,y)$
$\geqslant 1\, R \sqsubseteq C_i$	$R^{\mathcal{I}} \subseteq C_i^{\mathcal{I}} \times \Delta^{\mathcal{I}}$	$\forall x : Thing$ $\bullet\, minCardinality[R](x,1) \Rightarrow [\![C_i]\!](x)^a$
$\geqslant 1\, R \sqsubseteq A_i$	$R^{\mathcal{I}} \subseteq A_i^{\mathcal{I}} \times \Delta^{\mathcal{I}}$	**pred** $R : A_i \times Thing^b$
$\top \sqsubseteq \forall R.C_j$	$R^{\mathcal{I}} \subseteq \Delta^{\mathcal{I}} \times C_j^{\mathcal{I}}$	$\forall x : Thing$ $\bullet\, true \Rightarrow \forall y : Thing \bullet R(x,y) \Rightarrow [\![C_j]\!](y)^a$
$\top \sqsubseteq \forall R.A_j$	$R^{\mathcal{I}} \subseteq \Delta^{\mathcal{I}} \times A_j^{\mathcal{I}}$	**pred** $R : Thing \times A_j{}^b$
$R = R_0^-$	$R^{\mathcal{I}} = (R_0^{\mathcal{I}})^-$	$R(x,y) \Leftrightarrow R_0(y,x)$
$R = R^-$	$R^{\mathcal{I}} = (R^{\mathcal{I}})^-$	$R(x,y) \Leftrightarrow R(y,x)$
$\top \sqsubseteq \leqslant 1\, R^-$	$(R^{\mathcal{I}})^-$ is functional	$R(x,z) \wedge R(y,z) \Rightarrow x = y$
$Tr(R)$	$R^{\mathcal{I}} = (R^{\mathcal{I}})^+$	$R(x,y) \wedge R(y,z) \Rightarrow R(x,z)$
$R_1 = R_2$	$R_1^{\mathcal{I}} = R_2^{\mathcal{I}}$	$R_1(x,y) \Leftrightarrow R_2(x,y)$

[a] for C_i and/or C_j of form $A_1 \sqcup ... \sqcup A_n$ overloaded predicate profiles of all pairs are constructed; other description formulas for C_i and/or C_j yield argument restriction formulas in CASL-DL.

[b] if both domain and range are specified U gets type $A_i \times D$.

[c] D is either a datatype name or a named datatype range derived from $[\![D_i]\!]$.

[d] if both domain and range are specified R gets type $A_i \times A_j$.

Table 6. TBox Axioms for Functional Properties

$\mathcal{SHOIN}(\mathbf{D})$	Semantics	CASL-DL
$\top \sqsubseteq\, \leqslant 1\, U$	$U^{\mathcal{I}}$ is functional	**op** $U : Thing \to? D^a$
$\top \sqsubseteq\, \leqslant 1\, R$	$R^{\mathcal{I}}$ is functional	**op** $R : Thing \to? Thing$
$U \sqsubseteq U_i$	$U^{\mathcal{I}} \subseteq U_i^{\mathcal{I}}$	$\forall y : DATA \bullet U(x) = y \Rightarrow U_i(x,y)^b$
$\geqslant 1\, U \sqsubseteq C_i$	$U^{\mathcal{I}} \subseteq C_i^{\mathcal{I}} \times \Delta_D^{\mathcal{I}}$	$\forall x : Thing$ $\bullet\ minCardinality[U](x,1) \Rightarrow [\![C_i]\!](x)^{cd}$
$\geqslant 1\, U \sqsubseteq A_i$	$U^{\mathcal{I}} \subseteq A_i^{\mathcal{I}} \times \Delta_D^{\mathcal{I}}$	**op** $U : A_i \to? DATA^e$
$\top \sqsubseteq \forall U.D_i$	$U^{\mathcal{I}} \subseteq \Delta^{\mathcal{I}} \times D_i^{\mathcal{I}}$	**op** $U : Thing \to? D^{ae}$
$U = U_1$	$U^{\mathcal{I}} = U_1^{\mathcal{I}}$	$U(x) = y \Leftrightarrow U_1(x,y)^f$
$R \sqsubseteq R_i$	$R^{\mathcal{I}} \subseteq R_i^{\mathcal{I}}$	$R(x) = y \Rightarrow R_i(x,y)^b$
$\geqslant 1\, R \sqsubseteq C_i$	$R^{\mathcal{I}} \subseteq C_i^{\mathcal{I}} \times \Delta^{\mathcal{I}}$	$\forall x : Thing$ $\bullet\ minCardinality[R](x,1) \Rightarrow [\![C_i]\!](x)^{cd}$
$\geqslant 1\, R \sqsubseteq A_i$	$R^{\mathcal{I}} \subseteq A_i^{\mathcal{I}} \times \Delta^{\mathcal{I}}$	**op** $R : A_i \to? Thing^g$
$\top \sqsubseteq \forall R.C_j$	$R^{\mathcal{I}} \subseteq \Delta^{\mathcal{I}} \times C_j^{\mathcal{I}}$	$\forall x : Thing$ $\bullet\ true \Rightarrow def\, R(x) \Rightarrow [\![C_j]\!](R(x))^c$
$\top \sqsubseteq \forall R.A_j$	$R^{\mathcal{I}} \subseteq \Delta^{\mathcal{I}} \times A_j^{\mathcal{I}}$	**op** $R : Thing \to? A_j^g$
$R = R^-$	$R^{\mathcal{I}} = (R^{\mathcal{I}})^-$	$R(x) = y \Leftrightarrow R(y) = x$
$\top \sqsubseteq\, \leqslant 1\, R^-$	$(R^{\mathcal{I}})^-$ is functional	$R(x) = z \wedge R(y) = z \Rightarrow x = y$
$R = R_1$	$R^{\mathcal{I}} = R_1^{\mathcal{I}}$	$R(x) = y \Leftrightarrow R_1(x,y)^b$

[a] D is either a datatype name or a named datatype range derived from $[\![D_i]\!]$.

[b] if U_i or R_i is also declared as functional this formula is $\forall y : SD \bullet q(x) = y \Rightarrow$ $q_i(x) = y$ where SD is either $DATA$ or $Thing$ and q either U or R.

[c] for C_i and/or C_j of form $A_1 \sqcup ... \sqcup A_n$ overloaded function profiles of all pairs are constructed.

[d] where GENCARDINALITY is instantiated with **pred** $q(x : s_1 ; y : s_2) \Leftrightarrow q(x) = y$ where q is either R or U.

[e] if both domain and range are specified U gets type $A_i \to? D$.

[f] if U_1 or R_1 is also declared as functional the formula is $q(x) = y \Leftrightarrow q_1(x) = y$ where q is either U or R.

[g] if both domain and range are specified R gets type $A_i \to? A_j$.

Table 7. Axioms for the ABox

$\mathcal{SHOIN}(\mathbf{D})$	Semantics	CASL-DL
$o \in A$	$o^{\mathcal{I}} \in C_i^{\mathcal{I}}$	**op** $o : A$
$o \in C_i$	$o^{\mathcal{I}} \in C_i^{\mathcal{I}}$	$[\![C_i]\!](o)$
$\langle o, o_i \rangle \in R_i$	$\langle o^{\mathcal{I}}, o_i^{\mathcal{I}} \rangle \in R_i^{\mathcal{I}}$	$R_i(o, o_i)^a$
$\langle o, v_i \rangle \in U_i$	$\langle o^{\mathcal{I}}, v_i^{\mathcal{I}} \rangle \in U_i^{\mathcal{I}}$	$U_i(o, v_i)^a$
$o_1 = o_2$	$o_1^{\mathcal{I}} = o_2^{\mathcal{I}}$	$o_1 = o_2$
$o_1 \neq o_2$	$o_1^{\mathcal{I}} \neq o_2^{\mathcal{I}}$	$\neg o_1 = o_2$

[a] if R is declared as functional the translation result is $R_i(o) = o_i$.

[b] if U is declared as functional the translation result is $U_i(o) = v_i$.

Theoroidal Maps as Algebraic Simulations[*]

Narciso Martí-Oliet[1], José Meseguer[2], and Miguel Palomino[1]

[1] Departamento de Sistemas Informáticos,
Universidad Complutense de Madrid
[2] Computer Science Department,
University of Illinois at Urbana-Champaign
{narciso, miguelpt}@sip.ucm.es
meseguer@cs.uiuc.edu

Abstract. Computational systems are often represented by means of Kripke structures, and related using simulations. We propose rewriting logic as a flexible and executable framework in which to formally specify these mathematical models, and introduce a particular and elegant way of representing simulations in it: theoroidal maps. A categorical viewpoint is very natural in the study of these structures and we show how to organize Kripke structures in categories that afterwards are lifted to the rewriting logic's level. We illustrate the use of theoroidal maps with two applications: predicate abstraction and the study of fairness constraints.

1 Introduction

Formal reasoning about concurrent systems typically involves two levels of specification: (1) a *system specification* level, in which an explicit computational description of a concurrent system is given; and (2) a *property specification* level, in which different safety and liveness properties satisfied by the system are specified. A system specification typically determines a *mathematical model* (or set of models) about which we want to verify that some properties are satisfied. Frequently used mathematical models include transition systems, and Kripke structures—i.e., transition systems decorated with information about satisfaction of atomic predicates. For properties, different temporal and modal logics can be used; CTL* [5] is a common choice, because it contains the widely used LTL and CTL logics as special cases.

But how can such mathematical models be formally specified? There are many possibilities. In this paper we specify them by means of *rewrite theories*. This is a natural choice, because rewriting logic provides a flexible framework for specifying a wide range of concurrent systems at a high level [16, 14], yet in an executable way supported by languages such as Maude in which we can simulate and model check such systems [7, 8]. Essentially, system states are specified as elements of an initial algebra, and (parameterized) transitions as rewrite rules. Furthermore, it is then very easy to

---

[*] Research supported by ONR Grant N00014-02-1-0715, NSF Grant CCR-0234524, and by DARPA through Air Force Research Laboratory Contract F30602-02-C-0130; and by the Spanish projects MELODIAS TIC 2002-01167 and MIDAS TIC 2003–0100.

J.L. Fiadeiro, P. Mosses, and F. Orejas (Eds.): WADT 2004, LNCS 3423, pp. 126–143, 2005.
© Springer-Verlag Berlin Heidelberg 2005

equationally specify *atomic predicates* holding on the states in a theory extension. In this way, we can associate a Kripke structure $\mathcal{K}(\mathcal{R},k)_\Pi$ to a rewrite theory $\mathcal{R}$, a kind of states k, and atomic propositions Π. Given a CTL* formula φ, then the issue of whether the system specification satisfies the property φ becomes the question of verifying whether $\mathcal{K}(\mathcal{R},k)_\Pi \models \varphi$ holds.

However, it may be considerably easier to verify such a satisfaction relation using a *different* system specification $\mathcal{R}'$. For example, $\mathcal{K}(\mathcal{R},k)_\Pi$ may have infinitely many states, whereas $\mathcal{K}(\mathcal{R}',k)_\Pi$ may be a finite-state abstraction of $\mathcal{R}$ [19], so that we can use a model checker to verify $\mathcal{K}(\mathcal{R}',k)_\Pi \models \varphi$. From this we can infer that $\mathcal{K}(\mathcal{R},k)_\Pi \models \varphi$ holds, provided that $\mathcal{R}$ and $\mathcal{R}'$ can be related by an adequate *simulation map* $H : \mathcal{R} \longrightarrow \mathcal{R}'$. This of course suggests a categorical approach, and also exploring an adequate notion of theory morphism to define such simulations at a logical level. This is the goal of this paper. Specifically we:

- Define a category with objects Kripke structures and morphisms quite general "stuttering simulations," and show that properties specified by a natural subclass of CTL* formulas are reflected by such simulations.
- Show that those CTL* formulas, with Kripke structures as models and simulations as morphisms, form an *institution* [12].
- Explain the $\mathcal{K}(\mathcal{R},k)_\Pi$ construction in detail allowing us to specify Kripke structures by means of rewrite theories.
- Present a new notion of *partial theory morphism* which allows a more general and expressive way of relating theories than with ordinary theory morphisms.
- Define a category with rewrite theories (plus the specification of the kind of states and the state predicates) as objects, and suitable partial theory morphisms as morphisms, and show that they define a useful class of simulations between the underlying Kripke structures, which we call *theoroidal simulations*.
- Illustrate the usefulness of this notion in several areas, including predicate abstraction, and reasoning about temporal logic properties under fairness assumptions. Furthermore, theoroidal simulations greatly generalize *equational abstractions*, which were already shown to be very useful in [19].

An extended version of this paper with the missing proofs can be found in [15].

2 Prerequisites

2.1 Computational Systems

When reasoning about computational systems, it is usually convenient to abstract from as many details as possible by means of simple mathematical models that can be used to reason about them. For a state-based system we can represent its behavior by means of a *transition system*, which is a pair $\mathcal{A} = (A, \rightarrow_\mathcal{A})$ with A a set of states and $\rightarrow_\mathcal{A} \subseteq A \times A$ a binary relation called the transition relation.

A transition system, however, does not include any information about the relevant properties of the system. In order to reason about such properties it is necessary to add information about the atomic properties that hold in each state. In what follows, we assume a fixed set AP of atomic propositions and define a *Kripke structure* as a triple $\mathscr{A} = (A, \rightarrow_{\mathscr{A}}, L_{\mathscr{A}})$, where $(A, \rightarrow_{\mathscr{A}})$ is a transition system with $\rightarrow_{\mathscr{A}}$ a *total* relation, and $L_{\mathscr{A}} : A \rightarrow \mathscr{P}(AP)$ is a labeling function associating to each state the set of atomic propositions that hold in it. Note that the transition relation must be total [5]; given an arbitrary relation $\rightarrow$, we write $\rightarrow^{\bullet}$ for the total relation that extends $\rightarrow$ by adding a pair $a \rightarrow^{\bullet} a$ for each a such that there is no b with $a \rightarrow b$. A path in $\mathscr{A}$ is a function $\pi : \mathbb{N} \longrightarrow A$ such that, for each $i \in \mathbb{N}$, $\pi(i) \rightarrow_{\mathscr{A}} \pi(i+1)$.

To specify system properties we will use the logic $\text{ACTL}^*(AP)$, which is a sublogic of the branching-time temporal logic $\text{CTL}^*(AP)$ (see for example [5–Sect. 3.1]). There are two types of formulas in $\text{CTL}^*(AP)$: state formulas, denoted by $\text{State}(AP)$, and path formulas, denoted by $\text{Path}(AP)$. The semantics of the logic, specifying the satisfaction relations $\mathscr{A}, a \models \varphi$ and $\mathscr{A}, \pi \models \psi$ for a Kripke structure $\mathscr{A}$, an initial state $a \in A$, a state formula φ, a path π, and a path formula ψ, is defined as usual [5]. $\text{ACTL}^*(AP)$ is the restriction of $\text{CTL}^*(AP)$ to those formulas such that their negation-normal forms (with negations pushed to atoms) do not contain any existential path quantifiers. Sometimes, to avoid introducing existential quantifiers implicitly, it is more convenient to restrict ourselves to the negation-free fragment $\text{ACTL}^*\backslash\neg(AP)$ of $\text{ACTL}^*(AP)$, defined as follows:[1]

state formulas: $\varphi = p \in AP \mid \top \mid \bot \mid \varphi \vee \varphi \mid \varphi \wedge \varphi \mid \mathbf{A}\psi$

path formulas: $\psi = \varphi \mid \psi \vee \psi \mid \psi \wedge \psi \mid \mathbf{X}\psi \mid \psi \mathbf{U}\psi \mid \psi \mathbf{R}\psi \mid \mathbf{G}\psi \mid \mathbf{F}\psi.$

We write $\text{State}\backslash\neg(AP)$ and $\text{Path}\backslash\neg(AP)$ for the sets of state and path formulas in $\text{ACTL}^*\backslash\neg(AP)$, respectively.

2.2   Rewriting Logic

Rewriting logic [16] provides a very flexible framework for the system-level specification of concurrent systems. It is parameterized by an underlying equational logic, which we will also use to specify the system's properties; in this paper we use membership equational logic [17], whose main features we now review.

A *signature* in membership equational logic is a triple (K, Σ, S) (just Σ in the following), with K a set of *kinds*, $\Sigma = \{\Sigma_{k_1...k_n,k}\}_{(k_1...k_n,k) \in K^* \times K}$ a many-kinded signature, and $S = \{S_k\}_{k \in K}$ a pairwise disjoint K-kinded family of sets of *sorts*. The kind of a sort s is denoted by $[s]$. We write $T_{\Sigma,k}$ and $T_{\Sigma,k}(X)$ to denote respectively the set of ground Σ-terms with kind k and of Σ-terms with kind k over variables in X, where $X = \{x_1 : k_1, \ldots, x_n : k_n\}$ is a set of K-kinded variables. Intuitively, terms with a kind but without a sort represent undefined or error elements. An atomic formula is either an *equation* $t = t'$, where t and t' are Σ-terms of the same kind, or a *membership assertion* of the form $t : s$, where the term t has kind k and $s \in S_k$. Sentences are conditional formulas of the form $(\forall X) A_0$ if $A_1 \wedge \ldots \wedge A_n$, where each A_i is either an equation or a membership assertion, and X is a set of K-kinded variables containing all the variables in the A_i. A theory is a pair (Σ, E), where E is a set of sentences in membership equational logic over the signature Σ. We write $(\Sigma, E) \vdash \phi$, or just $E \vdash \phi$ if Σ is clear from the context, to

---

[1] $\mathbf{X}$, $\mathbf{G}$, and $\mathbf{F}$ stand for the classic *next* ($\bigcirc$), *henceforth* ($\square$), and *eventually* ($\lozenge$) LTL operators.

denote that (Σ, E) entails the sentence ϕ in the proof system of membership equational logic [17]. A theory (Σ, E) has an initial model $T_{\Sigma/E}$ whose elements are E-equivalence classes of terms $[t]$. Algebras over a signature are defined in a standard manner; we denote by A_f the interpretation of an operator f in the algebra A and by A_t that of a term t, and refer to [17] for a detailed presentation of the model theory.

Concurrent systems are axiomatized in rewriting logic by means of *rewrite theories* [16] of the form $\mathscr{R} = (\Sigma, E, R)$. The set of states is described by a membership equational theory (Σ, E) as the algebraic data type $T_{\Sigma/E,k}$ associated to the initial algebra $T_{\Sigma/E}$ of (Σ, E) by the choice of a kind k of states in Σ. The system's *transitions* are axiomatized by the *conditional rewrite rules R* which are of the form

$$\lambda : (\forall X)t \longrightarrow t' \text{ if } \bigwedge_{i \in I} p_i = q_i \wedge \bigwedge_{j \in J} w_j : s_j \wedge \bigwedge_{l \in L} t_l \longrightarrow t_l',$$

with λ a label, $p_i = q_i$ and $w_j : s_j$ atomic formulas in membership equational logic for $i \in I$ and $j \in J$, and for appropriate kinds k and $k_l, t, t' \in T_{\Sigma,k}(X)$, and $t_l, t_l' \in T_{\Sigma,k_l}(X)$ for $l \in L$. Under reasonable assumptions about E and R, rewrite theories are *executable*. Indeed, there are several rewriting logic language implementations, including CafeOBJ [11], ELAN [3], and Maude [7, 8]. Rewriting logic then has inference rules to infer all the possible concurrent computations in a system [16, 4], in the sense that, given two states $[u], [v] \in T_{\Sigma/E,k}$, we can reach $[v]$ from $[u]$ by some possibly complex concurrent computation iff we can prove $u \longrightarrow v$ in the logic; we denote this provability by $\mathscr{R} \vdash u \longrightarrow v$. In particular we can easily define the *one-step $\mathscr{R}$-rewriting relation*, which is a binary relation $\rightarrow^1_{\mathscr{R},k}$ on $T_{\Sigma,k}$ that holds between terms $u, v \in T_{\Sigma,k}$ iff there is a proof of $u \longrightarrow v$ in which only one rewrite rule in R is applied to a single subterm.

2.3   Computational Systems in Rewriting Logic

To associate a transition system to a rewrite theory we transfer the one-step rewriting relation $\rightarrow^1_{\mathscr{R},k}$ from terms in $T_{\Sigma,k}$ to states in $T_{\Sigma/E,k}$, by defining $[u] \rightarrow^1_{\mathscr{R},k} [v]$ iff $u' \rightarrow^1_{\mathscr{R},k} v'$ for some $u' \in [u]$, $v' \in [v]$. This definition determines a transition system $\mathscr{T}(\mathscr{R})_k = (T_{\Sigma/E,k}, (\rightarrow^1_{\mathscr{R},k})^\bullet)$ for each $k \in K$.

In order to associate temporal properties to a rewrite theory $\mathscr{R} = (\Sigma, E, R)$ we need to make explicit two things: the intended *kind k* of states in the signature Σ, and the relevant *state predicates*. Once the kind k is fixed, the transitions between states are given by $\mathscr{T}(\mathscr{R})_k$. In general, however, the state predicates need not be part of the system specification but only of the property specification. We assume that they have been defined by means of equations D in a *protecting* theory extension $(\Sigma', E \cup D)$ of (Σ, E); that is, the extension is conservative in the sense that the unique Σ-homomorphism $T_{\Sigma/E} \longrightarrow T_{\Sigma'/E \cup D}|_\Sigma$ should be bijective at each sort in Σ. We also assume that $(\Sigma', E \cup D)$ is a protecting theory extension of *BOOL*, the theory of Boolean values. Furthermore, we assume that the syntax defining the state predicates consists of a subsignature $\Pi \subseteq \Sigma'$ of operators, with each $p \in \Pi$ a state predicate symbol that can be *parameterized*, that is, p need not be a constant, but can in general be an operator $p : s_1 \ldots s_n \longrightarrow Prop$, with *Prop* the kind of propositions. If k is the kind of states, the *semantics* of the state predicates Π is defined with the help of an operator $\_ \models \_ : k \, Prop \longrightarrow Bool$ in Σ' and by equations $E \cup D$. By definition, given ground terms $u_1, \ldots, u_n$, we say that the state predicate $p(u_1, \ldots, u_n)$ *holds* in the state $[t]$ iff $E \cup D \vdash t \models p(u_1, \ldots, u_n) = true$.

Then, we associate to a rewrite theory $\mathscr{R} = (\Sigma, E, R)$ (with a selected kind k of states and with state predicates Π) a Kripke structure whose atomic propositions are specified by the set $AP_{\Pi} = \{\theta(p) \mid p \in \Pi, \theta$ ground substitution$\}$, where by convention we use the simplified notation $\theta(p)$ to denote the ground term $\theta(p(x_1, \ldots, x_n))$. We define $\mathscr{K}(\mathscr{R}, k)_{\Pi} = (T_{\Sigma/E,k}, (\rightarrow^{1}_{\mathscr{R},k})^{\bullet}, L_{\Pi})$, where $L_{\Pi}([t]) = \{\theta(p) \in AP_{\Pi} \mid \theta(p)$ holds in $[t]\}$.

3   Relating Systems

So far we have discussed how to mathematically capture the essential characteristics of computational systems and have proposed rewriting logic as a flexible framework in which to represent them. But we are not interested in computational systems in isolation. We would like to be able to study, for example, if a particular system is an abstraction, or an implementation, of another one. To do that, the concept of *simulation* is introduced.

3.1   Stuttering Simulations

Classically, a simulation $H : \mathscr{A} \longrightarrow \mathscr{B}$ of Kripke structures relates states that satisfy the same atomic propositions in such a way that to every path in $\mathscr{A}$ corresponds a path in $\mathscr{B}$. A key fact is that then every ACTL* formula that holds in $\mathscr{B}$ is also true in $\mathscr{A}$.

Our aim is to generalize the notion of simulation to give it a wider applicability. This generalization should satisfy the same two key properties of basic simulations: (i) be *compositional*, and (ii) *reflect* interesting properties. We achieve this goal by slightly restricting the logic; on the one hand, by forbidding negations (no real expressive power is lost) the condition that related states have to satisfy the same properties can be relaxed, and on the other, by forbidding the *next* operator **X** (see Section 3.2), we can allow paths to be simulated up to *stuttering* (which is what one really cares about most of the time).

Formally, for $\mathscr{A} = (A, \rightarrow_{\mathscr{A}})$ and $\mathscr{B} = (B, \rightarrow_{\mathscr{B}})$ transition systems and $H \subseteq A \times B$ a relation, we say that a path ρ in $\mathscr{B}$ *H-matches* a path π in $\mathscr{A}$ if there are strictly increasing functions $\alpha, \beta : \mathbb{N} \longrightarrow \mathbb{N}$ with $\alpha(0) = \beta(0) = 0$ such that, for all $i, j, k \in \mathbb{N}$, if $\alpha(i) \leq j < \alpha(i+1)$ and $\beta(i) \leq k < \beta(i+1)$, it holds that $\pi(j) H \rho(k)$. For example, the following diagram shows the beginning of two matching paths, where related elements are joined by dashed lines and $\alpha(0) = \beta(0) = 0$, $\alpha(1) = 2$, $\beta(1) = 3$, $\alpha(2) = 5$.

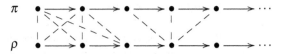

Definition 1. *Given transition systems $\mathscr{A}$ and $\mathscr{B}$, a* stuttering simulation *of transition systems $H : \mathscr{A} \longrightarrow \mathscr{B}$ is a binary relation $H \subseteq A \times B$ such that if aHb, then for each path π in $\mathscr{A}$ starting at a there is a path ρ in $\mathscr{B}$ starting at b that H-matches π. If H is a function we say that H is a* stuttering map *of transition systems. If both H and H^{-1} are stuttering simulations, then we call H a* stuttering bisimulation.*

Given Kripke structures $\mathscr{A} = (A, \rightarrow_{\mathscr{A}}, L_{\mathscr{A}})$ and $\mathscr{B} = (B, \rightarrow_{\mathscr{B}}, L_{\mathscr{B}})$ over AP, a stuttering AP-simulation *$H : \mathscr{A} \longrightarrow \mathscr{B}$ is a stuttering simulation of transition systems $H : (A, \rightarrow_{\mathscr{A}}) \longrightarrow (B, \rightarrow_{\mathscr{B}})$ such that if aHb then $L_{\mathscr{B}}(b) \subseteq L_{\mathscr{A}}(a)$. If H is a function we call H a* stuttering AP-map. *We call H a* stuttering AP-bisimulation *if H and H^{-1} are stuttering AP-simulations. We call H* strict *if aHb implies $L_{\mathscr{B}}(b) = L_{\mathscr{A}}(a)$.*

3.2   The Temporal Logic Institution

Simulations, as defined above, compose, and it is immediate to check that the identity function $1_{\mathscr{A}} : \mathscr{A} \longrightarrow \mathscr{A}$ is a simulation of transition systems and of Kripke structures. Therefore, transition systems together with their simulations define a category **STSys**, and similarly, for each set AP of atomic propositions there is a category $\mathbf{KSSim}_{AP}$ with a subcategory $\mathbf{KSMap}_{AP}$ of stuttering AP-maps. Note that if H is an isomorphism in $\mathbf{KSSim}_{AP}$ then it must be a map and a bisimulation. Note, finally, that the mapping $(A, \rightarrow_{\mathscr{A}}, L_{\mathscr{A}}) \mapsto (A, \rightarrow_{\mathscr{A}})$ extends to a forgetful functor $\mathbf{TS} : \mathbf{KSSim}_{AP} \longrightarrow \mathbf{STSys}$.

Although the main goal of this paper is the study of simulations and their representation in rewriting logic, we believe that a categorical viewpoint is indeed the most natural to understand these generalized simulations and hence consider worthwhile to devote the rest of this section to present some ideas in that context. In what follows we show how these categories can be neatly organized in an institution [12] for the logic ACTL*. Other institutions for temporal logics are discussed in [1], but their notions of signature morphism and of simulation (which roughly corresponds to our notion of bisimulation map) are more limited. As a side effect, we will also construct a Grothendieck category [20] which will allow us to relate Kripke structures over different sets of atomic propositions, further generalizing the notion of simulation.

Let us first define the category of signatures. A simple option would be to choose sets of atomic propositions as objects and functions between them as arrows, but we are aiming for the most general notion that still reflects satisfaction of suitable formulas.[2] For that, let $\text{State}\backslash\{\neg, \mathbf{X}\} : \mathbf{Set} \longrightarrow \mathbf{Set}$ be the functor mapping a set AP to $\text{State}\backslash\{\neg, \mathbf{X}\}(AP)$, the state formulas in $\text{ACTL}^* \backslash \neg(AP)$ that do not contain the *next* operator $\mathbf{X}$, and a function $\alpha : AP \longrightarrow AP'$ to its homomorphic extension

$$\overline{\alpha} : \text{State}\backslash\{\neg, \mathbf{X}\}(AP) \longrightarrow \text{State}\backslash\{\neg, \mathbf{X}\}(AP').$$

Then, the triple $\langle \text{State}\backslash\{\neg, \mathbf{X}\}, \eta, \mu \rangle$ is a monad [2], where $\eta : Id_{\mathbf{Set}} \Rightarrow \text{State}\backslash\{\neg, \mathbf{X}\}$ and $\mu : \text{State}\backslash\{\neg, \mathbf{X}\} \circ \text{State}\backslash\{\neg, \mathbf{X}\} \Rightarrow \text{State}\backslash\{\neg, \mathbf{X}\}$ are natural transformations such that $\eta_{AP}(p) = p$ and μ "unwraps" a formula into its basic atomic propositions. Our category of signatures will be $\mathbf{Set}_{\text{State}\backslash\{\neg, \mathbf{X}\}}$, the Kleisli category of the monad; its objects are just sets, and the morphisms $AP \longrightarrow AP'$ are functions $\alpha : AP \longrightarrow \text{State}\backslash\{\neg, \mathbf{X}\}(AP')$.

We also need a notion of a *reduct* of a Kripke structure, inspired by that of the reduct of an algebra. Given a function $\alpha : AP \longrightarrow \text{State}(AP')$ and a Kripke structure $\mathscr{A} = (A, \rightarrow_{\mathscr{A}}, L_{\mathscr{A}})$ over AP', we define the *reduct Kripke structure* $\mathscr{A}|_{\alpha} = (A, \rightarrow_{\mathscr{A}}, L_{\mathscr{A}|_{\alpha}})$ over AP, with labeling function $L_{\mathscr{A}|_{\alpha}}(a) = \{p \in AP \mid \mathscr{A}, a \models \alpha(p)\}$. We can now define the desired institution.

Definition 2. *The institution of Kripke structures,* $\mathscr{I}_{\mathbf{K}} = (\mathbf{Sign}_{\mathbf{K}}, sen_{\mathbf{K}}, \mathbf{Mod}_{\mathbf{K}}, \models)$, *is given by:*

---

[2] The simpler category, however, gives rise to a semiexact institution, which is not true for the one presented in the text; see [15] for more details.

- $\mathbf{Sign}_{\mathbf{K}} = \mathbf{Set}_{\mathrm{State}\setminus\{\neg,\mathbf{X}\}}$.
- $sen_{\mathbf{K}} : \mathbf{Set}_{\mathrm{State}\setminus\{\neg,\mathbf{X}\}} \longrightarrow \mathbf{Set}$ *is the functor mapping a set AP to* $\mathrm{State}\setminus\{\neg,\mathbf{X}\}(AP)$, *and a function* $\alpha : AP \longrightarrow \mathrm{State}\setminus\{\neg,\mathbf{X}\}(AP')$ *to its homomorphic extension* $\overline{\alpha} :$ $\mathrm{State}\setminus\{\neg,\mathbf{X}\}(AP) \longrightarrow \mathrm{State}\setminus\{\neg,\mathbf{X}\}(AP')$.
- $\mathbf{Mod}_{\mathbf{K}} : \mathbf{Set}_{\mathrm{State}\setminus\{\neg,\mathbf{X}\}} \longrightarrow \mathbf{Cat}^{\mathrm{op}}$ *is given by* $\mathbf{Mod}_{\mathbf{K}}(AP) = \mathbf{KSSim}_{AP}$ *and, for* $\alpha :$ $AP \longrightarrow AP'$ *in* $\mathbf{Set}_{\mathrm{State}\setminus\{\neg,\mathbf{X}\}}$, $\mathbf{Mod}_{\mathbf{K}}(\alpha)(\mathscr{A}) = \mathscr{A}|_{\alpha}$ *and* $\mathbf{Mod}_{\mathbf{K}}(\alpha)(H) = H$.
- *The satisfaction relation is defined as* $\mathscr{A} \models \varphi$ *iff* $\mathscr{A}, a \models \varphi$ *for all* $a \in A$.

Proposition 1. $\mathscr{I}_{\mathbf{K}}$ *is an institution.*

Now, having defined the indexed category $\mathbf{Mod}_{\mathbf{K}}$ allows us to construct the "flattened" category of Kripke structures over arbitrary sets of atomic propositions. Let us denote with $\mathbf{KSSim}$ the Grothendieck category [20] corresponding to $\mathbf{Mod}_{\mathbf{K}}$; spelling out the definition, this gives rise to our most general notion of simulation. A *stuttering simulation* $(\alpha, H) : (AP, \mathscr{A}) \longrightarrow (AP', \mathscr{B})$ in $\mathbf{KSSim}$ between a Kripke structure $\mathscr{A}$ over AP and another $\mathscr{B}$ over AP' consists of a function $\alpha : AP \longrightarrow \mathrm{State}\setminus\{\neg,\mathbf{X}\}(AP')$ together with an AP-simulation $H : \mathscr{A} \longrightarrow \mathscr{B}|_{\alpha}$. We say that (α, H) *reflects a state formula* φ if whenever aHb and $\mathscr{B}, b \models \overline{\alpha}(\varphi)$, then $\mathscr{A}, a \models \varphi$. Then, not only these generalized simulations still compose but they also reflect suitable ACTL* formulas.

Theorem 1. *Stuttering simulations always reflect satisfaction of* ACTL*$\setminus\{\neg,\mathbf{X}\}$ *formulas. In addition, strict stuttering simulations also reflect satisfaction of* ACTL* $\setminus \mathbf{X}$ *formulas.*

Note that by using different types of morphisms between Kripke structures and choosing as sentences those temporal formulas reflected by them, we can get different institutions and Grothendieck categories. For example, if we forget about stuttering and only allow simulations that preserve one-step transitions, and define the category of signatures through a functor State : $\mathbf{Set} \longrightarrow \mathbf{Set}$ mapping AP to State(AP), we get the institution of Kripke structures and classic simulations.

4   Theoroidal Maps

We have already noted that, in order to reason about computational systems, these can be abstractly described by means of transition systems and Kripke structures, and that rewriting logic can be used to specify both kinds of structures, as explained in the previous sections. Our goal now is to study how to relate different rewrite theories and how to lift to this specification level all the previous results about simulations of Kripke structures. For this, we consider four increasingly more general ways of defining simulations for rewrite theories specifying a concurrent system:

1. The easiest way of defining a simulation map for a rewrite theory (Σ, E, R) is by means of an *equational abstraction* [19], which consists in simply adding new equations, say E', to get a quotient system specified by $(\Sigma, E \cup E', R)$.
2. The previous method can be generalized by considering, instead of just theory inclusions $(\Sigma, E) \subseteq (\Sigma, E \cup E')$, arbitrary *theory interpretations* $H : (\Sigma, E) \longrightarrow (\Sigma', E')$ allowing arbitrary transformations on the data representation of states.

3. A third alternative consists in defining a simulation *map* between rewrite theories $\mathcal{R}$ and $\mathcal{R}'$ directly at the level of their associated Kripke structures by means of *equationally defined functions*.
4. Finally, the most general case is obtained by defining arbitrary simulations between rewrite theories $\mathcal{R}$ and $\mathcal{R}'$ by means of *rewrite relations*.

For each of the increasingly more general ways above of defining simulations, there are of course associated *correctness conditions* that must be verified. For equational abstractions they are considered in detail in [19]. Here we study the second case, that we call *theoroidal maps*; although not so general as the last two, there are still many interesting examples that can be explained with them, as we illustrate in Section 5. The remaining cases 3–4 will be treated elsewhere.

4.1   Generalized Theory Morphisms

The first thing to do is to make precise the meaning of *theory interpretation*. The idea is to use the standard concepts of signature and theory morphism. However, as we shall see in some of the examples below, the usual definition of signature morphism is sometimes not expressive enough. For this reason we introduce the following generalization of the concept of signature morphism in which a kind or an operator can be *erased*.

Definition 3. *Given two membership equational signatures $\Sigma = (K, \Sigma, S)$ and $\Sigma' = (K', \Sigma', S')$, a* generalized signature morphism $H : \Sigma \longrightarrow \Sigma'$ *is specified by:*

– *partial functions $H : K \longrightarrow K'$ and $H : S \longrightarrow S'$ such that, for all sorts $s \in \Sigma$, if $H(s)$ is defined so is $H([s])$ and $H([s]) = [H(s)]$.*
– *a partial function H assigning, to each $f \in \Sigma_{k_1 \ldots k_n, k}$ such that $H(k)$ is defined, a Σ'-term $H(f)$ of kind $H(k)$ such that $vars(H(f)) \subseteq \{x_{i_1} : H(k_{i_1}), \ldots, x_{i_m} : H(k_{i_m})\}$, where $k_{i_1}, \ldots, k_{i_m}$ is the (possibly empty) subsequence of $k_1, \ldots, k_n$ determined by those k_i such that $H(k_i)$ is defined. Otherwise, if $H(k)$ is undefined, so is $H(f)$.*

All standard constructions and results about signature morphisms apply to these generalized ones as well. Given $H : \Sigma \longrightarrow \Sigma'$ and a Σ'-algebra A, its reduct $U_H(A)$ over Σ is defined by:

– For each kind k, $U_H(A)_k = A_{H(k)}$ if $H(k)$ is defined; otherwise $U_H(A)_k = \{*\}$.
– For each sort s, $U_H(A)_s = A_{H(s)}$ if $H(s)$ is defined; otherwise $U_H(A)_s = \{*\}$.
– For each operator $f : k_1 \ldots k_n \longrightarrow k$, if $k_{i_1}, \ldots, k_{i_m}$ is the subsequence of those kinds in $k_1, \ldots, k_n$ for which H is defined,

$$U_H(A)_f(a_1, \ldots, a_n) = A_{H(f)}(a_{i_1}, \ldots, a_{i_m});$$

otherwise,

$$U_H(A)_f(a_1, \ldots, a_n) = *.$$

Given generalized signature morphisms $F : \Sigma \longrightarrow \Sigma'$ and $G : \Sigma' \longrightarrow \Sigma''$, their composition $G \circ F$ is defined for a kind k only if both $F(k)$ and $G(F(k))$ are defined, and then it is $(G \circ F)(k) = G(F(k))$; analogously for a sort s and an operator f.

Generalized signature morphisms can also be extended homomorphically to terms, but note that for t of kind k, if $H(k)$ is not defined then $H(t)$ is not defined either. This translation extends to formulas in the expected way, where by convention $H(t = t') = H(t : s) = \top$ if H is not defined for the kind of t (which is the same as that of t' and s). Our desired general notion of "theory interpretation" is then captured by the following:

Definition 4. *Given two membership equational theories* (Σ, E) *and* (Σ', E'), *a* generalized theory morphism *(resp. a* generalized theory morphism with initial semantics*)* $H : (\Sigma, E) \longrightarrow (\Sigma', E')$ *is a generalized signature morphism* $H : \Sigma \longrightarrow \Sigma'$ *such that for each* $\varphi \in E$, $E' \models H(\varphi)$ *(resp.* $T_{\Sigma'/E'} \models H(\varphi)$).

Note that, since $T_{\Sigma'/E'} \models E'$, each generalized theory morphism is a fortiori a generalized theory morphism with initial semantics, but not conversely. For example, if (Σ, E) is the theory with one sort, *Nat*, a binary operator $+$, and the equation $(\forall\{x, y : Nat\})\, x + y = y + x$, (Σ', E') is the usual equational definition of addition in Peano arithmetic, and H is the obvious signature inclusion, then we have $T_{\Sigma'/E'} \models (\forall\{x, y : Nat\})\, x + y = y + x$, but $E' \not\models (\forall\{x, y : Nat\})\, x + y = y + x$.

Again, generalized theory morphisms compose and together with membership equational theories give rise to a category $\mathbf{GTh}_{\mathrm{MEL}}$.

The new feature of generalized signature morphisms, which is inherited by generalized theory morphisms, is that kinds and operators can be removed. This could have been "implemented" using the standard notion of theory morphism in the following alternative manner:

Proposition 2. *A generalized theory morphism* $H : T \longrightarrow T'$ *is the same thing as an ordinary theory morphism* $H : T \longrightarrow T' \oplus ONE$, *where* $\oplus$ *denotes coproduct of theories, and ONE is a theory with a single kind* $[One]$ *and sort One, a constant* $*$ *of that kind, and the equation* $(\forall\{x : [One]\})\, x = *$.

Proof (sketch). Leaving a kind or sort undefined in a generalized signature morphism corresponds respectively to mapping it to $[One]$ or *One* in $T' \oplus ONE$, while leaving an operator undefined corresponds to mapping it to the term $*$. $\qquad\square$

Note that there is an equivalence of categories between the models of T' and those of $T' \oplus ONE$, because, even though we have introduced a new kind $[One]$, all its elements are collapsed by the equation $(\forall\{x : [One]\})\, x = *$ to the constant $*$ and can play no distinguished role.

Example. A special case of generalized theory morphisms are the projection functions from n-tuples to $(n - k)$-tuples. Consider a theory *3-TUPLE* for triples with kinds *3-Tuple, Elt@x, Elt@y, Elt@z*, an operator $\langle \_, \_, \_ \rangle : Elt@x\; Elt@y\; Elt@z \longrightarrow$ *3-Tuple*, projection operators p_1, p_2, and p_3, and the obvious equations. Similarly, the theory *2-TUPLE* has kinds *2-Tuple, Elt@x, Elt@z*, an operator $\langle \_, \_ \rangle : Elt@x\; Elt@z \longrightarrow$ *2-Tuple*, corresponding projection operators p_1 and p_2, and the equations for pairing. Projecting from a triple to a pair by projecting out the second component can be represented by the generalized theory morphism $H :$ *3-TUPLE* $\longrightarrow$ *2-TUPLE* mapping the kinds *Elt@x* and *Elt@z* to themselves, *3-Tuple* to *2-Tuple*, and the operator $\langle \_, \_, \_ \rangle$ to the term $\langle x_1 : Elt@x, x_3 : Elt@z \rangle$; the image of the kind *Elt@y* and the operator p_2 are left undefined.

4.2   Simulation Maps as Generalized Theory Morphisms

To be able to arrange rewrite theories specifying Kripke structures in a categorical way we need to consider a theory $BOOL_{\models}$ extending $BOOL$ with two new kinds, $State$ and $Prop$, and a new operator $\_ \models \_ : State\ Prop \longrightarrow Bool$.

We now have all the ingredients needed to define a category $\mathbf{SRWThHom}_{\models}$ in which stuttering maps are specified by theory interpretations. Objects in $\mathbf{SRWThHom}_{\models}$ are triples $(\mathscr{R}, (\Sigma', E \cup D), J)$ specifying, respectively, the transition relation, the atomic propositions, and the kind of the states. More precisely:

1. $\mathscr{R} = (\Sigma, E, R)$ is a rewrite theory specifying the transition system.
2. $(\Sigma, E) \subseteq (\Sigma', E \cup D)$ is a protecting theory extension, containing and protecting also the theory $BOOL$ of Booleans, that defines the atomic propositions satisfied by the states. We define $\Pi \subseteq \Sigma'$ as the subsignature of operators of coarity $Prop$.
3. $J : BOOL_{\models} \longrightarrow (\Sigma', E \cup D)$ is a membership equational theory morphism [17] that selects the distinguished kind of states $J(State)$, and such that: (i) it is the identity when restricted to $BOOL$, (ii) $J(Prop) = Prop$, and (iii) $J(\_ \models \_ : State\ Prop \rightarrow Bool) = \_ \models \_ : J(State)\ Prop \rightarrow Bool$.

Then, a morphism

$$H : (\mathscr{R}_1, (\Sigma'_1, E_1 \cup D_1), J_1) \longrightarrow (\mathscr{R}_2, (\Sigma'_2, E_2 \cup D_2), J_2)$$

in $\mathbf{SRWThHom}_{\models}$ is a *generalized signature morphism* $H : \Sigma_1 \cup \Pi_1 \longrightarrow \Sigma_2 \cup \Pi_2$ such that:

1. $H \circ J_1 = J_2$ (so that $BOOL$ is preserved and states in $\mathscr{R}_1$ are mapped to states in $\mathscr{R}_2$).
2. $H : (\Sigma_1, E_1) \longrightarrow (\Sigma_2, E_2)$ is a generalized morphism of membership equational theories with initial semantics, so that we have a unique Σ_1-homomorphism

$$\eta^H : T_{\Sigma_1/E_1} \longrightarrow U_H(T_{\Sigma_2/E_2}) : [t] \mapsto [H(t)].$$

3. (Preservation of transitions.) $\eta^H_{J_1(State)} : \mathscr{T}(\mathscr{R}_1)_{J_1(State)} \longrightarrow \mathscr{T}(\mathscr{R}_2)_{J_2(State)}$, the component corresponding to the kind $J_1(State)$ in η^H mapping $[t]$ to $[H(t)]$, is a stuttering map of transition systems.
4. (Preservation of predicates.) For each $t \in T_{\Sigma_1, J_1(State)}$ and state predicate $p(u_1, \ldots, u_n)$:

$$E_2 \cup D_2 \vdash H(t) \models H(p(u_1, \ldots, u_n)) = true \implies E_1 \cup D_1 \vdash t \models p(u_1, \ldots, u_n) = true.$$

We can analogously construct a subcategory $\mathbf{SRWThHom}^{str}_{\models}$ of strict maps. The definition is exactly the same except for item (4), where the implication must actually be an equivalence.

That H so constrained indeed gives rise to a map of Kripke structures is shown in Proposition 3 below. Let us define a functor $\mathscr{K} : \mathbf{SRWThHom}_{\models} \longrightarrow \mathbf{KSMap}$ as follows:

- for objects, $\mathscr{K}(\mathscr{R},(\Sigma',E\cup D),J) = \mathscr{K}(\mathscr{R},J(State))_\Pi$;
- for morphisms $H : (\mathscr{R}_1,(\Sigma'_1,E_1\cup D_1),J_1) \longrightarrow (\mathscr{R}_2,(\Sigma'_2,E_2\cup D_2),J_2)$, $\mathscr{K}(H) = (H|_{\Pi_1},\eta^H_{J_1(State)})$, where $H|_{\Pi_1}$ is the restriction of H to the state predicates Π_1.

Proposition 3. *With the above definitions,* $\mathscr{K}$: **SRWThHom**$_\models$ $\longrightarrow$ **KSMap** *is a functor with restriction* $\mathscr{K}$: **SRWThHom**$^{str}_\models$ $\longrightarrow$ **KSMap**^{str}.

Proof. $\mathscr{K}$ is well-defined on objects, and it is immediate to see that it preserves identities and composition of morphisms; the only thing we need to check is that, for all H, $\mathscr{K}(H)$ is indeed a map of Kripke structures. Let then $H : (\mathscr{R}_1,(\Sigma'_1,E_1\cup D_1),J_1) \longrightarrow (\mathscr{R}_2,(\Sigma'_2,E_2\cup D_2),J_2)$ be a morphism in **SRWThHom**$_\models$. By item (3) above, $\eta^H_{J_1(State)}$: $\mathscr{T}(\mathscr{R}_1)_{J_1(State)} \longrightarrow \mathscr{T}(\mathscr{R}_2)_{J_2(State)}$ is a stuttering map of transition systems. To show preservation of predicates, let $p(u_1,\dots,u_n) \in L_{\mathscr{K}(\mathscr{R}_2,J_2(State))_{\Pi_2}|_{H|_{\Pi_1}}}([H(t)])$. By definition of the reduct of a Kripke structure, $\mathscr{K}(\mathscr{R}_2,J_2(State))_{\Pi_2},[H(t)] \models H(p(u_1,\dots,u_n))$ which, by definition of $\mathscr{K}(\mathscr{R}_2,J_2(State))_{\Pi_2}$ and condition (4) in the definition of morphisms in **SRWThHom**$_\models$, implies that $p(u_1,\dots,u_n) \in L_{\mathscr{K}(\mathscr{R}_1,J_1(State))_{\Pi_1}}([t])$, as required. It is clear that if H belongs to **SRWThHom**$^{str}_\models$ the converse is also true and $\mathscr{K}(H)$ is a strict map. $\square$

An important consequence of this result and Theorem 1 is the following:

Theorem 2. *Given a morphism* $H : (\mathscr{R}_1,(\Sigma'_1,E_1\cup D_1),J_1) \longrightarrow (\mathscr{R}_2,(\Sigma'_2,E_2\cup D_2),J_2)$ *in* **SRWThHom**$_\models$ *or* **SRWThHom**$^{str}_\models$, *and a formula* φ *in* ACTL$^*\backslash\{\neg,\mathbf{X}\}(\Pi_1)$ *or* ACTL$^*\backslash\mathbf{X}(\Pi_1)$ *respectively, if* $H(\varphi)$ *holds in* $\mathscr{K}(\mathscr{R}_2,(\Sigma'_2,E_2\cup D_2),J_2)$ *then* φ *holds in* $\mathscr{K}(\mathscr{R}_1,(\Sigma'_1,E_1\cup D_1),J_1)$.

Similar constructions can be carried out when simulations are represented by means of equationally defined functions or rewrite relations (recall the introduction to this section), resulting in categories **SRWTh**$_\models$ and **SRelRWTh**$_\models$. Then, the lifting of Kripke structures to the framework of rewriting logic can be represented graphically with the following commutative diagram. In it, the horizontal arrows between categories associated to Kripke structures are inclusions, and those that map to categories associated to transition systems are the expected forgetful functors. (**SRWTh** is constructed analogously to **SRWThHom**$_\models$, but taking only the transitions into consideration.)

$$
\begin{array}{ccccccc}
\textbf{SRWThHom}_\models & \longrightarrow & \textbf{SRWTh}_\models & \longrightarrow & \textbf{SRelRWTh}_\models & \longrightarrow & \textbf{SRWTh} \\
\Big\downarrow{\scriptstyle\mathscr{K}} & & \Big\downarrow{\scriptstyle\mathscr{K}} & & \Big\downarrow{\scriptstyle\mathscr{K}} & & \Big\downarrow{\scriptstyle\mathscr{T}} \\
\textbf{KSMap} & \longrightarrow & \textbf{KSMap} & \longrightarrow & \textbf{KSSim} & \longrightarrow & \textbf{STSys}
\end{array}
$$

5 Applications

5.1 Predicate Abstraction

Simulations are useful to define abstractions that allow studying the properties of a complex system using a simpler one. A particular instance of the methodology of abstraction is *predicate abstraction* [13, 9]. Under this approach, the abstract domain is a Boolean algebra over a set of assertions and the abstraction function, typically as part of a Galois connection, is symbolically constructed as the conjunction of all expressions satisfying a certain condition, which is typically discharged using theorem proving. We now show how predicate abstractions can be understood as an instance of our notion of theoroidal map.

Let us first focus on the transition relation. Given a computational system, a set $\phi_1, \ldots, \phi_n$ of predicates over the states determines an abstraction function mapping a state S to the Boolean tuple $\langle \phi_1(S), \ldots, \phi_n(S) \rangle$. Let us assume that the transitions of the system are specified by a rewrite theory $\mathscr{R} = (\Sigma, E, R)$ whose kind of states is *State*. Then, if $\mathscr{R}$ is *State*-encapsulated with constructor $st : k_1 \ldots k_m \longrightarrow State$ (that is, among all operators in Σ the kind *State* only appears in the operator st, and only as its coarity), the above predicate abstraction can be represented in rewriting logic by means of a rewrite theory $\mathscr{R}_A = (\Sigma_A, E_A, R_A)$ where:

- Σ_A contains Σ and the signature of *BOOL*, together with a new kind *BState*, a new operator $bst : Bool^n \longrightarrow BState$ and, for each predicate ϕ_i, $1 \le i \le n$, an operator $p_i : State \longrightarrow Bool$ to represent it. We then have a signature morphism $H : \Sigma \longrightarrow \Sigma_A$ that maps the kind *State* to *BState*, the constructor st to the term

$$bst(p_1(st(x_1, \ldots, x_m)), \ldots, p_n(st(x_1, \ldots, x_m))),$$

  and is the identity everywhere else.
- E_A contains $H(E)$ and the equations in *BOOL*, together with equations for $p_1, \ldots, p_n$ specifying the predicates $\phi_1, \ldots, \phi_n$.
- $R_A = H(R)$.

By construction, then, $H : (\Sigma, E) \longrightarrow (\Sigma_A, E_A)$ is a theory morphism such that $t \rightarrow^1_{\mathscr{R}, State} t'$ implies $H(t) \rightarrow^1_{\mathscr{R}_A, BState} H(t')$, thus preserving the transition relation.

We can now turn our attention to the preservation of properties. Graphically, the relationship between the different theories involved is depicted in the following diagram,

$$
\begin{array}{ccc}
(\Sigma, E) & \lhook\joinrel\longrightarrow & (\Sigma', E \cup D) \\
H \downarrow & & \downarrow \\
(\Sigma_A, E_A) & \lhook\joinrel\longrightarrow & (\Sigma'_A, E_A \cup D_A)
\end{array}
$$

where $(\Sigma', E \cup D)$ is the equational theory specifying the properties of the given system, and $(\Sigma'_A, E_A \cup D_A)$ is the theory we have to associate to $\mathscr{R}_A$ defining its atomic propositions.

The syntax for the state predicates q (that we assume are constants) in the original system is given in a subsignature Π of Σ'. It is usually the case that for each of these q one of the predicates ϕ_i in the basis defining the abstraction has the meaning "the state S satisfies q." Let $q_1, \ldots, q_k$ be the state predicates in Π. We assume $k \le n$, and that each q_j, $1 \le j \le k$, corresponds to the predicate ϕ_j in the basis of the abstraction (but in general we may have $n > k$, with predicates $\phi_{k+1}, \ldots, \phi_n$ not having a counterpart in Π). That is, for a ϕ_j with a corresponding q_j in Π, its specification in E_A through $p_j(S)$ is thus essentially the same (modulo renaming) as that of $S \models q_j$ in D, so that $E \cup D \vdash (S \models q_j) = true \iff E_A \vdash p_j(S) = true$. Then, for the abstraction we use the same set of state predicates Π and they are specified in a theory extension $(\Sigma_A, E_A) \subseteq (\Sigma'_A, E_A \cup D_A)$, with $\Sigma'_A = \Sigma_A \cup \Sigma'$ and D_A containing, for each q_j in Π with associated ϕ_j, the equation

$$(\forall \{x_1, \ldots, x_n\}) (bst(x_1, \ldots, x_j, \ldots, x_n) \models q_j) = x_j .$$

Let us extend H to $\Sigma \cup \Pi$ by mapping each state predicate to itself. Thus, for all ground terms t of kind *State* and state predicates q_j, if $E_A \cup D_A \vdash (H(t) \models q_j) = true$ then, by the equation defining q_j in $E_A \cup D_A$ and since $H(t) = bst(p_1(t), \ldots, p_n(t))$, we have $E_A \cup D_A \vdash p_j(t) = true$ and even $E_A \vdash p_j(t) = true$ because p_j is completely specified in E_A. And hence, due to the relation between the equations defining $p_j(S)$ and $S \models q_j$, $E \cup D \vdash (t \models q_j) = true$ holds and preservation of predicates is guaranteed.

Finally, we can put all the pieces together and summarize the previous discussion as follows.

Theorem 3. *Let a concurrent system be specified as an object $(\mathscr{R}, (\Sigma', E \cup D), J)$ of* **SRWThHom**$_\models$, *where $\mathscr{R}$ is $J(State)$-encapsulated, and let $\phi_1, \ldots, \phi_n$ be a set of predicates over the kind $J(State)$, with each state predicate $q_j \in \Pi$ (we assume that all such q_j are constants) corresponding to a ϕ_j, $1 \leq j \leq k$. The result of applying predicate abstraction is the system given by $(\mathscr{R}_A, (\Sigma'_A, E_A \cup D_A), J_A)$, where $(\Sigma'_A, E_A \cup D_A)$ and $\mathscr{R}_A$ are defined as explained above, and where $J_A(State) = BState$. Then, with these definitions, $H : (\mathscr{R}, (\Sigma', E \cup D), J) \longrightarrow (\mathscr{R}_A, (\Sigma'_A, E_A \cup D_A), J_A)$ is an arrow in* **SRWThHom**$_\models$, *where H is the signature morphism $\Sigma \cup \Pi \longrightarrow \Sigma'_A \cup \Pi$.*

Let us illustrate these ideas by outlining how they apply to the bakery protocol. This is an infinite state protocol that achieves mutual exclusion between processes by dispensing a number to each process and serving them in sequential order according to the number they hold. For the case of two processes, the transitions can be specified in rewriting logic by a theory $\mathscr{R} = (\Sigma, E, R)$ such that:

- (Σ, E) contains declarations and equations specifying the natural numbers; in particular, the "equal to" (==) and "less than" (<) predicates are specified.
- States are constructed by an operator `st : Mode Nat Mode Nat -> State`. The first two components describe the status of the first process (the mode it is currently in, which can be `sleep`, `wait`, or `crit`, and its priority as given by the number according to which it will be served), and the last two components the status of the second process.
- R consists of eight rewrite rules, four for each process, describing all possible transitions. Among them, for example,

```
rl st(M, X, sleep, Y) => st(M, X, wait, s(X)) .
```

to represent that the second process can "awake" and move to `wait` mode, and

```
crl st(M, X, wait, Y) => st(M, X, crit, Y) if Y < X .
```

allowing the second process to move to the critical section if its counter is less than that of the first one.

The properties are defined in a theory extension $(\Sigma, E) \subseteq (\Sigma', E \cup D)$ that simply adds four constants 1wait, 1crit, 2wait, and 2crit to Σ to characterize when the first and second processes are in wait or crit mode, together with the obvious equations:

```
eq (st(wait, X, N, Y) |= 1wait) = true .
eq (st(sleep, X, N, Y) |= 1wait) = false .
eq (st(crit, X, N, Y) |= 1wait) = false .
...
```

For this protocol, we might be interested in verifying the following safety property: $\mathbf{AG}\neg(\text{1crit} \wedge \text{2crit})$.

We will use the following set of seven predicates to define the predicate abstraction:

$\phi_1(\text{st(M, X, N, Y)}) \Longleftrightarrow \text{M == wait}$ $\phi_5(\text{st(M, X, N, Y)}) \Longleftrightarrow \text{X == 0}$

$\phi_2(\text{st(M, X, N, Y)}) \Longleftrightarrow \text{M == crit}$ $\phi_6(\text{st(M, X, N, Y)}) \Longleftrightarrow \text{Y == 0}$

$\phi_3(\text{st(M, X, N, Y)}) \Longleftrightarrow \text{N == wait}$ $\phi_7(\text{st(M, X, N, Y)}) \Longleftrightarrow \text{X < Y}$

$\phi_4(\text{st(M, X, N, Y)}) \Longleftrightarrow \text{N == crit}$

Intuitively, we only care whether the processes are in wait or crit mode, whether their counters are equal to zero, and which counter is greater.

Note that the state predicates in the signature correspond to predicates 1–4. In terms of the notation used above, q_1 would be 1wait and it would be associated to ϕ_1, q_2 would be 1crit and would be associated to ϕ_2, and q_3 and q_4 would be 2wait and 2crit, associated to ϕ_3 and ϕ_4. Now, the abstract rewrite theory $\mathcal{R}_A = (\Sigma_A, E_A, R_A)$ is constructed by adding to $\mathcal{R}$:

- Operators p1 : State -> Bool,...,p7 : State -> Bool, together with a new kind BState and the constructor for abstract states

   ```
   op bst : Bool Bool Bool Bool Bool Bool Bool -> BState .
   ```

   This determines the signature morphism H, that maps the constructor operator st to the term

   ```
   bst(p1(st(M, X, N, Y)),...,p7(st(M, X, N, Y)))
   ```

- Equations associated to pi specifying ϕ_i for $i = 1, \ldots, 7$. Since predicates $\phi_1, \ldots, \phi_4$ correspond to the atomic propositions, their defining equations are "the same":

   ```
   eq p1(st(wait, X, N, Y)) = true .
   eq p1(st(sleep, X, N, Y)) = false .
   eq p1(st(crit, X, N, Y)) = false .
   eq p2(st(wait, X, N, Y)) = false .
   eq p2(st(sleep, X, N, Y)) = false .
   eq p2(st(crit, X, N, Y)) = true .
   ...
   ```

   The three remaining equations are also immediate:

   ```
   eq p5(st(M, X, N, Y)) = (X == 0) .
   eq p6(st(M, X, N, Y)) = (Y == 0) .
   eq p7(st(M, X, N, Y)) = (Y < X) .
   ```

– The translation of the rules in R by the signature morphism H. In particular, the two rules introduced before become:

```
rl bst(p1(st(M, X, sleep, Y)), ..., p7(st(M, X, sleep, Y))) =>
   bst(p1(st(M, X, wait, Y)), ..., p7(st(M, X, wait, s(X)))) .
crl bst(p1(st(M, X, wait, Y)), ..., p7(st(M, X, wait, Y))) =>
   bst(p1(st(M, X, crit, Y)), ..., p7(st(M, X, crit, Y)))
   if Y < X .
```

Finally, we have to write the equations in D_A defining the atomic propositions in the abstract model, which is straightforward.

```
eq (bst(B1, B2, B3, B4, B5, B6, B7) |= 1wait) = B1 .
eq (bst(B1, B2, B3, B4, B5, B6, B7) |= 1crit) = B2 .
eq (bst(B1, B2, B3, B4, B5, B6, B7) |= 2wait) = B3 .
eq (bst(B1, B2, B3, B4, B5, B6, B7) |= 2crit) = B4 .
```

By construction, this model is a predicate abstraction with respect to the basis $\phi_1, \ldots, \phi_7$ of the bakery protocol, in which the desired property can be model checked.

It is worth pointing out that this algebraic method of defining predicate abstractions cannot be expressed within the framework of [19], because the specification of the predicates ϕ_i requires, in general, to introduce auxiliary operators and thus a different signature $\Sigma_A \neq \Sigma$. Also, the resulting rewrite theory *is not executable* in general. This means that it cannot be directly used in a tool like the Maude model checker [10]. Predicate abstraction can be considered as a particular instance of our framework of algebraic simulations from a conceptual or foundational point of view, which is still quite useful because it provides a justification for the method within our framework. Current approaches to predicate abstraction do not work directly with the minimal transition relation (described in our account by $\mathscr{R}_A$). Instead, they compute a safe *approximation* of $\mathscr{R}_A$ by discharging some proof obligations. We are at present developing methods to compute such approximation within our framework using Maude's inductive theorem prover (ITP) [6] as the deductive engine to discharge such proof obligations.

5.2 A Fairness Example

We illustrate the use of theoroidal (bi)simulation maps to reason about fairness. The treatment can be made for very general classes of rewrite theories, and for quite flexible notions of fairness [18]. Here, we limit ourselves to illustrating some of the key ideas, including the use of theoroidal maps, by means of a simple communication protocol example. Note also that the same idea can be used for the representation and study of labeled transition systems in rewriting logic.

Consider a system consisting of a sender, a channel, and a receiver. The goal is to send a multiset of numbers (in arbitrary order) from the sender to the receiver through the channel. The channel can at any time contain several of these numbers. Besides the normal send and receive actions, the channel may stall an arbitrary number of times in sending some data. We can model the states of such a system by means of the signature

$$snd, ch, rcv : Nat \longrightarrow Conf$$
$$null : \longrightarrow Conf$$
$$\_\_ : Conf\ Conf \longrightarrow Conf$$

where the operator $\_\,\_$ (juxtaposition notation) denotes multiset union and satisfies the equations of associativity and commutativity, and has *null* as its identity element. For example, the term

$$snd(7)\,snd(3)\,snd(7)\,ch(2)ch(3)\,rcv(1)\,rcv(9)$$

describes a state in which 3 and two copies of 7 have not yet been sent, 2 and another copy of 3 are in the channel, and 1 and 9 have been received. The behavior of the system is specified by the following three rewrite rules:

$$send : snd(n) \longrightarrow ch(n)$$
$$stall : ch(n) \longrightarrow ch(n)$$
$$receive : ch(n) \longrightarrow rcv(n)$$

where n is a variable of sort *Nat*. Is this system terminating? Not without extra assumptions, since the *stall* rule could be applied forever. To make it terminating it is enough to assume the following "weak fairness" property about the *receive* rule, described by the formula

$$\textit{wf-receive} = \mathbf{FG}\,\textit{enabled-receive} \rightarrow \mathbf{GF}\,\textit{taken-receive};$$

that is, if eventually the *receive* rule becomes continuously enabled in a path, then it is taken infinitely often. Specifying the *enabled-receive* predicate equationally is quite easy (we just need to have some value in the channel) but the specification of the *taken-receive* predicate is more elusive. For example, does the *taken-receive* predicate hold of the state described above? We don't know; maybe the last action was receiving the value 1, in which case it would hold, but it could instead have been stalling on 3, or sending 2, and then it wouldn't. Here is where a theory transformation corresponding to a theoroidal map, and allowing us to define a bisimilar system where the *taken-receive* predicate can be defined, comes in. The new theory extends the above signature with the following new sorts and operators:

$$send, stall, receive, * : \longrightarrow Label$$
$$\{\_\,|\,\_\} : Conf\ Label \longrightarrow State$$

that is, a state now consists of a configuration-label pair, indicating the last rule that was applied. Since initially no rule has been applied, we add the label $*$ for all initial states. The rules of the transformed theory are now:

$$send : \{conf\ snd(n) \mid l\} \longrightarrow \{conf\ ch(n) \mid send\}$$
$$stall : \{conf\ ch(n) \mid l\} \longrightarrow \{conf\ ch(n) \mid stall\}$$
$$receive : \{conf\ ch(n) \mid l\} \longrightarrow \{conf\ rcv(n) \mid receive\}$$

where *conf* is a variable of sort *Conf*, and l a variable of sort *Label*. We can then define the predicates *enabled-send*, *enabled-receive*, and *taken-receive* by the equations

$$(\{conf\ snd(n) \mid l\} \models \textit{enabled-send}) = \textit{true}$$
$$(\{conf\ ch(n) \mid l\} \models \textit{enabled-receive}) = \textit{true}$$
$$(\{conf \mid receive\} \models \textit{taken-receive}) = \textit{true}$$

Then the fair termination property can be defined by the following formula, which indeed holds in the Kripke structure associated to this transformed theory for any initial state:

$$\mathbf{A}(\textit{wf-receive} \rightarrow \mathbf{F}(\neg\textit{enabled-send} \land \neg\textit{enabled-receive})).$$

Let $(\Sigma_{Comm}, E_{Comm})$ denote the underlying equational theory of our original rewrite theory, and let $(\Sigma_{LComm}, E_{Comm})$ denote that of the transformed theory (it has the same equations E_{Comm}). We can define a generalized theory morphism

$$H : (\Sigma_{LComm}, E_{Comm}) \longrightarrow (\Sigma_{Comm}, E_{Comm})$$

as follows. The sorts, implicit kinds, and operators in Σ_{Comm} are mapped identically to themselves; the sort *State* is mapped to *Conf*; and the sort *Label* is not mapped anywhere; the operator $\{\_ \mid \_\}$ is mapped to the variable *conf* of sort *Conf*; finally, the label constants are not mapped anywhere. Now, let Π_0 consist of the predicates *enabled-send* and *enabled-receive*, which in the original theory are defined by the equations

$$conf \; snd(n) \models \textit{enabled-send} = \textit{true}$$
$$conf \; ch(n) \models \textit{enabled-receive} = \textit{true}.$$

Then, if *Comm* and *LComm* denote our rewrite theories, H induces a theoroidal *bisimulation* (strict) map of Kripke structures

$$H : \mathcal{K}(LComm, [State])_{\Pi_0} \longrightarrow \mathcal{K}(Comm, [Conf])_{\Pi_0}.$$

Furthermore, in the case of *LComm* we can extend Π_0 to Π by adding the *taken-receive* predicate, so that fair termination can be properly specified and verified.

6   Conclusions

We have argued that a categorical approach to the study of Kripke structures and their generalized notion of simulation is very natural, and have shown this by neatly organizing them in an institution. Among the many ways that these Kripke structures and simulations can be formally specified we have proposed rewriting logic, which has proved to be a very flexible framework for this task. Simulations come in several flavors in rewriting logic and here we have focused on theoroidal maps; we have shown how they can be organized together with rewrite theories in a category that reflects that for Kripke structures, and how they apply to two interesting examples. An open line of research consists in the study of proof methods and the development of tool support to prove simulations correct; some preliminary results are reported in [15].

References

1. M. Arrais and J. L. Fiadeiro. Unifying theories in different institutions. In M. Haveraaen, O. Owe, and O.-J Dahl, editors, *Recent Trends in Data Type Specification, COMPASS/ADT, Selected Papers*, volume 1130 of *LNCS*, pages 81–101. Springer-Verlag, 1996.

2. M. Barr and C. Wells. *Category Theory for Computing Science.* Centre de Recherches Mathématiques, third edition, 1999.
3. P. Borovanský, C. Kirchner, H. Kirchner, and P.-E. Moreau. ELAN from a rewriting logic point of view. *Theoretical Computer Science,* 285(2):155–185, 2002.
4. R. Bruni and J. Meseguer. Generalized rewrite theories. In J. C. M. Baeten, J. K. Lenstra, J. Parrow, and G. J. Woeginger, editors, *Automata, Languages and Programming. ICALP 2003. Proceedings,* volume 2719 of *LNCS,* pages 252–266. Springer-Verlag, 2003.
5. E. M. Clarke, O. Grumberg, and D. A. Peled. *Model Checking.* MIT Press, 1999.
6. M. Clavel. The ITP Tool. http://geminis.sip.ucm.es/~clavel/itp, 2004.
7. M. Clavel, F. Durán, S. Eker, P. Lincoln, N. Martí-Oliet, J. Meseguer, and J. F. Quesada. Maude: Specification and programming in rewriting logic. *Theoretical Computer Science,* 285(2):187–243, 2002.
8. M. Clavel, F. Durán, S. Eker, P. Lincoln, N. Martí-Oliet, J. Meseguer, and C. Talcott. Maude manual (version 2.1). http://maude.cs.uiuc.edu/manual/, 2004.
9. M. A. Colón and T. E. Uribe. Generating finite-state abstractions of reactive systems using decision procedures. In A. J. Hu and M. Y. Vardi, editors, *Computer Aided Verification. CAV'98, Proceedings,* volume 1427 of *LNCS,* pages 293–304. Springer-Verlag, 1998.
10. S. Eker, J. Meseguer, and A. Sridharanarayanan. The Maude LTL model checker. In F. Gadducci and U. Montanari, editors, *Proceedings Fourth International Workshop on Rewriting Logic and its Applications, WRLA'02,* volume 71 of *ENTCS.* Elsevier, 2002.
11. K. Futatsugi and R. Diaconescu. *CafeOBJ Report.* World Scientific, AMAST Series, 1998.
12. J. Goguen and R. Burstall. Institutions: Abstract model theory for specification and programming. *Journal of the Association for Computing Machinery,* 39(1):95–146, 1992.
13. S. Graf and H. Saidi. Construction of abstract state graphs with PVS. In O. Grumberg, editor, *Computer Aided Verification. CAV'97, Proceedings,* volume 1254 of *LNCS,* pages 72–83. Springer-Verlag, 1997.
14. N. Martí-Oliet and J. Meseguer. Rewriting logic: Roadmap and bibliography. *Theoretical Computer Science,* 285(2):121–154, 2002.
15. N. Martí-Oliet, J. Meseguer, and M. Palomino. Algebraic simulations. http://maude.cs.uiuc.edu/papers/, 2004.
16. J. Meseguer. Conditional rewriting logic as a unified model of concurrency. *Theoretical Computer Science,* 96(1):73–155, 1992.
17. J. Meseguer. Membership algebra as a logical framework for equational specification. In F. Parisi-Presicce, editor, *Recent Trends in Algebraic Development Techniques, WADT'97, Selected Papers,* volume 1376 of *LNCS,* pages 18–61. Springer-Verlag, 1998.
18. J. Meseguer. Localized fairness: a rewriting semantics. Paper in preparation, 2004.
19. J. Meseguer, M. Palomino, and N. Martí-Oliet. Equational abstractions. In F. Baader, editor, *Automated Deduction - CADE-19. 19th International Conference on Automated Deduction, Proceedings,* volume 2741 of *LNCS,* pages 2–16. Springer-Verlag, 2003.
20. A. Tarlecki, R. M. Burstall, and J. A. Goguen. Some fundamental algebraic tools for the semantics of computation. Part 3: Indexed categories. *Theoretical Computer Science,* 91(2):239–264, 1991.

Behavioural Semantics of Algebraic Specifications in Arbitrary Logical Systems

Michał Misiak

Warsaw University, Faculty of Mathematics,
Informatics and Mechanics**

Abstract. Behavioural semantics for specifications plays a crucial role
in the formalization of the developments process, where a specification
need not to be implemented exactly but only so that the required system
behaviour is achieved. There are two main approaches to the definition
of behavioural semantics: the internal one (called behavioural semantics)
and external one (called abstractor semantics).

In this paper we present a notion of a behavioural concrete institution which is based on a notion of a concrete institution. The basic idea
to form a behavioural institution (i.e. to ensure the satisfaction condition holds) is adopted from [2]. The behavioural concrete institution is a
generalization of the COL-institution. In this work we also compare the
resulted behavioural semantics with the abstractor semantics.

1 Introduction

One of the problems of algebraic-style specification of software systems is that
the strict interpretation of a specification is often inadequate in practice. Typically a specification need not to be implemented exactly but only so that the
required system behaviour is achieved. To cope with this problem the semantics of specifications must be redefined resulting in the so called *behavioural* or
observational interpretation of specifications. There are two main approaches to
the definition of behavioural semantics of algebraic specifications. The internal
approach involves introducing an indistinguishability relation between elements
of models. The external approach is based on an equivalence relation between
models. These two approaches are related to each other and coincide in some
cases, see [4].

In this work we aim at a general definition of a behavioural semantics for
algebraic specifications in an arbitrary logical system. The key notion for this
purpose is the notion of institution, introduced in [6].

We propose a notion of a *behavioural concrete institution*. This framework is
based on the notion of concrete model category as introduced in [5]. The idea is to
equip the model categories of institutions considered with concretization functor,

---

** This work has been partially supported by KBN grant 7T11C 002 21 and European
AGILE project IST-2001-32747.

thus adding "carriers" to the models considered. Then, a *concrete institution* is just an ordinary institution in which all categories of models are concrete categories and each signature has a set of sorts.

To define a behavioural concrete institution we first need to define the behavioural satisfaction relation and so the behavioural semantics of flat specifications, and then extend it to form an institution. The behavioural semantics of flat specifications was introduced in [5], but this approach doesn't allow us to form an institution, since the satisfaction condition doesn't hold. Therefore we follow the idea presented in [2]. The behavioural concrete institution is a generalization of the COL-institution.

2   Basic Notions

An S-sorted set is a family $X = (X_s)_{s \in S}$ of sets. Most standard notions concerning sets can be generalized to S-sorted sets. For example, let $X = (X_s)_{s \in S}$, $Y = (Y_s)_{s \in S}$ be S-sorted sets:

- X is a subset of Y, written $X \subseteq Y$ if $X_s \subseteq Y_s$ for all $s \in S$;
- Cartesian product of X and Y is defined as $X \times Y = (X_s \times Y_s)_{s \in S}$;
- an S-sorted relation between elements of X and Y is $R \subseteq X \times Y$; if $x \in X_s$ and $y \in Y_s$ for some $s \in S$, then the fact that x is in relation R with y will be denoted $x \, R_s \, y$ or simply $x \, R \, y$;
- an S-sorted function from X to Y is $f = (f : X_s \to Y_s)_{s \in S}$;
- a kernel of an S-sorted function $f : X \to Y$ is $\ker(f) = (\ker(f_s))_{s \in S}$, where $\ker(f_s) = \{(x, x') \mid f_s(x) = f_s(x')\}$;

The subscript s will be often omitted, for example $x \in X_s$ will be written $x \in X$, for short.

A relation $\approx \subseteq X \times X$ is an equivalence if for all $s \in S$, $\approx_s$ is an equivalence. A quotient of X by an equivalence $\approx$ is defined $X/\approx = \{[x]_\approx \mid x \in X\}$ (and it is an S-sorted set), where $[x]_\approx = \{x' \in X \mid x \approx x'\}$.

S-sorted sets with S-sorted functions form the category $\mathbf{Set}^S$ of S-sorted sets.

Categories are denoted with bold faces, like $\mathbf{Set}^S$. Objects of a category $\mathbf{K}$ are denoted $|\mathbf{K}|$. The fact that A is an object of a category $\mathbf{K}$ is written $A \in |\mathbf{K}|$. If $f : A \to B$ is a morphism of $\mathbf{K}$ then it will be denoted $f \in \mathbf{K}$ or $f : A \to B \in \mathbf{K}$ (the latter brings an additional information about the source and target of the morphism f). The composition of morphisms $f : A \to B$ and $g : B \to C$ is denoted with ';' (semicolon) and written in the diagrammatic order, $f; g$.

Functors are also usually denoted with bold faces, $\mathbf{F} : \mathbf{K1} \to \mathbf{K2}$.

A notion of *institution* was introduced in [6], but in this paper we work with a slightly different definition. The definition we work with can be found, e.g in [7]. An institution $\mathbf{INS} = (\mathbf{Sign}, \mathbf{Sen}, \mathbf{Mod}, (\models_\Sigma)_{\Sigma \in |\mathbf{Sign}|})$ consists of:

- a category $\mathbf{Sign}$ of signatures;
- a *sentence functor* $\mathbf{Sen} : \mathbf{Sign} \to \mathbf{Set}$;
- a *model functor* $\mathbf{Mod} : \mathbf{Sign}^{op} \to \mathbf{Cat}$;

- for each signature $\Sigma \in |\mathbf{Sign}|$, a *satisfaction relation* $\models_\Sigma \subseteq |\mathbf{Mod}(\Sigma)| \times \mathbf{Sen}(\Sigma)$ such that for any signature morphism $\sigma : \Sigma \to \Sigma' \in \mathbf{Sign}$, Σ-sentence $\phi \in \mathbf{Sen}(\Sigma)$ and Σ'-model $M' \in |\mathbf{Mod}(\Sigma')|$:

$$M' \models_{\Sigma'} \mathbf{Sen}(\sigma)(\phi) \qquad iff \qquad \mathbf{Mod}(\sigma)(M') \models_\Sigma \phi.$$

The above condition is called *satisfaction condition*.

Throughout this paper the notation for $\mathbf{Sen}(\sigma)(\phi)$ and $\mathbf{Mod}(\sigma)(M')$ will be simplified, i.e. $\mathbf{Sen}(\sigma)(\phi)$ will be simply written as $\sigma(\phi)$ and $\mathbf{Mod}(\sigma)(M')$ will be denoted $M'|_\sigma$. The functor $\_|_\sigma : \mathbf{Mod}(\Sigma') \to \mathbf{Mod}(\Sigma)$ ($\mathbf{Mod}(\sigma)$) is called *reduct functor*.

An institution **INS** has the *amalgamation property* is for each pushout in the category of signatures **Sign**,

$$
\begin{array}{ccc}
\Sigma^1 & \xrightarrow{\ \sigma_1'\ } & \Sigma' \\
\sigma_1 \big\uparrow & & \big\uparrow \sigma_2' \\
\Sigma & \xrightarrow[\ \sigma_2\ ]{} & \Sigma^2,
\end{array}
$$

Σ^1-model $M_1 \in |\mathbf{Mod}(\Sigma^1)|$, Σ^2-model $M_2 \in |\mathbf{Mod}(\Sigma^2)|$ such that $M_1|_{\sigma_1} = M_2|_{\sigma_2}$ there exists a unique model $M' \in |\mathbf{Mod}(\Sigma')|$ such that $M|_{\sigma_1'} = M_1$ and $M'|_{\sigma_2'} = M_2$.

The semantics of a specification SP in any institution **INS** is a signature of this specification, Sig[SP] and a class of models of this specification, Mod[SP]. In each institution **INS** the following standard specification building operations are available:

- for $\Sigma \in |\mathbf{Sign}|$, $\Phi \subseteq \mathbf{Sen}(\Sigma)$, a basic specification (presentation), (Σ, Φ):
  - Sig[SP] $= \Sigma$,
  - Mod[SP] $= \{M \in |\mathbf{Mod}(\Sigma)| \mid M \models_\Sigma \Phi\}$;
- for any specification SP^1, SP^2 with the same signature Σ, their union $\mathrm{SP}^1 \cup \mathrm{SP}^2$:
  - Sig$[\mathrm{SP}^1 \cup \mathrm{SP}^2] = \Sigma$,
  - Mod$[\mathrm{SP}^1 \cup \mathrm{SP}^2] = \mathrm{Mod}[\mathrm{SP}^1] \cap \mathrm{Mod}[\mathrm{SP}^2]$;
- for a signature morphism $\sigma : \Sigma \to \Sigma'$ and a specification SP with the signature Σ, **translate** SP **by** σ:
  - Sig[**translate** SP **by** σ] $= \Sigma'$,
  - Mod[**translate** SP **by** σ] $= \{M' \in |\mathbf{Mod}(\Sigma')| \mid M'|_\sigma \in \mathrm{Mod}[\mathrm{SP}]\}$;
- for a signature morphism $\sigma : \Sigma \to \Sigma'$ and a specification SP$'$ with the signature Σ', **derive from** SP$'$ **by** σ:
  - Sig[**derive from** SP$'$ **by** σ] $= \Sigma$,
  - Mod[**derive from** SP$'$ **by** σ] $= \{M'|_\sigma \mid M' \in \mathrm{Mod}[SP']\}$.

Let $\sigma : \Sigma \to \Sigma'$ be a signature morphism. The reduct functor $\_|_\sigma$ is *isomorphic compatible* if for each Σ'-model $M' \in |\mathbf{Mod}(\Sigma')|$, Σ-model $N \in |\mathbf{Mod}(\Sigma)|$ that is isomorphic to $M'|_\sigma$ there exists a model $N' \in |\mathbf{Mod}(\Sigma')|$ isomorphic to M' such that $N'|_\sigma = N$. A specification SP *has isomorphic compatible reduct functors* if for each signature morphism used to build this specification, the corresponding reduct functor is isomorphic compatible.

3   Concrete Categories

The contents of this section is a selection of notions presented in [1] and [5]. The basic intuition to follow is that objects of a concrete category come equipped with carrier sets and morphisms can be though of as a functions between carrier sets that preserve the object structure. This additional structure of a category allows us to define many concepts from the universal algebra, like subobjects or quotients.

Definition 1. *An S-concrete category is a category* **K** *together with a concretization functor $|\_| : \mathbf{K} \to \mathbf{Set}^S$ that is faithful.*

The indicator S will be often omitted when dealing with S-concrete categories.

Concrete categories as defined above are similar to *constructs* in [1] but in this work we deal with many-sorted sets.

Concrete categories will be denoted simply by $|\_| : \mathbf{K} \to \mathbf{Set}^S$ instead of **K** together with $|\_| : \mathbf{K} \to \mathbf{Set}^S$ since in the concretization functors the whole information about the concrete category is included (i.e. the category **K** and the concretization functor itself).

Throughout this section, let $|\_| : \mathbf{K} \to \mathbf{Set}^S$ be a concrete category.

Proposition 1. *For any morphism $f : A \to B \in \mathbf{K}$:*

- *if $|f|$ is surjective then f is an epimorphism;*
- *if $|f|$ is injective then f is a monomorphism.*

Definition 2. *A concrete category $|\_| : \mathbf{K} \to \mathbf{Set}^S$ is* transportable *if for each object $A \in |\mathbf{K}|$ and a bijective function $i : |A| \to X$ there exists an object $B \in |\mathbf{K}|$ and an isomorphism $i' : A \to B$ such that $|i'| = i$ (and $|B| = X$).*

In [5] the notion of transportability is called *admitting of renaming of elements of objects.*

Definition 3. *An isomorphism $i : A \to B \in \mathbf{K}$ is* identity-carried *if $|i|$ is an identity. Two objects $A, B \in |\mathbf{K}|$ are* exactly isomorphic *if there exists an identity-carried isomorphism $i : A \to B$.*

Subobjects. A notion of a subobject can be found e.g. in [1], where it is called *initial subobject*, but in this work, to simplify matters, we use a slightly different definition.

Definition 4. *Let $A \in |\mathbf{K}|$ be an object of* **K**. *A* subobject *of A is an object $B \in |\mathbf{K}|$ together with a morphism $\iota_{B \hookrightarrow A} : B \to A$ such that $|\iota_{B \hookrightarrow A}| : |B| \to |A|$ is an inclusion and for each morphism $f : C \to A$ with $|f|(|C|) \subseteq |B|$[1] there exists a morphism $f' : C \to B$ such that $f'; \iota_{B \hookrightarrow A} = f$.*

---

[1] $|f|(|C|)$ is the image of the set $|C|$ under the function $|f|$.

If B is a subobject of A then there exists exactly one morphism $\iota_{B \hookrightarrow A} : B \to A$ such that $|\iota_{B \hookrightarrow A}|$ is an inclusion (follows from the faithfulness of $|\_|$). Moreover if $f : C \to A$ is a morphism with $|f|(|C|) \subseteq |B|$ then the morphism $f' : C \to B$ such that $f'; \iota_{B \hookrightarrow A} = f$ is unique.

The difference between the notion of a subobject presented here and the notion of an initial subobject from [1] is that the embedding ($\iota_{B \hookrightarrow A}$, see [1]) is required here to be an inclusion, not only an injection (like in [1]). That simplifies the definition of a generated subobject.

There is also a slight difference between the definition of a subobject presented here and in [5]. The subobjects defined here are called *full* in [5].

Proposition 2. *Let $A, B, C \in |\mathbf{K}|$. If B is a subobject of A and C is a subobject of B then C is a subobject of A.*

Definition 5. *Let $A \in |\mathbf{K}|$ be an object and $X \subseteq |A|$. A subobject of A generated by X is a subobject B of A such that $X \subseteq |B|$ and for any subobject C of A if $X \subseteq |C|$ then $|B| \subseteq |C|$.*

Proposition 3. *A subobject of A generated by $X \subseteq |A|$, if it exists, is unique up to an identity-carried isomorphism. Moreover, any object exactly isomorphic to a subobject of A generated by X is a subobject of A generated by X.*

The generated subobject of A by $X \subseteq |A|$ will be denoted $\langle X \rangle_A$ and the inclusion morphism, $\iota_{\langle X \rangle_A \hookrightarrow A} : \langle X \rangle_A \to A$, will be written $\iota_{X \hookrightarrow A}$, for short.

Definition 6. *A concrete category $|\_| : \mathbf{K} \to \mathbf{Set}^S$ has generated subobjects if for each $A \in |\mathbf{K}|$ and $X \subseteq |A|$ there exists the subobject of A generated by X.*

Quotients

Definition 7. *Let $A \in |\mathbf{K}|$ be an object of $\mathbf{K}$. A quotient of A is an object $B \in |\mathbf{K}|$ together with an epimorphism $\pi_{A/B} : A \to B$ such that for any morphism $f : A \to C$ with $\ker(|\pi_{A/B}|) \subseteq \ker(|f|)$ there exists a morphism $f' : B \to C$ such that $\pi_{A/B}; f' = f$.*
A quotient B of A is final if $|\pi_{A/B}|$ is surjective.

If B is a quotient of A then the morphism $\pi_{A/B}$ is called the *quotient projection*.

The notion of a final quotient as defined above comes from [1]. In [5] final quotients are called *surjective quotients*.

Definition 8. *Let $A \in |\mathbf{K}|$ be an object of $\mathbf{K}$. An equivalence relation $\approx \subseteq |A| \times |A|$ is a congruence on A if there exists a morphism $f : A \to B$ such that $\ker(|f|) = \approx$.*
A quotient of A by $\approx$ is a quotient B of A with $\pi_{A/B} : A \to B$ such that $\ker(|\pi_{A/B}|) = \approx$.

Proposition 4. *A quotient of A by a congruence $\approx \subseteq |A| \times |A|$, if it exists, is unique up to an isomorphism. Moreover, any object isomorphic to a quotient of A by $\approx$ is a quotient of A by $\approx$.*

The quotient of A by a congruence $\approx\, \subseteq |A| \times |A|$ will be denoted $A/\approx$ and the morphism $\pi_{A/(A/\approx)}$ will be simply written as $\pi_{A/\approx} : A \to A/\approx$.

Let $A \in |\mathbf{K}|$ be an object of K and $\approx\, \subseteq |A| \times |A|$ be a congruence. If B is a subobject of A then the quotient (if it exists) of B by $\approx \cap |B| \times |B|$ will be simply denoted $B/\approx$ instead of $B/(\approx \cap |B| \times |B|)$. Notice that if $\approx$ is a congruence on A then $\approx$ (more precisely $\approx \cap |B| \times |B|$) is also a congruence on B.

Definition 9. *A concrete category $|\_| : \mathbf{K} \to \mathbf{Set}^S$ has (final) quotients if for each object $A \in |\mathbf{K}|$ and a congruence $\approx\, \subseteq |A| \times |A|$ on A there exists a (final) quotient of A by $\approx$.*

Definition 10. *In a concrete category $|\_| : \mathbf{K} \to \mathbf{Set}^S$ subobjects are compatible with quotients if for each object $A \in |\mathbf{K}|$, its subobject $\iota_{B \hookrightarrow A} : B \to A$ and a congruence $\approx\, \subseteq |A| \times |A|$ if quotients $A/\approx$ and $B/\approx$ exist then $B/\approx$ is a subobject of $A/\approx$ (formally there exist an object $C \in |\mathbf{K}|$ isomorphic to $B/\approx$ which is a subobject of A) and the following diagram commute:*

$$
\begin{array}{ccc}
B & \xrightarrow{\;\iota_{B \hookrightarrow A}\;} & A \\[4pt]
{\scriptstyle \pi_{B/\approx}}\big\downarrow & & \big\downarrow{\scriptstyle \pi_{A/\approx}} \\[6pt]
B/\approx & \xrightarrow[\;\iota_{B/\approx \hookrightarrow A/\approx}\;]{} & A/\approx,
\end{array}
$$

i.e. $\iota_{B \hookrightarrow A}; \pi_{A/\approx} = \pi_{B/\approx}; \iota_{B/\approx \hookrightarrow A/\approx}$.

4 Concrete Institutions

In this section we follow the ideas of the previous section and define a *concrete institution* which is an extension of the notion of the institution introduced in [6]. A concrete institution is an institution in which for each signature a set of sorts of this signature is available and each category of models is a concrete category.

The notation for the category of S-sorted sets, $\mathbf{Set}^S$, can be extended to denote the functor: $\mathbf{Set}^{(-)} : \mathbf{Set}^{op} \to \mathbf{Cat}$. For a set S, $\mathbf{Set}^S$ is the category of S-sorted sets. For a function $\sigma : S \to S'$, $\mathbf{Set}^\sigma$ is the reduct functor, $\mathbf{Set}^\sigma : \mathbf{Set}^{S'} \to \mathbf{Set}^S$ defined: $\mathbf{Set}^\sigma((X_s)_{s \in S'}) = (Y_s)_{s \in S}$ with $Y_s = X_{\sigma(s)}$ and similarly for S'-sorted functions.

If $\approx'\, \subseteq A' \times B'$ is an S'-sorted relation then $\mathbf{Set}^\sigma(\approx')$ is well defined since, in fact, $\approx'$ is an S'-sorted set (and the result is an S-sorted relation between elements of $\mathbf{Set}^\sigma(A)$ and $\mathbf{Set}^\sigma(B)$).

Definition 11. *A concrete institution $\mathbf{INS_c}$ based on an institution $\mathbf{INS} = (\mathbf{Sign}, \mathbf{Sen}, \mathbf{Mod}, (\models_\Sigma)_{\Sigma \in |\mathbf{Sign}|})$ consists of $\mathbf{INS}$ together with a functor sorts : $\mathbf{Sign} \to \mathbf{Set}$ and a natural transformation $|\_| : \mathbf{Mod} \to sorts^{op}; \mathbf{Set}^{(-)}$ between functors from $\mathbf{Sign}^{op}$ to $\mathbf{Cat}$.*

Thus, a concrete institution is a tuple:

$$\mathbf{INS_c} = (\mathbf{Sign}, \mathbf{Sen}, \mathbf{Mod}, (\models_\Sigma)_{\Sigma \in |\mathbf{Sign}|}, sorts, |\_|).$$

The functor $sorts : \textbf{Sign} \to \textbf{Set}$ yields, for each signature Σ, a set of sorts of this signature. The natural transformation $|\_| : \textbf{Mod} \to sorts^{op}; \textbf{Set}^{(-)}$ is a family of concretization functors, $(|\_|_\Sigma : \textbf{Mod}(\Sigma) \to \textbf{Set}^{sorts(\Sigma)})_{\Sigma \in |\textbf{Sign}|}$. The naturality of this transformation ensures that the following diagram commutes:

$$
\begin{array}{ccc}
\Sigma & \textbf{Mod}(\Sigma) \xrightarrow{\ |\_|_\Sigma\ } & \textbf{Set}^{sorts(\Sigma)} \\[2mm]
\sigma \downarrow \quad & \ _{-|_\sigma} \uparrow \quad\quad & \quad\quad \uparrow \textbf{Set}^{sorts(\sigma)} \\[2mm]
\Sigma' & \textbf{Mod}(\Sigma') \xrightarrow[\ |\_|_{\Sigma'}\ ]{} & \textbf{Set}^{sorts(\Sigma')},
\end{array}
$$

where $\sigma : \Sigma \to \Sigma' \in \textbf{Sign}$.

The commutativity of the above diagram allows us to simplify the notation. The functor $\textbf{Set}^{sorts(\sigma)} : \textbf{Set}^{sorts(\Sigma')} \to \textbf{Set}^{sorts(\Sigma)}$ will be denoted $\_|_\sigma : \textbf{Set}^{sorts(\Sigma')} \to \textbf{Set}^{sorts(\Sigma)}$.

Let $\textbf{INS}_\textbf{c} = (\textbf{Sign}, \textbf{Sen}, \textbf{Mod}, (\models_\Sigma)_{\Sigma \in |\textbf{Sign}|}, sorts, |\_|)$ be a concrete institution, fixed throughout this section.

Definition 12. *A satisfaction relation* $\models_\Sigma \subseteq |\textbf{Mod}(\Sigma)| \times \textbf{Sen}(\Sigma)$, *where* $\Sigma \in |\textbf{Sign}|$, *is* isomorphism compatible *if for all isomorphic models* $A, B \in |\textbf{Mod}(\Sigma)|$ *and* Σ-*sentence* $\phi \in \textbf{Sen}(\Sigma)$ *the following holds:* $A \models_\Sigma \phi$ *iff* $B \models_\Sigma \phi$.

A concrete institution $\textbf{INS}_\textbf{c} = (\textbf{Sign}, \textbf{Sen}, \textbf{Mod}, (\models_\Sigma)_{\Sigma \in |\textbf{Sign}|}, sorts, |\_|)$ *has* isomorphic compatible satisfaction relations *if for each signature* $\Sigma \in |\textbf{Sign}|$ *the satisfaction relation* $\models_\Sigma$ *is isomorphic compatible.*

Definition 13. *A reduct functor* $\_|_\sigma$, *where* $\sigma : \Sigma \to \Sigma' \in \textbf{Sign}$, *preserves* subobjects *if for each* Σ'-*model* $A' \in \textbf{Mod}(\Sigma')$ *and its subobject* B' *the reduct* $B'|_\sigma$ *is a subobject of* $A'|_\sigma$.

In a concrete institution $\textbf{INS}_\textbf{c} = (\textbf{Sign}, \textbf{Sen}, \textbf{Mod}, (\models_\Sigma)_{\Sigma \in |\textbf{Sign}|}, sorts, |\_|)$ *reduct functors preserve subobjects if for each signature morphism* $\sigma : \Sigma \to \Sigma' \in \textbf{Sign}$, *the reduct functor* $\_|_\sigma$ *preserves subobjects.*

Definition 14. *A reduct functor* $\_|_\sigma$, *where* $\sigma : \Sigma \to \Sigma' \in \textbf{Sign}$, *preserves* quotients *if for each* Σ'-*model* $A' \in \textbf{Mod}(\Sigma')$ *and its quotient* $\pi_{A'/B'} : A' \to B'$ *the reduct* $\pi_{A'/B'}|_\sigma : A'|_\sigma \to B'|_\sigma$ *is a quotient of* $A'|_\sigma$.

In a concrete institution $\textbf{INS}_\textbf{c} = (\textbf{Sign}, \textbf{Sen}, \textbf{Mod}, (\models_\Sigma)_{\Sigma \in |\textbf{Sign}|}, sorts, |\_|)$ *reduct functors preserve quotients if for each signature morphism* $\sigma : \Sigma \to \Sigma' \in \textbf{Sign}$, *the reduct functor* $\_|_\sigma$ *preserves quotients.*

A concrete institution $\textbf{INS}_\textbf{c} = (\textbf{Sign}, \textbf{Sen}, \textbf{Mod}, (\models_\Sigma)_{\Sigma \in |\textbf{Sign}|}, sorts, |\_|)$

- is transportable;
- has generated subobjects;
- has (final) quotients;
- subobjects are compatible with quotients

if for each signature $\Sigma \in |\textbf{Sign}|$, the concrete category $|\_|_\Sigma : \textbf{Mod}(\Sigma) \to \textbf{Set}^{sorts(\Sigma)}$ has the corresponding property.

5   Behavioural Concrete Institutions

The extra structure of a concrete institution allows us to redefine the satisfaction relation to obtain its behavioural version.

Let $\mathbf{INS_c} = (\mathbf{Sign}, \mathbf{Sen}, \mathbf{Mod}, (\models_\Sigma)_{\Sigma \in |\mathbf{Sign}|}, sorts, |\_|)$ be a concrete institution, fixed throughout this section. We assume that $\mathbf{INS_c}$

- is transportable,
- has generated subobjects,
- has final quotients,
- subobjects are compatible with quotients,
- has isomorphic compatible satisfaction relations,
- reduct functors preserve subobjects and quotients.

5.1   Behavioural Satisfaction Relation

Reachability is an important concept of system specifications. A *reachability structure* on a model is a subset of the carrier set of this models. It contains the elements which are of interest from the user's point of view. In this work we follow the ideas introduced in [2] and do not require a reachability structure to be a subobject of a model considered contrary e.g. to [4] (where a reachability structure is implicitly incorporated into a notion of a partial congruence).

Definition 15. *A* reachability structure *over a signature* $\Sigma \in |\mathbf{Sign}|$ *is a family* $\mathcal{R} = (\mathcal{R}_M)_{M \in |\mathbf{Mod}(\Sigma)|}$ *of sorts*(Σ)*-sorted sets such that* $\mathcal{R}_M \subseteq |M|_\Sigma$ *for each* $M \in |\mathbf{Mod}(\Sigma)|$.

Another important aspect of system specifications is the concept of observability. In this work we generalize the notion of observational equality from [2] which we call here an *observability structure*. An observability structure on a model is an equivalence relation on the carrier set of this model. Unlike the approach presented in [4] we do not impose any further restrictions on an observability structure. The idea comes from [2].

Definition 16. *An* observability structure *over a signature* $\Sigma \in |\mathbf{Sign}|$ *is a family* $\approx = (\approx_M)_{M \in |\mathbf{Mod}(\Sigma)|}$ *of equivalence relations such that* $\approx_M \subseteq |M|_\Sigma \times |M|_\Sigma$ *(i.e.* $\approx_M$ *is an equivalence relation on* $|M|_\Sigma$).

Usually a reachability structure is determined by a distinguished set of constructor operations and an observability structure by a distinguished set of observer operations, see [2], but it is not the purpose of this work to present how those structures can be defined. The problem here is more complicated since in an arbitrary (concrete) institution the notion of an operation is not available. In this work we only present the way of defining the behavioural semantics of specifications given arbitrary reachability and observability structures.

A pair $(\mathcal{R}, \approx)$, where $\mathcal{R}$ is a reachability structure over a signature Σ and $\approx$ is an observability structure over the signature Σ, is called a *behavioural structure* over the signature Σ.

Definition 17. *A behavioural signature $\Sigma_{\text{Beh}} = (\Sigma, \mathcal{R}, \approx)$ consists of:*

- *a signature $\Sigma \in |\mathbf{Sign}|$;*
- *a reachability structure $\mathcal{R}$ over the signature Σ;*
- *an observability structure $\approx$ over the signature Σ.*

Following the ideas from [2], since no restrictions were imposed on reachability and observability structures, we introduce two kinds of constraints on the class of models: the reachability and the observability constraint. The former is a well-known constraint which expresses the property that the only admissible models are those on which the reachability structure is a subobject of the model considered (intuitively it is closed under operations). But if we deal both with the reachability and observability concepts such a requirement is too strong, since from the user's point of view this is not different from allowing the elements of the submodel generated by its reachability structure to be indistinguishable from some elements in this reachability structure.

The latter (the observability constraint) simply states that the observability structure on a model must be a congruence on the subobject of this model generated by the reachability structure.

Definition 18. *Let $\Sigma_{\text{Beh}} = (\Sigma, \mathcal{R}, \approx)$ be a behavioural signature. A Σ-model $M \in |\mathbf{Mod}(\Sigma)|$ satisfies*

- *the* reachability constraint *if for each $a \in |\langle \mathcal{R}_M \rangle_M|_\Sigma$ there exists $b \in \mathcal{R}_M$ such that $a \approx_M b$;*
- *the* observability constraint *if $\approx_M$ is a congruence on $\langle \mathcal{R}_M \rangle_M$ (more precisely if $\approx \cap |\langle \mathcal{R}_M \rangle_M|_\Sigma \times |\langle \mathcal{R}_M \rangle_M|_\Sigma$ is a congruence on $\langle \mathcal{R}_M \rangle_M$).*

Definition 19. *Let $\Sigma_{\text{Beh}} = (\Sigma, \mathcal{R}, \approx)$ be a behavioural signature. A Σ-model $M \in |\mathbf{Mod}(\Sigma)|$ is called* behavioural *if it satisfies the reachability and the observability constraints.*

The class of all behavioural models over a behavioural signature Σ_{Beh} will be denoted $\text{Mod}_{\text{Beh}}(\Sigma_{\text{Beh}})$.

The standard way of defining a *behavioural satisfaction relation* independently on the logical system is by the notion of a *behaviour* of a model, see [4] or [5]. For the logical systems in which the satisfaction relation is based on an equality between terms this approach (this definition of a behavioural satisfaction relation) is equivalent to the approach which involves changing the semantics of equality, see [4] or [2].

Definition 20. *Let $\Sigma_{\text{Beh}} = (\Sigma, \mathcal{R}, \approx)$ be a behavioural signature. The* behaviour $\mathcal{B}_{\Sigma_{\text{Beh}}}(M)$ *of a behavioural model $M \in \text{Mod}_{\text{Beh}}(\Sigma_{\text{Beh}})$ is defined:*

$$\mathcal{B}_{\Sigma_{\text{Beh}}}(M) = \langle \mathcal{R}_M \rangle_M / \approx_M.$$

Definition 21. *Let* $\Sigma_{\mathrm{Beh}} = (\Sigma, \mathcal{R}, \approx)$ *be a behavioural signature. A behavioural model* $M \in \mathrm{Mod}_{\mathrm{Beh}}(\Sigma_{\mathrm{Beh}})$ *behaviourally satisfies a sentence* $\phi \in \mathbf{Sen}(\Sigma)$,

$$M \models_{\Sigma_{\mathrm{Beh}}} \phi$$

if its behaviour satisfies the sentence ϕ *in the original sense,*

$$\mathcal{B}_{\Sigma_{\mathrm{Beh}}}(M) \models_{\Sigma} \phi.$$

5.2   Behavioural Concrete Institutions

The above section covers only the case of flat specifications, sometimes called *satisfaction frame*, which is only a single fibre of an institution. In this section we present how the notions of the previous section can be used to form a *behavioural concrete institution*.

The first step is to impose additional requirements of the signature morphisms of the original institution $\mathbf{INS_c}$, to eliminate those that violate the satisfaction condition for behavioural satisfaction relation and behavioural models.

Definition 22. *A* behavioural signature morphism $\sigma : \Sigma_{\mathrm{Beh}} \to \Sigma'_{\mathrm{Beh}}$, *where* $\Sigma_{\mathrm{Beh}} = (\Sigma, \mathcal{R}, \approx)$ *and* $\Sigma'_{\mathrm{Beh}} = (\Sigma', \mathcal{R}', \approx')$, *is a signature morphism* $\sigma : \Sigma \to \Sigma'$ *such that it preserves the reachability structure and the observability structure, i.e. if for each* Σ'-*model* $M' \in |\mathbf{Mod}(\Sigma')|$ *the following holds:* $\mathcal{R}'_{M'}|_{\sigma} = \mathcal{R}_{M'|_{\sigma}}$ *and* $\approx'_{M'}|_{\sigma} = \approx_{M'|_{\sigma}}$.

Now, given the notion of a behavioural morphism we can define the *category of all behavioural signatures*. This category contains all behavioural signatures and morphisms of this category are all behavioural signature morphisms.

Definition 23. *The* category of all behavioural signatures, **ASign**, *consists of:*

- *objects are all behavioural signatures* $\Sigma_{\mathrm{Beh}} = (\Sigma, \mathcal{R}, \approx)$ *such that* $\Sigma \in |\mathbf{Sign}|$ *and* $(\mathcal{R}, \approx)$ *is a behavioural structure over* Σ;
- *morphisms are all behavioural signature morphisms.*

The functor from the category of all behavioural signatures **ASign** to the category of signatures **Sign** which simply "forgets" about behavioural structures is called the *forgetful functor*, $\mathbf{AF} : \mathbf{ASign} \to \mathbf{Sign}$. It is defined: $\mathbf{AF}(\Sigma_{\mathrm{Beh}}) = \Sigma$ for $\Sigma_{\mathrm{Beh}} = (\Sigma, \mathcal{R}, \approx) \in |\mathbf{ASign}|$ and $\mathbf{AF}(\sigma) = \sigma$ for $\sigma \in \mathbf{ASign}$.

Theorem 1. *Let* $\sigma : \Sigma_{\mathrm{Beh}} \to \Sigma'_{\mathrm{Beh}} \in \mathbf{ASign}$ *be a behavioural signature morphism,* $\Sigma_{\mathrm{Beh}} = (\Sigma, \mathcal{R}, \approx)$ *and* $\Sigma'_{\mathrm{Beh}} = (\Sigma', \mathcal{R}', \approx')$. *Then for any behavioural model* $M' \in \mathbf{Mod}(\Sigma')$ *the reduct of this model,* $M'|_{\sigma}{}^2$, *is behavioural.*

---

[2] Formally, it should be written $M'|_{\mathbf{AF}(\sigma)}$, but to simplify matters, since it doesn't throw into confusion it will be denoted like above (i.e. $M'|_{\sigma}$).

The above theorem allows us to define the *behavioural model functor*, which for each behavioural signature $\Sigma_{\mathrm{Beh}} = (\Sigma, \mathcal{R}, \approx)$ yields the category of all behavioural models over the signature Σ and for each behavioural signature morphism $\sigma : \Sigma_{\mathrm{Beh}} \to \Sigma'_{\mathrm{Beh}}$ it yields the restriction of the reduct functor $\_|_\sigma$ to the category of all behavioural models over the signature Σ' ($\Sigma'_{\mathrm{Beh}} = (\Sigma', \mathcal{R}', \approx')$). Th. 1 states that this definition is correct, i.e. the reduct of a behavioural model over the signature Σ' is a behavioural model over the signature Σ (w.r.t $(\mathcal{R}, \approx)$, where $\Sigma_{\mathrm{Beh}} = (\Sigma, \mathcal{R}, \approx)$).

Definition 24. *The* behavioural model functor $\mathbf{AMod} : \mathbf{ASign}^{op} \to \mathbf{Cat}$ *is defined:*

- *for* $\Sigma_{\mathrm{Beh}} = (\Sigma, \mathcal{R}, \approx) \in |\mathbf{ASign}|$, $\mathbf{AMod}(\Sigma_{\mathrm{Beh}}) = \mathrm{Mod}_{\mathrm{Beh}}(\Sigma_{\mathrm{Beh}})$ *is the full subcategory of* $\mathbf{Mod}(\Sigma)$;
- *for* $\sigma : \Sigma_{\mathrm{Beh}} \to \Sigma'_{\mathrm{Beh}} \in \mathbf{ASign}$, *where* $\Sigma_{\mathrm{Beh}} = (\Sigma, \mathcal{R}, \approx)$ *and* $\Sigma'_{\mathrm{Beh}} = (\Sigma', \mathcal{R}', \approx')$, $\mathbf{AMod}(\sigma) = \_|_\sigma$ *is the restriction of the reduct functor* $\_|_\sigma : \mathbf{Mod}(\Sigma') \to \mathbf{Mod}(\Sigma)$ *to the category* $\mathbf{AMod}(\Sigma'_{\mathrm{Beh}})$.

Theorem 2. *For each behavioural signature morphism* $\sigma : \Sigma_{\mathrm{Beh}} \to \Sigma'_{\mathrm{Beh}}$, *where* $\Sigma_{\mathrm{Beh}} = (\Sigma, \mathcal{R}, \approx)$ *and* $\Sigma'_{\mathrm{Beh}} = (\Sigma', \mathcal{R}', \approx')$, *for each behavioural* Σ'-*model* $M' \in |\mathbf{AMod}(\Sigma'_{\mathrm{Beh}})|$ *and* Σ-*sentence* $\phi \in \mathbf{Sen}(\Sigma)$ *the following holds:*

$$M'|_\sigma \models_{\Sigma_{\mathrm{Beh}}} \phi \qquad iff \qquad M' \models_{\Sigma'_{\mathrm{Beh}}} \sigma(\phi).$$

Definition 25. *The tuple*

$$\mathbf{AINS_c} = (\mathbf{ASign}, \mathbf{ASen}, \mathbf{AMod}, (\models_{\Sigma_{\mathrm{Beh}}})_{\Sigma_{\mathrm{Beh}} \in |\mathbf{ASign}|}, asorts, |\_|^a)$$

is the behavioural concrete institution *also called the* concrete institution of behavioural logic *based on the concrete institution* $\mathbf{INS_c}$, *where*

- $\mathbf{ASign}$ *is the category of all behavioural signatures,*
- $\mathbf{ASen} : \mathbf{ASign} \to \mathbf{Set}$ *is the behavioural sentence functor, defined:* $\mathbf{ASen} = \mathbf{AF}; \mathbf{Sen}$,
- $\mathbf{AMod} : \mathbf{ASign}^{op} \to \mathbf{Cat}$ *is the behavioural model functor,*
- *for each* $\Sigma_{\mathrm{Beh}} \in |\mathbf{ASign}|$, $\models_{\Sigma_{\mathrm{Beh}}}$ *is the behavioural satisfaction relation,* $\models_{\Sigma_{\mathrm{Beh}}} \subseteq |\mathbf{AMod}(\Sigma)| \times \mathbf{ASen}(\Sigma)$,
- $asorts : \mathbf{ASign} \to \mathbf{Set}$ *is the behavioural sorts functor, defined:* $asorts = \mathbf{AF}; sorts$,
- $|\_|^a : \mathbf{AMod} \to asorts^{op}; \mathbf{Set}^{(-)}$ *is a natural transformation between functors from* $\mathbf{ASign}^{op}$ *to* $\mathbf{Cat}$, *defined: for a behavioural signature* $\Sigma_{\mathrm{Beh}} \in |\mathbf{ASign}|$, $|\_|^a_{\Sigma_{\mathrm{Beh}}} : \mathbf{AMod}(\Sigma) \to \mathbf{Set}^{asorts(\Sigma_{\mathrm{Beh}})}$ *is the restriction of the functor* $|\_|_\Sigma : \mathbf{Mod}(\Sigma) \to \mathbf{Set}^{sorts(\mathbf{AF}(\Sigma_{\mathrm{Beh}}))}$ *to the category of all behavioural models over* Σ_{Beh}.

The superscript 'a' in the natural transformation $|\_|^a : \mathbf{AMod} \to asorts^{op}; \mathbf{Set}^{(-)}$ will be omitted.

The institution $\mathbf{AINS_c}$ is a rather "large" institution. The category of signatures of this institution contains all behavioural signatures $\Sigma_{\mathrm{Beh}} = (\Sigma, \mathcal{R}, \approx)$

such that Σ is a signature from the original category of signatures **Sign** and $(\mathcal{R}, \approx)$ is an arbitrary behavioural structure. Such a freedom is usually inadequate in practice (when defining a behavioural semantics of a specification language based on the original semantics). Interesting cases are when behavioural structures are determined for example by given sets of constructor and observer operations (see [2]). The institution **AINS$_c$** was introduced for technical reasons, to express some properties concerning behavioural structures. Therefore we introduce a new behavioural concrete institution **BINS$_c$** in which the category of signatures **BSign** contains only some behavioural signatures and behavioural signature morphisms. In other words **BSign** is a subcategory of **ASign**. Formally **BINS$_c$** is a tuple:

$$\mathbf{BINS_c} = (\mathbf{BSign}, \mathbf{BSen}, \mathbf{BMod}, (\models_{\Sigma_{\mathrm{Beh}}})_{\Sigma_{\mathrm{Beh}} \in |\mathbf{BSign}|}, bsorts, |\_|^b).$$

The other components (apart from **BSign**) of **BINS$_c$** are defined in exactly the same way as in the institution **AINS$_c$**. The forgetful functor $\mathbf{BF} : \mathbf{BSign} \to$ **Sign** can also be easily defined.

An institution **BINS$_c$** can be thought of as a "subinstitution" of **AINS$_c$** with a smaller category of signatures. Note that **AINS$_c$** is a special case of **BINS$_c$**.

5.3   Properties of Behavioural Concrete Institutions

Let **INS$_c$** be a concrete institution that satisfies all the properties mentioned in the beginning of this section and **BINS$_c$** be an arbitrary behavioural concrete institution based on **INS$_c$**. Of course there also exists the behavioural concrete institution **AINS$_c$** based on **INS$_c$** in which the category of signatures contains all behavioural signatures and all behavioural signature morphisms.

In this subsection we assume that the functor $sorts : \mathbf{Sign} \to \mathbf{Set}$ is cocontinuous.

Proposition 5. *If the category* **Sign** *of signatures if cocomplete then so is the category of all behavioural signatures* **ASign** *and the forgetful functor* $\mathbf{AF} :$ **ASign** $\to$ **Sign** *is cocontinuous.*

One important property of an institution is the amalgamation property. Unfortunately a behavioural concrete institution doesn't have the amalgamation property even if the concrete institution on which it is based on has the amalgamation property. The counterexample can be found in [3], where the constructor based observational logic institution, which is a special case of a behavioural concrete institution, is presented. However there are some conditions under which the amalgamation union of two models exists. These conditions are generalization of the conditions for amalgamation from [3].

Proposition 6. *Let*

$$
\begin{array}{ccc}
\Sigma^1_{\mathrm{Beh}} = (\Sigma^1, \mathcal{R}^1, \approx^1) & \xrightarrow{\ \sigma'_1\ } & \Sigma'_{\mathrm{Beh}} = (\Sigma', \mathcal{R}', \approx') \\[4pt]
{\scriptstyle \sigma_1}\big\uparrow & & \big\uparrow{\scriptstyle \sigma'_2} \\[4pt]
\Sigma_{\mathrm{Beh}} = (\Sigma, \mathcal{R}, \approx) & \xrightarrow[\ \sigma_2\ ]{} & \Sigma^2_{\mathrm{Beh}} = (\Sigma^2, \mathcal{R}^2, \approx^2)
\end{array}
$$

be a pushout in the category **BSign** *of behavioural signatures such that, the image of this diagram under the forgetful functor* **BF**,

$$
\begin{array}{ccc}
\Sigma^1 & \xrightarrow{\;\sigma_1'\;} & \Sigma' \\[4pt]
\sigma_1 \big\uparrow & & \big\uparrow \sigma_2' \\[4pt]
\Sigma & \xrightarrow[\;\sigma_2\;]{} & \Sigma^2,
\end{array}
$$

is a pushout in the category **Sign**. *Assume that* **INS$_c$** *has the amalgamation property on this pushout, i.e. for any* $N_1 \in |\mathbf{Mod}(\Sigma^1)|$, $N_2 \in |\mathbf{Mod}(\Sigma^2)|$ *such that* $N_1|_{\sigma_1} = N_2|_{\sigma_2}$ *there exists the unique amalgamation* $N' \in |\mathbf{Mod}(\Sigma')|$ *of* N_1 *and* N_2 *(i.e.* $N'|_{\sigma_1'} = N_1$ *and* $N'|_{\sigma_2'} = N_2$*). Now, let* $M_1 \in |\mathbf{BSign}(\Sigma^1)|$, $M_2 \in |\mathbf{BSign}(\Sigma^2)|$ *be behavioural models such that* $M_1|_{\sigma_1} = M_2|_{\sigma_2}$. *If* $\langle \mathcal{R}_{M_1}^1 \rangle_{M_1}|_{\sigma_1} = \langle \mathcal{R}_{M_1|_{\sigma_1}} \rangle_{M_1|_{\sigma_1}}$ *and* $\langle \mathcal{R}_{M_2}^2 \rangle_{M_2}|_{\sigma_2} = \langle \mathcal{R}_{M_2|_{\sigma_2}} \rangle_{M_2|_{\sigma_2}}$ *then there exists the unique amalgamation* $M' \in |\mathbf{BSign}(\Sigma')|$ *of* M_1 *and* M_2.

5.4   Behavioural Specifications

Given a behavioural concrete institution **BINS$_c$** all standard specification building operations are available:

- for $\Sigma_{\mathrm{Beh}} \in |\mathbf{BSign}|$, $\Phi \subseteq \mathbf{BSen}(\Sigma_{\mathrm{Beh}})$, a basic specification (presentation) $(\Sigma_{\mathrm{Beh}}, \Phi)$:
  - $\mathrm{Sig}[(\Sigma_{\mathrm{Beh}}, \Phi)] = \Sigma_{\mathrm{Beh}}$,
  - $\mathrm{Mod}[(\Sigma_{\mathrm{Beh}}, \Phi)] = \{ M \in |\mathbf{BMod}(\Sigma_{\mathrm{Beh}})| \mid M \models_{\Sigma_{\mathrm{Beh}}} \Phi \}$;
- for any specifications $\mathrm{SP}^1_{\mathrm{Beh}}$, $\mathrm{SP}^2_{\mathrm{Beh}}$ with the same signature Σ_{Beh}, their union $\mathrm{SP}^1_{\mathrm{Beh}} \cup \mathrm{SP}^2_{\mathrm{Beh}}$:
  - $\mathrm{Sig}[\mathrm{SP}^1_{\mathrm{Beh}} \cup \mathrm{SP}^2_{\mathrm{Beh}}]$,
  - $\mathrm{Mod}[\mathrm{SP}^1_{\mathrm{Beh}} \cup \mathrm{SP}^2_{\mathrm{Beh}}] = \mathrm{Mod}[\mathrm{SP}^1_{\mathrm{Beh}}] \cap \mathrm{Mod}[\mathrm{SP}^2_{\mathrm{Beh}}]$;
- for a behavioural signature morphism $\sigma : \Sigma_{\mathrm{Beh}} \to \Sigma'_{\mathrm{Beh}}$ and a specification $\mathrm{SP}_{\mathrm{Beh}}$ with the signature Σ_{Beh}, **translate** $\mathrm{SP}_{\mathrm{Beh}}$ **by** σ:
  - $\mathrm{Sig}[\textbf{translate } \mathrm{SP}_{\mathrm{Beh}} \textbf{ by } \sigma] = \Sigma'_{\mathrm{Beh}}$,
  - $\mathrm{Mod}[\textbf{translate } \mathrm{SP}_{\mathrm{Beh}} \textbf{ by } \sigma] = \{ M' \in |\mathbf{BMod}(\Sigma'_{\mathrm{Beh}})| \mid M'|_\sigma \in \mathrm{Mod}[\mathrm{SP}_{\mathrm{Beh}}] \}$;
- for a signature morphism $\sigma : \Sigma_{\mathrm{Beh}} \to \Sigma'_{\mathrm{Beh}}$ and a specification $\mathrm{SP}'_{\mathrm{Beh}}$ with the signature Σ'_{Beh}, **derive from** $\mathrm{SP}'_{\mathrm{Beh}}$ **by** σ:
  - $\mathrm{Sig}[\textbf{derive from } \mathrm{SP}'_{\mathrm{Beh}} \textbf{ by } \sigma] = \Sigma_{\mathrm{Beh}}$,
  - $\mathrm{Mod}[\textbf{derive from } \mathrm{SP}'_{\mathrm{Beh}} \textbf{ by } \sigma] = \{ M'|_\sigma \mid M' \in \mathrm{Mod}[\mathrm{SP}'_{\mathrm{Beh}}] \}$.

For each behavioural specification $\mathrm{SP}_{\mathrm{Beh}}$ (i.e. a specification in the institution **BINS$_c$**) there exists a standard specification SP (i.e. in the institution **INS$_c$**) built in the same way as $\mathrm{SP}_{\mathrm{Beh}}$, which corresponds to this behavioural specification with the following property: $\mathrm{Sig}[\mathrm{SP}] = \mathbf{BF}(\mathrm{Sig}[\mathrm{SP}_{\mathrm{Beh}}])$. This correspondence can be easily defined by the induction on the structure of specifications. For example, if $\mathrm{SP}_{\mathrm{Beh}} = (\Sigma_{\mathrm{Beh}}, \Phi)$ is a basic specification, where $\Sigma_{\mathrm{Beh}} = (\Sigma, \mathcal{R}, \approx)$, then the corresponding standard specification is defined: $SP = (\Sigma, \Phi)$. The definition of the correspondence for the others specification building operations are obvious, e.g. if $SP_{Beh} = \textbf{derive from } SP'_{Beh} \textbf{ by } \sigma$ then the corresponding standard specification is defined: $SP = \textbf{derive from } SP' \textbf{ by } \sigma$, where SP' is a standard specification that corresponds to SP'_{Beh}.

5.5   Examples

The described above notion of a behavioural concrete institution covers many institution of interest: institution of standard algebras, partial algebras with strong homomorphisms and CASL -institution with a slightly changed notion of a homomorphism between models.

6   Relating Behavioural and Abstractor Semantics

In this section we present relations between the behavioural and abstractor semantics.

Let $\textbf{INS}_c = (\textbf{Sign}, \textbf{Sen}, \textbf{Mod}, (\models_\Sigma)_{\Sigma \in |\textbf{Sign}|}, sorts, |\_|)$ be a concrete institution that satisfies all the assumptions presented in the beginning of the previous section (which allow us to define a behavioural concrete institution based on $\textbf{INS}_c$), fixed throughout this section.

6.1   Abstractor Specifications

Let us now briefly focus on the abstractor semantics. More information about it can be found for example in [5] or [4].

In fact the additional structure available in concrete institutions is not needed to define the abstractor semantics (i.e. the notion of the standard institution is sufficient for that purpose).

Let $\Sigma \in |\textbf{Sign}|$ be a signature of a concrete institution $\textbf{INS}_c$. An *abstractor equivalence* over the signature Σ is an equivalence relation between Σ-models, $\equiv \subseteq |\textbf{Mod}(\Sigma)| \times |\textbf{Mod}(\Sigma)|$. An abstractor equivalence $\equiv$ is called *isomorphism protecting* if all isomorphic models $M, N \in |\textbf{Mod}(\Sigma)|$ are equivalent, $M \equiv N$.

For any class of Σ-models $\mathcal{M} \subseteq |\textbf{Mod}(\Sigma)|$, the *abstractor closure* of $\mathcal{M}$ is the closure of this class under the abstractor equivalence, $Abs_\equiv(\mathcal{M}) = \{M \in |\textbf{Mod}(\Sigma)| \mid M \equiv N \text{ for some } N \in \mathcal{M}\}$.

The notion of an abstractor closure allows us to define the abstractor semantics of specifications. Let SP be a specification with the signature Σ. The class of models which behaviourally (up to the abstractor equivalence $\equiv$) satisfy the specification SP is the abstractor closure of the class of models which satisfy the specification SP literally, i.e. $\text{Mod}[\textbf{abstract } SP \textbf{ wrt } \equiv] = Abs_\equiv(\text{Mod}[SP])$.

6.2   Behavioural Specifications and Behavioural Closure Operator

Let $\textbf{BINS}_c = (\textbf{BSign}, \textbf{BSen}, \textbf{BMod}, (\models_{\Sigma_{Beh}})_{\Sigma_{Beh} \in |\textbf{BSign}|}, bsorts, |\_|)$ be a behavioural concrete institution based on $\textbf{INS}_c$.

A similar condition to isomorphism protecting can be expressed for behavioural structures. It is called isomorphism compatibility and it differs slightly from the one introduced in [4] since in this framework not for each model the behaviour is defined.

Definition 26. *A behavioural signature* $\Sigma_{\text{Beh}} = (\Sigma, \mathcal{R}, \approx) \in |\textbf{BSign}|$ *is isomorphism compatible if for each isomorphic Σ-models $M, N \in |\textbf{Mod}(\Sigma)|$ if M is behavioural then N is behavioural and in that case their behaviours are isomorphic,* $\mathcal{B}_{\Sigma_{\text{Beh}}}(M) \cong \mathcal{B}_{\Sigma_{\text{Beh}}}(N)$.

The notion of a fully abstract model can be found in [4] or [5]. A model is fully abstract if the behavioural structure on this model is trivial, i.e. the reachability structure on such a model is the whole carrier set of this model and the observability structure is an identity relation.

Definition 27. *Let $\Sigma_{\text{Beh}} = (\Sigma, \mathcal{R}, \approx) \in |\textbf{BSign}|$ be a behavioural signature. A model $M \in |\textbf{Mod}(\Sigma)|$ is fully abstract if $\mathcal{R}_M = |M|_{\Sigma_{\text{Beh}}}$ and $\approx_M = id_{|M|_{\Sigma_{\text{Beh}}}}$. If $\mathcal{M} \subseteq |\textbf{Mod}(\Sigma)|$ is a class of Σ-models then,* $\text{FA}_{\Sigma_{\text{Beh}}}(\mathcal{M})$ *denotes the class of all fully abstract models in $\mathcal{M}$,* $\text{FA}_{\Sigma_{\text{Beh}}}(\mathcal{M}) = \{M \in \mathcal{M} \mid M \text{ is fully abstract}\}$.

Note that if a model $M \in |\textbf{Mod}(\Sigma)|$ is fully abstract then it is behavioural. Therefore $\text{FA}_{\Sigma_{\text{Beh}}}(\mathcal{M})$ is a class of behavioural models (even if in $\mathcal{M}$ there are non-behavioural models).

The regularity of a behavioural structure (signature) is also an important property, see [4]. It express the idempotency of the behaviour operator.

Definition 28. *A behavioural signature $\Sigma_{\text{Beh}} \in |\textbf{BSign}|$ is called:*

- weakly regular *if for each behavioural model $M \in |\textbf{BMod}(\Sigma_{\text{Beh}})|$ its behaviour is behavioural and it is isomorphic to the behaviour of the behaviour of this model, i.e.* $\mathcal{B}_{\Sigma_{\text{Beh}}}(M) \cong \mathcal{B}_{\Sigma_{\text{Beh}}}(\mathcal{B}_{\Sigma_{\text{Beh}}}(M))$;
- regular *if for each behavioural model $M \in |\textbf{BMod}(\Sigma_{\text{Beh}})|$ its behaviour is a fully abstract model.*

Regularity implies weak regularity. If a behavioural signature Σ_{Beh} is weakly regular and isomorphic compatible then the behavioural satisfaction relation $\models_{\Sigma_{\text{Beh}}}$ is isomorphism compatible.

The above conditions which should be satisfied by any reasonable behavioural concrete institution (by all signatures in the behavioural concrete institution) will allow us to express relations between the internal approach and the external approach to the definition of behavioural semantics.

Definition 29. *Let $\Sigma_{\text{Beh}} = (\Sigma, \mathcal{R}, \approx) \in |\textbf{BSign}|$ be a behavioural signature, $\mathcal{M} \subseteq |\textbf{Mod}(\Sigma)|$ be a class of Σ-models. The behavioural closure of the class M is a class* $\text{Beh}_{\Sigma_{\text{Beh}}}(\mathcal{M}) = \{M \in |\textbf{BMod}(\Sigma_{\text{Beh}})| \mid \mathcal{B}_{\Sigma_{\text{Beh}}}(M) \in \mathcal{M}\}$.

Note that, similarly as for the operator which yields the class of fully abstract models, even if there are some non-behavioural models in $\mathcal{M}$, the behavioural closure of this class contains only behavioural models.

Corollary 1. *Let* $SP_{Beh} = (\Sigma_{Beh}, \Phi)$ *be a behavioural specification and* SP *be its corresponding standard specification. Then* $Mod[SP_{Beh}] = Beh_{\Sigma_{Beh}}(Mod[SP])$.

The following lemma is useful to prove relations between the behavioural and abstractor semantics.

Lemma 1. *Let* SP_{Beh} *be a behavioural specification and* SP *its corresponding standard specification. If* SP *has isomorphic compatible reduct functors then* $Mod[SP_{Beh}] \subseteq Beh_{\Sigma_{Beh}}(Mod[SP])$, *where* $\Sigma_{Beh} = Sig[SP_{Beh}]$.

The proof of the above lemma is by the induction on the structure of specifications.

The opposite inclusion doesn't hold in general (i.e. $Beh_{\Sigma_{Beh}}(Mod[SP]) \nsubseteq Mod[SP_{Beh}]$), even in the case of standard algebras and equational logic.

6.3 Relations

The crucial notion for expressing relations between the two approaches to the definition of the behavioural semantics is the notion of *factorizability*, introduced in [4].

Definition 30. *Let* $\Sigma_{Beh} = (\Sigma, \mathcal{R}, \approx)$ *be a behavioural signature and* $\equiv$ *be an abstractor equivalence over* Σ. *The abstractor equivalence* $\equiv$ *is called* factorizable *by* Σ_{Beh} *(or by the behavioural structure* $(\mathcal{R}, \approx)$*) if the following two conditions hold:*

- *for all behavioural models* $M, N \in |\mathbf{BMod}(\Sigma_{Beh})|$, $M \equiv N$ *iff* $\mathcal{B}_{\Sigma_{Beh}}(M) \cong \mathcal{B}_{\Sigma_{Beh}}(N)$;
- *for each behavioural model* $M \in |\mathbf{BMod}(\Sigma_{Beh})|$ *and* $N \in |\mathbf{Mod}(\Sigma)|$ *if* $M \equiv N$ *then* N *is behavioural,* $N \in |\mathbf{BMod}(\Sigma_{Beh})|$.

The second condition in the above definition states that the class of behavioural models over Σ (w.r.t $(\mathcal{R}, \approx)$) is closed under the abstractor equivalence $\equiv$. The first condition is standard, i.e. it comes from the original definition of factorizability in [4].

The abstractor equivalence $\equiv_{Obs,In}$ from [4] is factorizable by Σ_{COL} w.r.t the above definition, if Obs is the set of observable sorts of Σ_{COL} (i.e. S_{Obs}) and In is the set of loose sorts of Σ_{COL} (i.e. S_{Loose}), see [2] for the definition of observable and loose sorts of a COL-signature.

Throughout the rest of this section we assume that all behavioural signatures $\Sigma_{Beh} \in |\mathbf{BSign}|$ are isomorphic compatible and all abstractor equivalences considered are isomorphic protecting.

Lemma 2. *Let* $\Sigma_{Beh} = (\Sigma, \mathcal{R}, \approx)$ *be a behavioural signature that is weakly regular and* $\equiv$ *is an abstractor equivalence over* Σ, *factorizable by* Σ_{Beh}. *Then for any class of models* $\mathcal{M} \subseteq |\mathbf{Mod}(\Sigma)|$ *the following holds:* $Beh_{\Sigma_{Beh}}(\mathcal{M}) \subseteq Abs_{\equiv}(\mathcal{M})$. *If moreover* $\mathcal{M}$ *is closed under isomorphism and* $\mathcal{M} \subseteq Beh_{\Sigma_{Beh}}(\mathcal{M})$ *(behavioural consistency) then* $Beh_{\Sigma_{Beh}}(\mathcal{M}) = Abs_{\equiv}(\mathcal{M})$.

Note that in the above lemma, the class $\mathcal{M}$ is not required to be a class of behavioural models. But if $\mathcal{M}$ is behaviourally consistent ($\mathcal{M} \subseteq \mathrm{Beh}_{\Sigma_{\mathrm{Beh}}}(\mathcal{M})$) then it implies that $\mathcal{M}$ contains only behavioural models.

Proposition 7. *Let $\Sigma_{\mathrm{Beh}} = (\Sigma, \mathcal{R}, \approx)$ be a behavioural signature that is weakly regular and $\equiv$ be an abstractor equivalence over Σ, factorizable by Σ_{Beh}. Let also $\mathrm{SP}_{\mathrm{Beh}}$ be a behavioural specification with $\mathrm{Sig}[\mathrm{SP}_{\mathrm{Beh}}] = \Sigma_{\mathrm{Beh}}$ and SP be a standard specification which corresponds to the behavioural specification $\mathrm{SP}_{\mathrm{Beh}}$. We assume that SP has isomorphic compatible reduct functors. Then $\mathrm{Mod}[\mathrm{SP}_{\mathrm{Beh}}] \subseteq \mathrm{Mod}[\textbf{abstract SP wrt} \equiv]$.*

The opposite inclusion doesn't hold in general. Consider a basic specification $\mathrm{SP}_{\mathrm{Beh}}$ with an empty set of axioms. If not all models over the signature $\mathrm{Sig}[\mathrm{SP}_{\mathrm{Beh}}]$ are behavioural then $\mathrm{Mod}[\textbf{abstract SP wrt} \equiv] \not\subseteq \mathrm{Mod}[\mathrm{SP}_{\mathrm{Beh}}]$ (SP is the standard specification corresponding to $\mathrm{SP}_{\mathrm{Beh}}$), since on the left-hand side of the inclusion there is the whole class of models over $\mathrm{Sig}[\mathrm{SP}]$ and on the right-hand side only behavioural models.

The last fact in this subsection concerns relations between behavioural specifications and the abstractor closure of the class of fully abstract models.

Lemma 3. *Let $\Sigma_{\mathrm{Beh}} = (\Sigma, \mathcal{R}, \approx)$ be a behavioural signature that is regular and $\equiv$ is an abstractor equivalence over Σ, factorizable by Σ_{Beh}. Then for any class of models closed under isomorphism $\mathcal{M} \subseteq |\mathbf{Mod}(\Sigma)|$ the following holds: $\mathrm{Beh}_{\Sigma_{\mathrm{Beh}}}(\mathcal{M}) = \mathrm{Abs}_{\equiv}(\mathrm{FA}_{\Sigma_{\mathrm{Beh}}}(\mathcal{M}))$.*

7 Final Remarks

In this paper we attempted to define a behavioural semantics for specifications built in an arbitrary logical system formalized as an institution. Although the presented framework covers many institutions of interest it doesn't cover, for example the institution of continuous algebras (an institution of behavioural logic for continuous algebras can be defined if we deal only with the observability concepts). In fact, all assumptions needed to form a behavioural concrete institution from an ordinary concrete institution are quite numerous.

A technical tool used in this work are standard techniques of concrete categories. However, an interesting issue, for further work is to define a behavioural institution based on an institution (without the additional structure of concrete institutions) by a given behaviour operator.

Another important issue for further work is to find a proof system for behavioural specifications basing on a given proof system for ordinary specifications.

References

1. J. Adámek, H. Herrlich, and G. E. Strecker. *Abstract and Concrete Categories*. Pure and Applied Mathematics. Wiley-Interscience Publication, 1990.

2. M. Bidoit and R. Hennicker. On the integration of observability and reachability concepts. In *Proc. 5th Int. Conf. Foundations of Software Science and Computation Structures (FOSSACS 2002)*, volume 2303 of *Lecture Notes in Computer Science*, pages 21–36. Springer, 2002.
3. M. Bidoit and R. Hennicker. The constructor-based observational logic. Research Report, LSV-03-9, 2003.
4. M. Bidoit, R. Hennicker, and M. Wirsing. Behavioural and abstractor specifications. In *Science of Computer Programming*, pages 149–186, 1995.
5. M. Bidoit and A. Tarlecki. Behavioural satisfaction and equivalence in concrete model categories. In *Proc. 21st Int. Coll. Trees in Algebra and Programming*, volume 1059 of *Lecture Notes in Computer Science*, pages 241–256. Springer, 1996.
6. R. Burstall and J. Goguen. Institutions: abstract model theory for computer science. *Journal of the Assoc. for Computing Machinery*, 39:95–146, 1992.
7. A. Tarlecki. Institutions: An abstract framework for formal specifications. In *Algebraic foundations of system specification*, chapter 3, pages 105–130. Springer, 1999.

A Simple Refinement Language for CASL*

Till Mossakowski[1], Donald Sannella[2], and Andrzej Tarlecki[3]

[1] BISS, Department of Computer Science, University of Bremen
[2] LFCS, School of Informatics, University of Edinburgh, Edinburgh, UK
[3] Institute of Informatics, Warsaw University and
Institute of Computer Science, PAS, Warsaw, Poland

Abstract. We extend CASL architectural specifications with a simple refinement language that allows the formalization of developments as refinement trees. The essence of the extension is to allow refinements of unit specifications in CASL architectural specifications.

1 Introduction

The standard development paradigm of algebraic specification [1] postulates that the development begins with a formal *requirement specification* SP_0 (extracted from a software project's informal requirements) that fixes only expected properties but ideally says nothing about implementation issues; this is to be followed by a number of *refinement* steps that fix more and more details of the design, until a specification SP_n is obtained that is detailed enough that its conversion into a program P is relatively straightforward:

$$SP_0 \rightsquigarrow SP_1 \rightsquigarrow \cdots \rightsquigarrow SP_n \cdots\!\!> P$$

Actually, this picture is too simple in practice: for complex software systems, it is necessary to reduce complexity by introducing branching points into the chain of refinement steps, so that the resulting implementation tasks can be resolved independently, e.g. by different developers. CASL architectural specifications [3, 8] have been designed for this purpose, based on the insight that structuring of implementations is different from structuring specifications [9].

However, CASL architectural specifications allow for the specification of individual branching points only. In this work, we extend CASL with a simple refinement language that adds the means to formalize whole developments in the form of *refinement trees*.

As it stands, this is not a formal proposal for an extension to CASL and some of the details of syntax etc. are rather tentative. It is intended as a basis for further discussion and experimentation, that may eventually lead to such a proposal.

---

* This work has been partially supported by KBN grant 7T11C 002 21, European projects AGILE IST-2001-32747 (AT) and MRG IST-2001-33149 (DS), British–Polish Research Partnership Programme (DS, AT), and German DFG project KR 1191/5-2 (TM).

J.L. Fiadeiro, P. Mosses, and F. Orejas (Eds.): WADT 2004, LNCS 3423, pp. 162–185, 2005.

The paper is organized as follows. Section 2 recalls CASL, Sect. 3 introduces simple refinements and Sect. 4 branching refinements. These are related to constructor implementations in Sect. 5, which leads to the question of how programs are modelled in CASL. This is addressed in Sect. 6. Sections 7 and 8 describe the syntax and semantics of our proposed refinement language; to facilitate understanding, we first deal with a simpler version in Sect. 7 before exposing the full complexity of the proposed refinement language and its semantics in Sect. 8. Finally, Sect. 9 sketches a larger example and Sect. 10 concludes the paper.

2   CASL Preliminaries

CASL [2, 8] consists of several major *layers*, which are quite independent and may be understood and used separately:

Basic specifications are written in many-sorted first-order logic, extended by subsorting, partial functions and induction axioms for datatypes. Indeed, the details are quite irrelevant here as long as a basic specification determines a *signature* together with a set of *axioms*. The semantics of a basic specification is then given by the signature and the class of all *models* that satisfy the axioms. Formally, $[\![\langle \Sigma, \Phi \rangle]\!] = \{M \in \mathbf{Mod}(\Sigma) \mid M \models \Phi\}$, where $\mathbf{Mod}(\Sigma)$ is the class of all CASL models over the signature Σ.

Structured specifications allow specifications to be built from basic specifications by the use of translation, reduction, union, and extension, as well as generic (parametrized) and named specifications; semantics of structured specifications is given in terms of signatures and model classes, as for basic specifications.

Architectural specifications describe the structure of an implementation by defining how it may be constructed out of software components (*units*) that satisfy given specifications. These *unit specifications* describe self-contained units (models, as above), or generic units (corresponding to parametrized programs) mapping such models to other models.

CASL admits a clean separation of the layer of basic specifications from the other layers. Any logic can be used in the basic specification layer, as long as it is formalized as an *institution* [10]. The semantics of the other layers is defined for an arbitrary institution. The architectural specification layer is also independent of the details of the features used for building structured specifications.

3   Simple Refinements

The simplest form of refinement is just model class inclusion. Consider the following standard specification of monoids:

spec MONOID =
    sort *Elem*
    ops *0* : *Elem*;
        $\_\_ + \_\_$: *Elem* × *Elem* → *Elem*, *assoc, unit 0*

This specification is rather loose. It can be refined in many different ways, e.g. into the natural numbers. We first specify the natural numbers inductively in terms of zero and successor, then define addition, and finally hide the successor operation:

> **spec** NATWITHSUC =
>> **free type** $Nat ::= 0 \mid suc(Nat)$
>> **op** $1 : Nat = suc(0)$
>> **op** $\_\_ + \_\_ : Nat \times Nat \to Nat,\, unit\ 0$
>>> $\forall x, y : Num \bullet x + suc(y) = suc(x + y)$

> **spec** NAT =
>> NATWITHSUC **hide** suc

The refinement between the two specifications can now be stated as follows:

> **refinement** R1 =
>> MONOID **refined via** $Elem \mapsto Nat$ **to** NAT

Correctness of this refinement means that given any NAT-model, its reduct along $Elem \mapsto Nat$ yields a MONOID-model, formally

$$M|_\sigma \in [\![\text{MONOID}]\!] \text{ for each } M \in [\![\text{NAT}]\!]$$

where σ maps $Elem$ to Nat (σ is generated from the symbol map $Elem \mapsto Nat$ in a straightforward way, see [8]). Of course, this just states that the natural numbers with addition form a monoid, or, in other words, that $\sigma : \text{MONOID} \to \text{NAT}$ is a specification morphism. Specification morphisms arise in CASL already as *views*, used for instantiating generic specifications. For that application it is useful to allow them to be generic themselves, and this leads to certain complications. Here specification morphisms are used for a different purpose where the complications of generic views are irrelevant and distracting.

The specification NAT can be taken as a realisation of the natural numbers, but quite an inefficient one. It is far more efficient to use a binary representation (++ is binary addition with carry):

> **spec** NATBIN =
>> **generated type** $Bin ::= 0 \mid 1 \mid \_\_0(Bin) \mid \_\_1(Bin)$
>> **ops** $\_\_ + \_\_ ,\_\_ ++ \_\_ : Bin \times Bin \to Bin$
>> $\forall x, y : Bin$
>>> - $0\ 0 = 0$
>>> - $\neg\, (0 = 1)$
>>> - $\neg\, (x\ 0 = y\ 1)$
>>> - $0 + 0 = 0$
>>> - $x\ 0 + y\ 0 = (x + y)\ 0$
>>> - $x\ 0 + y\ 1 = (x + y)\ 1$
>>> - $x\ 1 + y\ 0 = (x + y)\ 1$
>>> - $x\ 1 + y\ 1 = (x ++ y)\ 0$
>>
>>> - $0\ 1 = 1$
>>> - $x\ 0 = y\ 0 \Rightarrow x = y$
>>> - $x\ 1 = y\ 1 \Rightarrow x = y$
>>> - $0 ++ 0 = 1$
>>> - $x\ 0 ++ y\ 0 = (x + y)\ 1$
>>> - $x\ 0 ++ y\ 1 = (x ++ y)\ 0$
>>> - $x\ 1 ++ y\ 0 = (x ++ y)\ 0$
>>> - $x\ 1 ++ y\ 1 = (x ++ y)\ 1$

We now have a further refinement:

> **refinement** R2 =
>> NAT **refined via** $Nat \mapsto Bin$ **to** NATBIN

Note that it is quite typical that the target specification of the refinement adds auxiliary operations, which are forgotten by reducing along the signature morphism.

The two refinements can be combined into a *chain* of refinements:

refinement R3 =
   MONOID **refined via** *Elem* ↦ *Nat* **to**
      NAT **refined via** *Nat* ↦ *Bin* **to** NATBIN

which can be depicted as follows:

$$\text{MONOID} \overset{\sigma}{\rightsquigarrow} \text{NAT} \overset{\theta}{\rightsquigarrow} \text{NATBIN}$$

Here, σ and θ are the specification morphisms associated to the refinements, and the correctness conditions for the individual refinements guarantee that the chain is also a correct refinement in the following sense:

$$M|_{\sigma;\theta} = (M|_\theta)|_\sigma \in [\![\text{MONOID}]\!] \text{ for each } M \in [\![\text{NATBIN}]\!]$$

If we want to save some typing, we can also write:

refinement R3′ =
   MONOID **refined via** *Elem* ↦ *Nat* **to** R2

or equivalently

refinement R3″ =
   MONOID **refined via** *Elem* ↦ *Nat* **to** NAT **then** R2

which can be rewritten as

refinement R3‴ = R1 **then** R2

4   Branching Refinements

Suppose that we want to implement not only NAT, but NATWITHSUC, i.e. also the successor function. Now, while the presence of the successor function enables an easy *specification* of the natural numbers, it may be a little distracting in achieving an efficient *implementation*. So we can help the implementor and impose (via a CASL architectural specification) that the natural numbers should be implemented with addition, and the successor function should only be implemented afterwards, *in terms of* addition:

arch spec ADDITION_FIRST =
   units N : NAT;
         M: { **op** $suc(n : Nat) : Nat = n + 1$ } **given** N
   result M

We thus have chosen to split the implementation of NATWITHSUC into two independent subtasks: the implementation of NAT, and the implementation of a generic program, that given *any* NAT-model will realise the successor function on top of it. The generic program is then applied once to the implementation N of NAT. A version making this genericity explicit is the following:

> **arch spec** ADDITION_FIRST_GENERIC =
>   **units** N : NAT;
> F : NAT $\rightarrow$ { **op** $suc(n : Nat) : Nat = n + 1$ };
> $M = F[N]$
>   **result** M

Here, F is a generic program unit, that is, a parametrized program, that may be applied to any program unit matching its parameter specification, not only to N. The specification

$$\text{NAT} \ \rightarrow \ \{ \textbf{op} \ suc(n : Nat) : Nat = n + 1 \}$$

is a so-called (generic) *unit specification*. It denotes the class of all functions F mapping NAT-models to models of

$$\text{NAT} \ \textbf{then} \ \{ \textbf{op} \ suc(n : Nat) : Nat = n + 1 \}$$

in such a way that the argument model (unit) is preserved, i.e. $F(N)|_{\text{NAT}} = N$ for any NAT-model N.

The term $F[N]$ is a *unit term* (in this case: a unit application) computing a unit out of the (generic and non-generic) units introduced earlier. $M = F[N]$ is a *unit definition* which defines the unit M to be exactly the (value of the) unit term $F[N]$. The unit M is then used as the *result* unit term. In general, the result unit may be given by an arbitrary unit term involving the units declared or defined within the architectural specification. If the result unit is itself to be generic, the unit term has to preceded by a λ-abstraction (this is one form of *unit expression*).

We can express that ADDITION_FIRST is a refinement of NATWITHSUC as follows:

> **refinement** R4 =
>   NATWITHSUC **refined to arch spec** ADDITION_FIRST

This time, we have left out the signature morphism, since it is just the identity. Since the refined specification is an architectural specification, we use the keywords **arch spec** before the refined specification.

If we want to combine this design decision with the decision to implement NAT with NATBIN, we can write a refinement directly after the specification of the unit in question:

> **arch spec** ADDITION_FIRST_WITH_BIN =
>   **units** N : NAT **refined via** $Nat \mapsto Bin$ **to** NATBIN;
> F : NAT $\rightarrow$ { **op** $suc(n : Nat) : Nat = n + 1$ };
> $M = F[N]$
>   **result** M

or, more briefly, using the refinement of NAT into NATBIN named R2 above:

arch spec ADDITION_FIRST_WITH_BIN' =
    units N : R2;
        F : NAT $\rightarrow$ { **op** $suc(n : Nat) : Nat = n + 1$ };
        $M = F[N]$
    result M

5 Refinement: Constructor Implementations

Semantically, all the types of refinements introduced so far can be seen as *constructor implementations* in the sense of [15]. Constructor implementations are written as

$$SP \rightsquigarrow_{\kappa} SP'$$

Here, a constructor κ is a function mapping models to models; formally κ : $\mathbf{Mod}(Sig[SP']) \rightarrow \mathbf{Mod}(Sig[SP])$. Such a constructor implementation is *correct* if

$$\text{for all } A' \in [\![SP']\!], \, \kappa(A') \in [\![SP]\!].$$

In our proposed refinement language, constructors are induced by specification morphisms $\sigma \colon SP \longrightarrow SP'$, that is, signature morphisms from $Sig[SP]$ to $Sig[SP']$ with $[\![SP']\!]|_{\sigma} \subseteq [\![SP]\!]$. The constructor is just the reduct functor induced by σ, and correctness is equivalent to σ being a specification morphism.

Constructors correspond to *generic program modules* in programming languages, such as generic packages in Ada or functors in Extended ML:

```
functor K(X:SP'):SP = ... code ...
```

In the framework of [15], a specification is implemented via a sequence of refinement steps, until ultimately the empty specification EMPTY is reached:

$$SP_0 \rightsquigarrow_{\kappa_1} SP_1 \rightsquigarrow_{\kappa_2} \cdots \rightsquigarrow_{\kappa_n} SP_n = \text{EMPTY}$$

If all these steps are correct, the combination of the constructors (starting with the trivial model *empty* of the EMPTY specification) yields a model of the original specification SP_0:

$$\kappa_1(\kappa_2(\cdots(\kappa_n(empty))\cdots)) \in [\![SP_0]\!]$$

Architectural specifications introduce *branching*: a specification may be refined to *several* specifications, which requires n-ary constructors:

$$SP \rightsquigarrow_{\kappa} \begin{cases} SP_1 \\ \vdots \\ SP_n \end{cases}$$

As expected, correctness here means that

$$\text{for all } A_1 \in [\![SP_1]\!], \, \ldots, \, A_n \in [\![SP_n]\!], \, \kappa(A_1, \ldots, A_n) \in [\![SP]\!].$$

The corresponding CASL architectural specification is written

> **arch spec**
>   **units** U_1: SP_1
>   . . .
> U_n: SP_n
>   **result** UT

where UT is a unit term describing the constructor κ, which may involve the unit names $U_1, \ldots, U_n$. Unit terms are built by renaming of units, hiding parts thereof, amalgamation of units, application of generic units to arguments, as well as local unit definitions (introducing local names for unit terms). The requirements imposed by the semantics on the result unit terms ensure that the induced constructors are always defined for the relevant argument units.

Analogously to the linear situation, once we have a tree of correct refinement steps with leaves being empty specifications, as follows:

$$SP \rightsquigarrow_\kappa \begin{cases} SP_1 \rightsquigarrow_{\kappa_1} \text{EMPTY} \\ \quad \vdots \\ SP_n \rightsquigarrow_{\kappa_n} \begin{cases} SP_{n1} \rightsquigarrow_{\kappa_{n1}} \begin{cases} SP_{n11} \rightsquigarrow_{\kappa_{n11}} \text{EMPTY} \\ \cdots \end{cases} \\ SP_{nm} \rightsquigarrow_{\kappa_{nm}} \text{EMPTY} \end{cases} \end{cases}$$

we can construct a model of the original requirement specification by successively applying the constructors, starting with the trivial model *empty*:

$$\kappa(\kappa_1(empty),$$
$$\cdots$$
$$\kappa_n(\kappa_{n1}(\kappa_{n11}(empty)),$$
$$\cdots$$
$$\kappa_{nm}(empty))) \quad \in [\![SP]\!]$$

Note that this whole section applies not only to the refinement of ordinary program units (models), but also to generic units (functions from models to models). We will come back to this later.

6   Programs in CASL

One problem with the approach described so far is that the constructors provided by specification morphisms and architectural specifications in CASL do not suffice for implementing specifications. In a sense, these constructors only provide means to *combine* or *modify* existing program units—but there is no way to build program units *from scratch*. That is, CASL lacks a notion of *program*.

An obvious way out of this situation is to add more unit operations that can be used for unit terms (or unit expressions) in architectural specifications.

Concerning construction of datatypes, one could add a simple version of free extension, giving a model of a datatype that is determined uniquely up to isomorphism, and that corresponds to an algebraic datatype in a functional programming language. For the construction of operations on top of these datatypes, one could use reducts along derived signature morphisms. Derived signature morphisms may map an operation to a term or to a recursive definition by means of equations, like function definitions in a functional programming language. See [16, Chap. 4] for a more detailed account of this approach.

Note that this approach is necessarily no longer institution independent. The details of the kind of free extensions that actually correspond to datatypes in a programming language depend both on the institution and the programming language at hand. The same remark holds for the definition of derived signature morphisms.

An alternative is to approximate the institution independent essence of programs by considering *monomorphic specifications*. A unit specification is monomorphic if the result specification is a monomorphic extension of the argument specifications. This means that it is a construction that is unique up to isomorphism. Ultimately, monomorphic unit specifications need to be translated to (parametrized) programs in some programming language. As above, this process obviously depends on both the institution and the programming language in question. The difference is that the specification language itself remains institution independent, since the translation to a programming language is not part of the specification language.

In some cases it is possible to perform the translation automatically, for unit specifications that obey certain syntactic restrictions. For functional programming languages such as Haskell and ML, one would require that all sorts are given as free types, and all functions are defined by means of recursive equations in such a way that termination is provable. Indeed, the translation of a parametrized program then provides a construction that is unique, not only unique up to isomorphism. See [14] for details, and [6] for a translation of a subset of CASL to OCAML. Using free extensions, it is also possible to capture partial recursive functions, see [6, 14]. Moreover, with Haskell (and its type class *Eq*) as target language, generated types with explicitly given equality can also be used. For ML and Haskell, there is also a direct correspondence at the level of CASL unit terms, see Fig. 1.

For other programming languages, the translation between monomorphic specifications and programs might be much less straightforward. In general, it may be necessary to translate manually, and prove that the resulting program is a correct realization of the specification. There may also be a mismatch between the constructs that are available for combining modules in the programming language and the constructs that CASL provides for combining unit terms. Then, one possibility would be to view unit terms in architectural specifications as prescriptions for the composition and transformation of the component units, and carry these out manually using the constructs that the programming language provides. (This may be automated by devising operations on program

CASL	ML	Haskell
non-generic unit	structure	module
generic unit	functor	multi-parameter type class in a module
monomorphic unit specification with free types and recursive definitions	structure with datatypes and recursive definitions	module with datatypes and recursive definitions
unit application	functor application	type class instantiation
unit amalgamation	combination of structures	combination of modules
unit hiding	restriction to subsignature	hiding
unit renaming	redefinition	redefinition
architectural specification	structure/functor using other structures/functors	module using other modules

Fig. 1. Unit term constructs in ML and Haskell

texts corresponding to unit term constructs.) Alternatively, one might take the target programming language into account in the refinement process and simply avoid in unit terms any use of the constructs that have no counterpart in the programming language at hand.

With this approach, the use of a parametrized program κ in a constructor implementation $SP \rightsquigarrow_\kappa SP'$ is expressed as

> **arch spec**
>   **unit** $K : SP' \rightarrow SP$ **refined to** USP
>   **result** K

where USP is a monomorphic specification of κ from which the corresponding parameterized program may be obtained directly. Such a constructor can also be used in the context of another refinement. For example, the refinement

$$SP \rightsquigarrow_\kappa SP' \rightsquigarrow_\kappa SP''$$

is expressed as

> **refinement** R5 =
> SP **refined to**
>   **arch spec**
>   { **units**
> $K : SP' \rightarrow SP$ **refined to** USP
> $A' : SP'$ **refined to arch spec**
>   { **units**
> $K' : SP'' \rightarrow SP'$ **refined to** USP'
> $A'' : SP''$
>   **result** $K'[A'']$ }
>   **result** $K[A']$ }

where USP and USP' are monomorphic specifications of κ and κ', respectively.

7   Refinements in CASL

Let us now come to a more systematic treatment of the refinement language that we propose.

The grammar below extends the grammar for the concrete syntax of CASL given in the CASL Reference Manual [8]. The new parts of the grammar are marked in *italics*, while removed parts are ~~crossed out~~. The details here are formulated in terms of concrete syntax, in contrast to [8], to make the presentation more accessible to readers who are not intimately familiar with the details of the CASL design. A corresponding abstract syntax is given in the appendix.

The central notion of the refinement language is specification refinement. These can take various forms, all of which have already been discussed along with concrete examples above.

```
SPEC-REF      ::= SPEC-NAME
              | UNIT-SPEC
              | UNIT-SPEC refined via SYMB-MAP-ITEMS* to SPEC-REF
              | UNIT-SPEC refined to SPEC-REF
              | arch spec ARCH-SPEC
              | SPEC-REF then SPEC-REF
```

Like ordinary specifications and unit specifications, specification refinements can be named:

```
SPEC-REF-DEFN ::= refinement SPEC-NAME = SPEC-REF end
```

Here, the notation <u>end</u> stands for optional **end**.

The syntax of declarations of units within architectural specifications is relaxed: arbitrary specification refinements are allowed, not only unit specifications:

```
UNIT-DECL     ::= UNIT-NAME : SPEC-REF
```

To avoid additional complexity but mainly for methodological reasons, we leave out refinements of unit specifications with imports (the "**given**" clause in ADDITION_FIRST in Sect. 4). See Sect. 10 for justification and discussion.

Finally, since we allow for coercion of architectural specifications to specification refinements, there is no need for coercing them to unit specifications:

```
UNIT-SPEC     ::= SPEC
              | SPEC *...* SPEC -> SPEC
              | arch spec ARCH-SPEC
              | closed UNIT-SPEC
```

As with the rest of CASL, the semantics is given in two parts: the static semantics and the model semantics. In the semantics below, we ignore global environments which store the meanings of global names; consequently, we also omit the case of named specification refinements. Details, which are straightforward, follow a similar pattern as in [8, III:5 and III:6].

The static semantics of specification refinements is given in Fig. 2. The judgements are of the form $\vdash SPR \triangleright (U\Sigma, U\Sigma')$. Here, $U\Sigma$ is the unit signature for

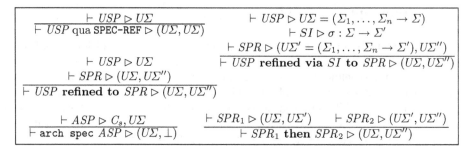

Fig. 2. Static semantics of specification refinements

units of the specification being refined and $U\Sigma'$ is the unit signature for units of the specification after refinement. A unit signature consists of a tuple of argument signatures (which is empty and may be omitted for non-generic units) and a result signature. Further details can be found in [8, III:5]. For instance, we have

$$\vdash \text{Monoid } \textbf{refined via } Elem \mapsto Nat \textbf{ to } \text{Nat} \rhd (\Sigma_{\text{Monoid}}, \Sigma_{\text{Nat}}),$$

where Σ_{Monoid} and Σ_{Nat} are the signatures of Monoid and Nat respectively. Since so far we don't allow for further refinement of architectural specifications, only of their units, if SPR is an architectural specification we mark this by putting $U\Sigma' = \bot$.

The model semantics of specification refinements is given in Fig. 3. The judgements are of the form $\vdash SPR \Rightarrow \mathcal{R}$. If $\vdash SPR \rhd (U\Sigma, U\Sigma')$ then $\mathcal{R}$ is a class of pairs (U, U') such that U and U' are units over unit signatures $U\Sigma$ and $U\Sigma'$ respectively and $\mathcal{R}^{-1}$ is a partial function mapping $U\Sigma'$-units to $U\Sigma$-units. $\mathcal{R}^{-1}$ is the constructor involved in the refinement and its domain is the class of models of the specification after refinement. A unit is either a model (when it is non-generic) or a unit function, mapping compatible tuples of argument models to result models. Further details can be found in [8, III:5]. For instance, we have

$$\vdash \text{Monoid } \textbf{refined via } Elem \mapsto Nat \textbf{ to } \text{Nat} \Rightarrow \{(N|_\sigma, N) \mid N \in [\![\text{Nat}]\!]\}$$

where σ maps $Elem$ to Nat. Again, this takes a special form when SPR is an architectural specification: the second component of each pair is then $\bot$.

Both static semantics and model semantics rules rely on the semantics of unit specifications [8, III:5], symbol mappings [8, III:4] and architectural specifications [8, III:5]. We just recall that the static semantics of an architectural specification consists of a static unit context (which we ignore here, but see Sect. 8) and a result unit signature. An architectural model consists of a unit environment (again, ignored here, but see Sect. 8) and a result unit. Note that the semantics of architectural specifications has to be adjusted as well, since its unit declarations may now involve arbitrary specification refinements rather than only unit specifications. Luckily, going from the semantics of the former to the plain Casl semantics of the latter is very easy here. In the static semantics,

$$\frac{\vdash USP \Rightarrow \mathcal{U}}{\vdash USP \textbf{ qua SPEC-REF} \Rightarrow \{(U, U) \mid U \in \mathcal{U}\}}$$

$$\frac{\vdash USP \Rightarrow \mathcal{U} \qquad \vdash SPR \Rightarrow \mathcal{R}}{\vdash USP \textbf{ refined to } SPR \Rightarrow \mathcal{R}}$$
$$U' \in \mathcal{U}, \text{ for all } (U', U'') \in \mathcal{R}$$

$$\frac{\vdash USP \Rightarrow \mathcal{U} \qquad \vdash SI \rhd \sigma : \Sigma \to \Sigma' \qquad \vdash SPR \Rightarrow \mathcal{R}}{\vdash USP \textbf{ refined via } SI \textbf{ to } SPR \Rightarrow \mathcal{R}'}$$
$$U'|_\sigma \in \mathcal{U}, \text{ for all } (U', U'') \in \mathcal{R}$$
$$\mathcal{R}' = \{(U'|_\sigma, U'') \mid (U', U'') \in \mathcal{R}\}$$

$$\frac{\vdash ASP \Rightarrow \mathcal{AM}}{\vdash \textbf{arch spec } ASP \Rightarrow \{(U, \bot) \mid (E, U) \in \mathcal{AM}\}}$$

$$\frac{\vdash SPR_1 \Rightarrow \mathcal{R}_1 \qquad \vdash SPR_2 \Rightarrow \mathcal{R}_2}{\vdash SPR_1 \textbf{ then } SPR_2 \Rightarrow \mathcal{R}}$$
$$\text{for all } (U', U'') \in \mathcal{R}_2, (U, U') \in \mathcal{R}_1 \text{ for some } U$$
$$\mathcal{R} = \{(U, U'') \mid (U, U') \in \mathcal{R}_1, (U', U'') \in \mathcal{R}_2 \text{ for some } U'\}$$

Fig. 3. Model semantics of specification refinements

from the semantics of specification refinements we just take the first component (the unit signature of the specification being refined). In the model semantics, we project the semantics of specification refinements onto the first component, thus taking the class of all units that may arise as results of the construction involved in the refinement.

The semantics of specification refinements relies on the simplifying assumption that the parameter specifications of generic unit specifications do not change under refinement. This allows us to freely use the reduct notation $U|_\sigma$, even when U is a generic unit; in this case, the notation denotes the unit function obtained by reducing the result via σ after applying U. In practice, this restriction is not troublesome, since we always can write an architectural specification that adjusts the parameter specification as required. Namely, given unit specifications $SP \to SP'$ and $SP_1 \to SP_1'$ with a specification morphism $\sigma : SP_1 \to SP$, the following is a correct specification refinement[1]

$$SP \to SP' \textbf{ refined via } \tau \textbf{ to arch spec}$$
$$\{\textbf{unit } F{:}SP_1 \to SP_1'$$
$$\textbf{result } \lambda X{:}SP.F[X \textbf{ fit } \sigma]\}$$

---

[1] Assuming that all symbols shared between SP_1' and SP originate in SP_1, as imposed by CASL rules for application of generic units.

where τ is a specification morphism from SP_1 to the pushout specification $SP'_1 \oplus SP$ in the following diagram:

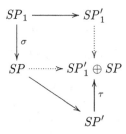

In the terminology of [11], (σ, τ) is a *first-order morphism* from $SP \to SP'$ to $SP_1 \to SP'_1$.

A crucial property of development trees as are now captured by architectural specifications with specification refinements is that adding correct refinements to unit specifications in an architectural specification, and thus expanding the development tree by additional refinement steps at its leaves, preserves correctness of the entire development. In particular, the semantics of architectural specifications with new correct refinements remains well-defined. This holds by [3, Thm. 2] (a technical assumption necessary there holds trivially in the absence of imports).

8   Component Refinements

Refinements introduced in Sect. 7 do not allow the user to refine architectural specifications as such. Only refinements for individual units are allowed, and they must be inserted into the architectural specification, directly into unit declarations. Consider for instance the following example from Sect. 4, where a refinement for unit N in the architectural specification ADDITION_FIRST_GENERIC was captured as follows:

> **arch spec** ADDITION_FIRST_WITH_BIN$'$ =
>   **units** N : R2;
> F : NAT $\rightarrow$ { **op** $suc(n : Nat) : Nat = n + 1$ };
> $M = F[N]$
>   **result** M

This is not very convenient: given an already defined architectural specification (in this case, ADDITION_FIRST_GENERIC), one would like to avoid rewriting it when indicating that specifications of some of the units (N here) are to be refined (using R2 here). Instead, one would rather refer to the architectural specification as given, and indicate refinements that are to follow, in this case writing:

> **refinement** R =
>   **arch spec** ADDITION_FIRST_GENERIC **then** {N to R2}

where $\{N$ **to** $R2\}$ is a refinement of an architectural specification having a unit named N. The refinement R then consists of the architectural specification AD-DITION_FIRST_GENERIC with N further refined according to R2. The need for such syntax is perhaps even more visible in complex examples involving nested refinements, like refinement R5 at the end of Sect. 6, which one would prefer to restructure as follows:

refinement $R5' =$
> SP **refined to arch spec** $\{$**units**
>> $K : SP' \to SP;$
>> $A' : SP'$
>> **result** $K[A']\}$
>
> **then** $\{K$ **to** $USP,$
>> A' **to arch spec** $\{$**units**
>>> $K' : SP'' \to SP';$
>>> $A'' : SP''$
>>> **result** $K'[A'']\}$
>> **then** $\{K'$ **to** $USP'\}\}$

or even:

refinement $R5'' =$
> SP **refined to arch spec** $\{$**units**
>> $K : SP' \to SP;$
>> $A' : SP'$
>> **result** $K[A']\}$
>
> **then** $\{K$ **to** $USP,$
>> A' **to arch spec** $\{$**units**
>>> $K' : SP'' \to SP';$
>>> $A'' : SP''$
>>> **result** $K'[A'']\}\}$
> **then** $\{A'$ **to** $\{K'$ **to** $USP'\}\}$

Of course, all the architectural specifications used here, as well as the refinements, would typically be defined earlier and then referred to by their names when refined further.

To capture such possibilities we extend the syntax for refinements introduced in Sect. 7, adding a new form:

```
SPEC-REF  ::= ...
          | {UNIT-NAME_1 to SPEC-REF_1, ..., UNIT-NAME_n to SPEC-REF_n}
```

However, this apparently simple change considerably increases the conceptual (and then semantic) complexity here, since in fact we are now dealing with three kinds of refinements:

- *unit specification refinements* which lead from a unit specification to another unit specification;
- *branching specification refinements* which generalise unit refinements by additionally allowing the target specification to be architectural; and

$$\frac{\vdash USP \rhd U\Sigma}{\vdash USP \text{ qua } \textbf{SPEC-REF} \rhd (U\Sigma, U\Sigma)}$$

$$\frac{\vdash USP \rhd U\Sigma = (\Sigma_1, \ldots, \Sigma_n \to \Sigma)}{\vdash SI \rhd \sigma : \Sigma \to \Sigma'}$$
$$\frac{\vdash SPR \rhd ((\Sigma_1, \ldots, \Sigma_n \to \Sigma'), B\Sigma'')}{\vdash USP \text{ \textbf{refined via} } SI \text{ \textbf{to} } SPR \rhd (U\Sigma, B\Sigma'')}$$

$$\frac{\vdash USP \rhd U\Sigma \qquad \vdash SPR \rhd (U\Sigma, B\Sigma'')}{\vdash USP \text{ \textbf{refined to} } SPR \rhd (U\Sigma, B\Sigma''')}$$

$$\frac{\vdash SPR_1 \rhd R\Sigma_1 \qquad \vdash SPR_2 \rhd R\Sigma_2 \qquad R\Sigma = R\Sigma_1 ; R\Sigma_2}{\vdash SPR_1 \text{ \textbf{then} } SPR_2 \rhd R\Sigma}$$

$$\frac{\vdash ASP \rhd RstC, U\Sigma}{\vdash \textbf{arch spec } ASP \rhd (U\Sigma, \pi_2(RstC))}$$

$$\frac{UN_1, \ldots, UN_n \text{ are distinct} \qquad \vdash SPR_1 \rhd R\Sigma_1 \quad \cdots \quad \vdash SPR_n \rhd R\Sigma_n}{\vdash \{UN_1 \text{ to } SPR_1, \ldots, UN_n \text{ to } SPR_n\} \rhd \{UN_1 \mapsto R\Sigma_1, \ldots, UN_n \mapsto R\Sigma_n\}}$$

Fig. 4. Static semantics of extended refinements

- *component specification refinements* which name units whose specifications are to be further refined as indicated.

The corresponding semantic concepts come now in three flavours as well. For the static semantics, we introduce *refinement signatures*, $R\Sigma$, which take one of the following forms:

- *unit refinement signatures* $(U\Sigma, U\Sigma')$ which consist of two unit signatures (this corresponds to the typical case in the static semantics of Sect. 7);
- *branching refinement signatures* $(U\Sigma, B\Sigma')$ which consist of a unit signature $U\Sigma$ and a *branching signature* $B\Sigma'$, which is either a unit signature $U\Sigma'$ (in which case the branching refinement signature is a unit refinement signature) or a *branching static context* $BstC'$, which is in turn a (finite) map assigning branching signatures to unit names. Note that therefore all static contexts as used in the plain CASL semantics [8, III:5] are branching static contexts, but not vice versa;
- *component refinement signatures* which are (finite) maps $\{UN_i \mapsto R\Sigma_i\}_{i \in \mathcal{J}}$ from unit names to refinement signatures. When all $R\Sigma_i$, $i \in \mathcal{J}$, in such a map are branching refinement signatures, we refer to it as a *refined-unit static context*. Any refined-unit static context $RstC = \{UN_i \mapsto (U\Sigma_i, B\Sigma_i)\}_{i \in \mathcal{J}}$ can be naturally coerced to the static context $\pi_1(RstC) = \{UN_i \mapsto U\Sigma_i\}_{i \in \mathcal{J}}$ of the plain CASL semantics, as well as to the branching static context $\pi_2(RstC) = \{UN_i \mapsto B\Sigma_i\}_{i \in \mathcal{J}}$.

New rules for the static semantics of refinements are given in Fig. 4. The first three rules are essentially inherited from Sect. 7, with a minor change to allow for the target signature to be branching. The new rule for architectural refinements allows for their further refinement by replacing the $\perp$ mark by the branching static context that emerges from the semantics of the architectural specification (see the end of this section for a brief discussion of the new semantics for architectural specifications). The rule for individual component refinements is straightforward: it just stores the refinement signatures obtained from the refinements attached to unit names. The extra complexity is hidden in the rule for refinement composition using an auxiliary partial composition operation on refinement signatures. Given refinement signatures $R\Sigma_1$ and $R\Sigma_2$, their *composition* $R\Sigma_1 ; R\Sigma_2$ is defined inductively depending on the form of the first argument:

- $R\Sigma_1 = (U\Sigma, U\Sigma')$: then $R\Sigma_1 ; R\Sigma_2$ is defined only if $R\Sigma_2$ is a branching refinement signature of the form $(U\Sigma', B\Sigma'')$. Then $R\Sigma_1 ; R\Sigma_2 = (U\Sigma, B\Sigma'')$.
- $R\Sigma_1 = (U\Sigma, BstC')$: then $R\Sigma_1 ; R\Sigma_2$ is defined only if $R\Sigma_2$ is a component refinement signature such that $R\Sigma_2$ *matches* $BstC'$, that is, $dom(R\Sigma_2) \subseteq dom(BstC')$ and for each $UN \in dom(R\Sigma_2)$,
  - either $BstC'(UN)$ is a unit signature and then $R\Sigma_2(UN) = (U\Sigma', B\Sigma'')$ with $U\Sigma' = BstC'(UN)$, or
  - $BstC'(UN)$ is a branching static context and then $R\Sigma_2(UN)$ matches $BstC'(UN)$.

  Then $R\Sigma_1 ; R\Sigma_2 = (U\Sigma, BstC'[R\Sigma_2])$, where given any branching static context $BstC'$ and component refinement signature $R\Sigma_2$ that matches $BstC'$, $BstC'[R\Sigma_2]$ modifies $BstC'$ on each $UN \in dom(R\Sigma_2)$ as follows:
  - if $BstC'(UN)$ is a unit signature then $BstC'[R\Sigma_2](UN) = B\Sigma''$ where $R\Sigma_2(UN) = (BstC'(UN), B\Sigma'')$,
  - if $BstC'(UN)$ is a branching static context then $BstC'[R\Sigma_2](UN) = BstC'(UN)[R\Sigma_2(UN)]$.
- $R\Sigma_1$ is a component refinement signature: then $R\Sigma_1 ; R\Sigma_2$ is defined only if $R\Sigma_2$ is a component refinement signature too, and moreover, for all $UN \in dom(R\Sigma_1) \cap dom(R\Sigma_2)$, $R\Sigma_{UN} = R\Sigma_1(UN) ; R\Sigma_2(UN)$ is defined. Then $R\Sigma_1 ; R\Sigma_2$ modifies the (ill-defined) union of $R\Sigma_1$ and $R\Sigma_2$ by putting $(R\Sigma_1 ; R\Sigma_2)(UN) = R\Sigma_{UN}$ for $UN \in dom(R\Sigma_1) \cap dom(R\Sigma_2)$.

The complexity of the model semantics for refinements increases similarly. Given a refinement signature $R\Sigma$, *refinement relations*, $\mathcal{R}$, are classes of *assignments*, R, which take the following forms:

- *unit assignments*, for $R\Sigma = (U\Sigma, U\Sigma')$, are pairs (U, U') of units over unit signatures $U\Sigma$ and $U\Sigma'$, respectively;
- *branching assignments*, for $R\Sigma = (U\Sigma, B\Sigma')$, are pairs (U, BM'), where U is a unit over the unit signature $U\Sigma$ and BM' is a *branching model* over the branching signature $B\Sigma'$, which is either a unit over $B\Sigma'$ when $B\Sigma'$ is a unit signature (in which case the branching assignment is a unit assignment), or a *branching environment* BE' that fits $B\Sigma'$ when $B\Sigma'$ is a branching static context. Branching environments are (finite) maps assigning branching models to unit names, with the obvious requirements to ensure compatibility with the branching signatures indicated in the corresponding branching static context. Note that therefore all unit environments as used in the plain CASL semantics [8, III:5] are branching environments, but not vice versa.

$$\frac{\vdash \mathit{USP} \Rightarrow \mathcal{U}}{\vdash \mathit{USP}\ \text{qua}\ \texttt{SPEC-REF} \Rightarrow \{(U,U) \mid U \in \mathcal{U}\}}$$

$$\frac{\begin{array}{c}\vdash \mathit{USP} \Rightarrow \mathcal{U} \qquad \vdash \mathit{SPR} \Rightarrow \mathcal{R} \\ U' \in \mathcal{U},\ \text{for all } (U', BM'') \in \mathcal{R}\end{array}}{\vdash \mathit{USP}\ \textbf{refined to}\ \mathit{SPR} \Rightarrow \mathcal{R}}$$

$$\frac{\begin{array}{c}\vdash \mathit{USP} \Rightarrow \mathcal{U} \qquad \vdash \mathit{SI} \rhd \sigma : \Sigma \to \Sigma' \qquad \vdash \mathit{SPR} \Rightarrow \mathcal{R} \\ U'|_\sigma \in \mathcal{U},\ \text{for all } (U', BM'') \in \mathcal{R} \\ \mathcal{R}' = \{(U'|_\sigma, BM'') \mid (U', BM'') \in \mathcal{R}\}\end{array}}{\vdash \mathit{USP}\ \textbf{refined via}\ \mathit{SI}\ \textbf{to}\ \mathit{SPR} \Rightarrow \mathcal{R}'}$$

$$\frac{\vdash \mathit{ASP} \Rightarrow \mathcal{AM}}{\vdash \texttt{arch spec}\ \mathit{ASP} \Rightarrow \{(U, \pi_2(\mathit{RE})) \mid (\mathit{RE}, U) \in \mathcal{AM}\}}$$

$$\frac{\vdash \mathit{SPR}_1 \Rightarrow \mathcal{R}_1 \quad \cdots \quad \vdash \mathit{SPR}_n \Rightarrow \mathcal{R}_n}{\begin{array}{r}\vdash \{\mathit{UN}_1\ \texttt{to}\ \mathit{SPR}_1, \ldots, \mathit{UN}_n\ \texttt{to}\ \mathit{SPR}_n\} \Rightarrow \{R \mid \mathit{dom}(R) = \{\mathit{UN}_1, \ldots, \mathit{UN}_n\}, \\ R(\mathit{UN}_1) \in \mathcal{R}_1, \ldots, R(\mathit{UN}_n) \in \mathcal{R}_n\}\end{array}}$$

$$\frac{\vdash \mathit{SPR}_1 \Rightarrow \mathcal{R}_1 \qquad \vdash \mathit{SPR}_2 \Rightarrow \mathcal{R}_2 \qquad \mathcal{R} = \mathcal{R}_1 \,;\mathcal{R}_2}{\vdash \mathit{SPR}_1\ \textbf{then}\ \mathit{SPR}_2 \Rightarrow \mathcal{R}}$$

Fig. 5. Model semantics of extended refinements

- *component assignments*, for $R\Sigma = \{UN_i \mapsto R\Sigma_i\}_{i \in \mathcal{J}}$, are (finite) maps $\{UN_i \mapsto R_i\}_{i \in \mathcal{J}}$ from unit names to assignments over the respective refinement signatures. When $R\Sigma$ is a refined-unit static context (and so each R_i, $i \in \mathcal{J}$, is a branching assignment) we refer to $RE = \{UN_i \mapsto (U_i, BM_i)\}_{i \in \mathcal{J}}$ as a *refined-unit environment.* Any such refined-unit environment can be naturally coerced to a unit environment $\pi_1(RE) = \{UN_i \mapsto U_i\}_{i \in \mathcal{J}}$ of the plain CASL semantics, as well as to a branching environment $\pi_2(RE) = \{UN_i \mapsto BM_i\}_{i \in \mathcal{J}}$.

New rules for the model semantics of refinements are given in Fig. 5. As with the static semantics, the non-trivial change is hidden in the rule for refinement composition using the auxiliary partial operation to compose refinement relations. Given two refinement relations $\mathcal{R}_1, \mathcal{R}_2$ over refinement signatures $R\Sigma_1, R\Sigma_2$, respectively, such that the composition $R\Sigma = R\Sigma_1 \,; R\Sigma_2$ is defined, the *composition* $\mathcal{R}_1 \,; \mathcal{R}_2$ is defined as a refinement relation over $R\Sigma$ as follows:

- $R\Sigma_1 = (U\Sigma, U\Sigma')$, $R\Sigma_2 = (U\Sigma', B\Sigma'')$: then $\mathcal{R}_1 \,; \mathcal{R}_2$ is defined only if for all $(U', BM'') \in \mathcal{R}_2$ we have $(U, U') \in \mathcal{R}_1$ for some U. Then
$$\mathcal{R}_1 \,; \mathcal{R}_2 = \{(U, BM'') \mid (U, U') \in \mathcal{R}_1, (U', BM'') \in \mathcal{R}_2 \text{ for some } U'\}$$

– $R\Sigma_1 = (U\Sigma, BstC')$ and $R\Sigma_2$ is a component refinement signature that matches $BstC'$: then $\mathcal{R}_1 ; \mathcal{R}_2$ is defined only if for each $R_2 \in \mathcal{R}_2$ there is $(U, BE') \in \mathcal{R}_1$ such that R_2 *matches* BE', that is, for each $UN \in dom(R_2)$,
  • either $BstC'(UN)$ is a unit signature and then $R_2(UN) = (U'', BM'')$ with $U'' = BE'(UN)$, or
  • $BstC'(UN)$ is a branching static context and then $R_2(UN)$ matches $BE'(UN)$.

Then
$$\mathcal{R}_1 ; \mathcal{R}_2 = \{(U, BE'[R_2]) \mid (U, BE') \in \mathcal{R}_1, R_2 \in \mathcal{R}_2, R_2 \text{ matches } BE'\}$$

where given any branching environment BE' that fits $BstC'$ and assignment R_2 that matches BE', $BE'[R_2]$ modifies BE' on each $UN \in dom(R_2)$ as follows:
  • if $BstC'(UN)$ is a unit signature then $BE'[R_2](UN) = BM''$ where $R_2(UN) = (BE'(UN), BM'')$;
  • if $BstC'(UN)$ is a branching static context then we put $BE'[R_2](UN) = BE'(UN)[R_2(UN)]$.

– $R\Sigma_1$ and $R\Sigma_2$ are component refinement signatures such that for all $UN \in dom(R\Sigma_1) \cap dom(R\Sigma_2)$, $R\Sigma_{UN} = R\Sigma_1(UN) ; R\Sigma_2(UN)$ is defined then $\mathcal{R}_1 ; \mathcal{R}_2$ is defined only if for each $R_2 \in \mathcal{R}_2$ there is $R_1 \in \mathcal{R}_1$ such that R_1 *transfers to* R_2, that is, for each $UN \in dom(R_1) \cap dom(R_2)$,
  • either $R\Sigma_1(UN)$ is a unit refinement signature $(U\Sigma, U\Sigma')$, and then $R_1(UN) = (U, U'_1)$ and $R_2(UN) = (U'_2, BM'')$ with $U'_1 = U'_2$, or
  • $R\Sigma_1(UN)$ is a branching refinement signature $(U\Sigma, BstC')$, and then $R_1(UN) = (U, BE')$ and $R_2(UN)$ is an assignment that matches BE', or
  • $R\Sigma_1(UN)$ is component refinement signature, and then $R_1(UN)$ transfers to $R_2(UN)$.

Then
$$\mathcal{R}_1 ; \mathcal{R}_2 = \{R_1 ; R_2 \mid R_1 \in \mathcal{R}_1, R_2 \in \mathcal{R}_2, R_1 \text{ transfers to } R_2\}$$

where given any assignments R_1, R_2 over $R\Sigma_1$, $R\Sigma_2$, respectively, such that R_1 transfers to R_2, $R_1 ; R_2$ is the assignment that modifies the (ill-defined) union of R_1 and R_2 on each $UN \in dom(R_1) \cap dom(R_2)$ as follows:
  • if $R\Sigma_1(UN) = (U\Sigma, U\Sigma')$, $R_1(UN) = (U, U'_1)$ and $R_2(UN) = (U'_2, BM'')$ (hence $U'_1 = U'_2$) then $(R_1 ; R_2)(UN) = (U, BM'')$;
  • if $R\Sigma_1(UN) = (U\Sigma, BstC')$, $R_1(UN) = (U, BE')$ (hence $R_2(UN)$ is an assignment that matches BE') then $(R_1 ; R_2)(UN) = (U, BE'[R_2(UN)])$;
  • if $R\Sigma_1(UN)$ is a component refinement signature (hence $R_1(UN)$ and $R_2(UN)$ are component assignments such that $R_1(UN)$ transfers to $R_2(UN)$) then $(R_1 ; R_2)(UN) = R_1(UN) ; R_2(UN)$.

We also have to consider the necessary changes to the semantics of architectural specifications in [8, III:5]. Most visibly, as sketched above, we have to modify the semantic concepts for architectural specifications to work with refined-unit static contexts and refined-unit environments rather than unit static contexts and unit environments. This would alter most of the rules only formally. At crucial places, where units are used, the easy modification relies on the π_1 coercions from refined-unit static contexts to static contexts and from refined-unit environments to unit environments for static and model semantics, respectively.

Further straightforward modification concerns the semantics of unit declarations, now with arbitrary specification refinements in place of unit specifications. The new static semantics imposes the restriction that only branching specification refinements (so: no component specification refinements) are allowed here[2], and stores the appropriate branching refinement signature for the declared unit name in the refined-unit static context. Then, the model semantics produces the context that consists of all refined-unit environments that map the declared unit name to a branching assignment in the semantics of the branching specification refinement used in the declaration.

Finally, the semantics of unit definitions involves additional unit refinement signatures and assignments with the $\perp$ mark as the second component to indicate that unit definitions cannot be further refined.

9   The Steam Boiler Example

So far, we have illustrated the refinement language by means of toy examples. A discussion of realistic examples would exceed the space limitations of this paper. However, the CASL User Manual [2, Chap. 13] contains a specification of an industrial case study, namely a steam boiler control system that serves to control the water level in a steam boiler. Reference [2, Sect. 13.10] contains several architectural specifications explaining how to decompose the steam boiler control system into subsystems, using e.g. a specification VALUE for physical values, a specification SBCS_STATE for the specification of the state of the steam boiler, a specification PU_PREDICTION for prediction of the pump behaviour, etc. There is no space here to repeat the details of this example, so we refer the reader to [2, Chap. 13] and only use the additional linguistic features introduced in Sects. 7 and 8 to present specification refinements that formally capture the development described there.

The development in [2, Sect. 13.10] begins by indicating the initial architectural design for the overall requirement specification of the system:

> **arch spec** ARCH_SBCS =
>   **units** P: VALUE $\rightarrow$ PRELIMINARY;
> S: PRELIMINARY $\rightarrow$ SBCS_STATE;
> A: SBCS_STATE $\rightarrow$ SBCS_ANALYSIS;
> C: SBCS_ANALYSIS $\rightarrow$ STEAM_BOILER_CONTROL_SYSTEM
>   **result** λV : VALUE • $C[A[S[P[V]]]]$

---

[2] The (abstract) syntax of specification refinements may be massaged so that some of the restrictions imposed by the static semantics on the composability and use of specification refinements are incorporated in the (context-free) grammar.

We may record this initial refinement now:

refinement REF_SBCS =
STEAM_BOILER_CONTROL_SYSTEM **to** ARCH_SBCS

In [2, Sect. 13.10], specifications refining the individual units of the above architectural specification are given. We extend the refinement REF_SBCS to capture them as well:

refinement REF_SBCS' =
REF_SBCS **then** {P **to arch spec** ARCH_PRELIMINARY,
S **to** UNIT_SBCS_STATE,
A **to arch spec** ARCH_ANALYSIS,
C **to** UNIT_SBCS_SYSTEM   }

The resulting specification for the unit S, UNIT_SBCS_STATE, is monomorphic:

unit spec UNIT_SBCS_STATE =
PRELIMINARY $\rightarrow$ SBCS_STATE_IMPL

Development within CASL stops at this point, the last step being the passage to a program in a programming language. This also holds for the component C, even though the corresponding unit specification UNIT_SBCS_SYSTEM is not explicitly provided in [2, Chap. 13].

The architectural specification ARCH_ANALYSIS used in the refinement above is given in [2, Sect. 13.10] as follows:

arch spec ARCH_ANALYSIS =
units FD : SBCS_STATE $\rightarrow$ FAILURE_DETECTION;
PR : FAILURE_DETECTION $\rightarrow$ PU_PREDICTION;
ME : PU_PREDICTION $\rightarrow$ MODE_EVOLUTION [PU_PREDICTION];
MTS : MODE_EVOLUTION [PU_PREDICTION] $\rightarrow$ SBCS_ANALYSIS
result λS : SBCS_STATE $\bullet$ $MTS\,[ME\,[PR\,[FD\,[S]]]]$

As remarked in [2, Sect. 13.10], the specifications for the components ME and MTS are simple enough to be directly implemented, so we stop their development at this point. For the other two units, we record the corresponding refinements from [2, Sect. 13.10]:

refinement REF_ARCH_ANALYSIS =
{FD **to arch spec** ARCH_FAILURE_DETECTION,
PR **to arch spec** ARCH_PREDICTION}

Finally, we put the above together and capture the overall development sketched in [2, Sect. 13.10]:

refinement REF_SBCS'' =
REF_SBCS' **then** {A **to** REF_ARCH_ANALYSIS}

10   Conclusion and Future Work

The issue of refinement has been discussed in many specification frameworks, starting with [12] and [13], and some frameworks provide methods for proving correctness of refinements. But this is normally regarded as a "meta-level" issue and specification languages have typically not included syntactic constructs for formally stating such relationships between specifications that are analogous to those presented here for CASL. A notable exception is Specware [17], where specifications (and implementations) are structured using specification diagrams, and refinements correspond to specification morphisms for which syntax is provided. This, together with features for expanding specification diagrams, provides sufficient expressive power to capture our branching specification refinements. A difference is that Specware does not include a distinction between structured specifications and CASL-like architectural specifications, and refinements are required to preserve specification structure.

One point of this proposal that requires further work is the treatment of *shared subcomponents*, such as S in the following:

$$\textbf{arch spec}\ \ \text{ASP} = \textbf{units}\ \ S : USP$$

$$\begin{array}{ll}A_1 : & \textbf{arch spec} \\ & \{\,\textbf{units} \\ & \quad\quad B_1 : USP'_1 \\ & \quad\quad \ldots \\ & \quad\quad B_m : USP'_m \\ & \quad\quad \textbf{result}\ \ldots B_1 \ldots S \ldots B_m \ldots\} \\ & \ldots \\ A_2 : & \textbf{arch spec} \\ & \{\,\textbf{units} \\ & \quad\quad C_1 : USP'_1 \\ & \quad\quad \ldots \\ & \quad\quad C_p : USP'_p \\ & \quad\quad \textbf{result}\ \ldots C_1 \ldots S \ldots C_p \ldots\} \\ & \textbf{result}\ \ldots A_1 \ldots A_2 \ldots \end{array}$$

This requires a relatively straightforward modification to the semantics of CASL architectural specifications to make declared units visible within architectural specifications for further units.

We have not provided a treatment of refinements of unit specifications with imports, as was pointed out in Sect. 7. A formal account of imports would add considerably to the complexity of the semantics, see [8, III:5]. However, they can be regarded as implicit formal parameters which are instantiated only once, as in the specification ADDITION_FIRST_GENERIC. And moreover, this seems to be the appropriate view when refinements are considered. The ultimate target of refinement of such a specification will necessarily involve a parametrized program, and at some point in the refinement process this needs to be made explicit. Thus we regard the lack of treatment of imports as methodologically sound rather than merely a convenient simplification. That said, given modified visibility rules as

sketched above, we could allow for refinements from a specification of the form
SP **given** UT to an architectural specification of the form

arch spec
   {**units** F: $SP_{par} \to SP'$
   result $F[UT]$}

which would be correct provided UT : SP_{par} and $[\![SP]\!] \supseteq [\![SP_{par}$ **and** $SP']\!]$.
Notice that here UT typically refers to units from the level of the unit that is
specified by SP **given** UT; this is the reason why the modified visibility rules
are necessary.

Finally, we have not discussed *behavioural refinement*, corresponding to *abstractor implementations* in [15]. Often, a refined specification does not satisfy
the initial requirements literally, but only up to some sort of behavioural equivalence: for example, if stacks are implemented as arrays-with-pointer, then two
arrays-with-pointer differing only in their "junk" entries (that is, those that are
"above" the pointer) exhibit the same behaviour in terms of the stack operations, and hence correspond to the same abstract stack. This can be taken
into account by re-interpreting unit specifications to include models that are behaviourally equivalent to literal models, see [4, 5]; then specification refinements
as considered here become behavioural.

Acknowledgments: Our thanks to the anonymous referees, and to Michel
Bidoit, whose suggestions encouraged us to make some important improvements.

References

1. E. Astesiano, H.-J. Kreowski, and B. Krieg-Brückner. *Algebraic Foundations of
   Systems Specification.* Springer, 1999.
2. M. Bidoit and P.D. Mosses. CASL *User Manual.* LNCS Vol. 2900 (IFIP Series).
   Springer, 2004.
3. M. Bidoit, D. Sannella, and A. Tarlecki. Architectural specifications in CASL.
   Formal Aspects of Computing, 13:252–273, 2002.
4. M. Bidoit, D. Sannella, and A. Tarlecki. Global development via local observational
   construction steps. In *Proc. 27th Intl. Symp. on Mathematical Foundations of
   Computer Science,* LNCS Vol. 2420, pages 1–24. Springer, 2002.
5. M. Bidoit, D. Sannella, and A. Tarlecki. Observational interpretation for CASL
   specifications. In preparation, 2004.
6. T. Brunet. Génération automatique de code à partir de spécifications formelles.
   Master's thesis, Université de Poitiers, 2003.
7. CoFI. The Common Framework Initiative for algebraic specification and development, electronic archives. Notes and Documents accessible from `http:
   //www.cofi.info/`.
8. CoFI (The Common Framework Initiative). CASL *Reference Manual.* LNCS
   Vol. 2960 (IFIP Series). Springer, 2004.
9. J. Fitzgerald and C. Jones. Modularizing the formal description of a database
   system. In *Proc. VDM'90 Conference,* LNCS Vol. 428. Springer, 1990.

10. J. Goguen and R. Burstall. Institutions: Abstract model theory for specification and programming. *Journal of the Association for Computing Machinery*, 39:95–146, 1992.

11. J. Goguen and K. Lin. Morphisms and semantics for higher order parameterized programming. Available from http://www.cs.ucsd.edu/users/goguen/pps/shom.ps, August 2002.

12. C.A.R. Hoare. Correctness of data representations. *Acta Informatica*, 1:271–281, 1972.

13. R. Milner. An algebraic definition of simulation between programs. In *Proc. 2nd Intl. Joint Conf. on Artificial Intelligence*, pages 481–489. British Computer Society, 1971.

14. T. Mossakowski. Two "functional programming" sublanguages of CASL. Note L-9, in [7], March 1998.

15. D. Sannella and A. Tarlecki. Toward formal development of programs from algebraic specifications: implementations revisited. *Acta Informatica*, 25:233–281, 1988.

16. D. Sannella and A. Tarlecki. *Foundations of Algebraic Specifications and Formal Program Development*. Cambridge University Press, 2005, to appear. See http://homepages.inf.ed.ac.uk/dts/book/index.html.

17. D. Smith. Designware: Software development by refinement. In *Proc. Conference on Category Theory and Computer Science, CTCS'99*, volume 29 of *Electronic Notes in Theoretical Computer Science*. Elsevier, 2000.

A   Abstract Syntax for CASL Architectural Specifications

The grammar extends the grammar given in the CASL Reference Manual [8]. The new parts of the grammar are marked in *italics*, while removed parts are ~~crossed out~~.

```
ARCH-SPEC-DEFN     ::= arch-spec-defn ARCH-SPEC-NAME ARCH-SPEC
ARCH-SPEC          ::= basic-arch-spec UNIT-DECL-DEFN+ RESULT-UNIT
                     | ARCH-SPEC-NAME
UNIT-DECL-DEFN     ::= UNIT-DECL | UNIT-DEFN

UNIT-DECL          ::= unit-decl UNIT-NAME SPEC-REF UNIT-IMPORTED
UNIT-IMPORTED      ::= unit-imported UNIT-TERM*
UNIT-DEFN          ::= unit-defn UNIT-NAME UNIT-EXPRESSION

UNIT-SPEC-DEFN     ::= unit-spec-defn SPEC-NAME UNIT-SPEC
UNIT-SPEC          ::= UNIT-TYPE | SPEC-NAME
                     | arch-unit-spec ARCH-SPEC
                     | closed-unit-spec UNIT-SPEC
UNIT-TYPE          ::= unit-type SPEC* SPEC

SPEC-REF-DEFN      ::= ref-unit-spec-defn SPEC-NAME SPEC-REF
SPEC-REF           ::= SPEC-NAME
                     | unit-spec UNIT-SPEC
                     | refinement UNIT-SPEC SYMB-MAP-ITEMS* SPEC-REF
                     | arch-unit-spec ARCH-SPEC
                     | compose-ref SPEC-REF SPEC-REF
                     | component-ref UNIT-REF*

UNIT-REF           ::= unit-ref UNIT-NAME SPEC-REF

RESULT-UNIT        ::= result-unit UNIT-EXPRESSION
UNIT-EXPRESSION    ::= unit-expression UNIT-BINDING* UNIT-TERM
UNIT-BINDING       ::= unit-binding UNIT-NAME UNIT-SPEC
UNIT-TERM          ::= unit-translation UNIT-TERM RENAMING
                     | unit-reduction UNIT-TERM RESTRICTION
                     | amalgamation UNIT-TERM+
                     | local-unit UNIT-DEFN+ UNIT-TERM
                     | unit-appl UNIT-NAME FIT-ARG-UNIT*
FIT-ARG-UNIT       ::= fit-arg-unit UNIT-TERM SYMB-MAP-ITEMS*

ARCH-SPEC-NAME     ::= SIMPLE-ID
UNIT-NAME          ::= SIMPLE-ID
```

A SPEC-NAME can be a SPEC-REF either directly, or indirectly via UNIT-SPEC. This ambiguity is solved by looking up the SPEC-NAME in the global environment, which is expected to keep information about UNIT-SPECs and SPEC-REFs separately.

A Distributed and Mobile Component System Based on the Ambient Calculus

Nikos Mylonakis and Fernando Orejas

Departament LSI,
Universitat Politècnica de Catalunya

Abstract. In this paper we present a new component concept equivalent to the one of [2] which is more appropriate for distributed applications. After that, we present the notion of component system and define a set of operations of component systems, some of which are used to define an ambient calculus [1] with component systems. Finally we present an example.

1 Introduction

The final aim of this work is the study of the modelling and development of component-based distributed applications with mobile processes in the internet. In particular, we believe that the concept of component can play an important role in the development of such applications where components can be defined as independent units with a specific task or functionality. These systems are in general heterogeneous which means that they can be described or implemented using different formalisms. To develop these systems, we use a uniform notion of component which is, to a certain extent, independent of the formalism which is used. The framework which we define is based on a generic notion of transformation or refinement which is used to define the semantics of components and their interconnection. In particular, in [2], the conditions which must satisfy these transformations in order to instantiate the given framework to a concrete formalism are presented. Additionally, a simple notion of composition of components is given. In our work, we define a new semantics of components which is a variation of the one defined in [2] which make possible the definition of more complex forms of composition.

In this work we also introduce the notion of component system. A component system is a set of components where every component can have several import interfaces and several export interfaces. Additionally, we can connect an import interface of a component with an export interface of another component. A component system is in general a graph. We define also different operations on component systems, including a non-deterministic operation to establish some or all the possible connections between two component systems. In particular, when a component moves into a new context, it may establish different kinds of connections through its interfaces with several interfaces from the other components. In this situation we assume that some of these connections may be

J.L. Fiadeiro, P. Mosses, and F. Orejas (Eds.): WADT 2004, LNCS 3423, pp. 186–200, 2005.

established nondeterministically. Moreover, we assume that this interconnection can vary over time. The intuition behind this operation is that mobile components are autonomous agents that, on a given situation, may choose which connections to establish.

In order to explicitly deal with mobility we decided to integrate our generic approach with a specific well-established calculus for distribution and mobility like the ambient calculus [1]. This is a first step to use our component framework in distributed and mobile applications although the extension is not as general as we would have liked to. A possible future work which would be a more general extension than the one which we propose, would be to incorporate component systems in bigraphs [3].

In order to extend component systems with the ambient calculus, it is necessary to define a forest of hierarchies of component systems. Thus, an ambient expression will denote a forest of hierarchies of component systems and after making a move, a sub-hierarchy of a hierarchy will be associated to another hierarchy in the forest. Every component system in the hierarchy can have associated a forest of sub-hierarchies. The main modification in the calculus is the possibility to have a component system as a process and the obligation to have a component system associated to the name of an ambient.

Finally, we present an example of a server with a firewall together with two clients trying to obtain a software component to finish an application. To present the example, we instantiate our component system with a formalism based on algebra transformation systems.

The structure of this paper is the following: in the first section we present our component concept reviewing the notion of the transformation framework in [2]. In the next section we present our component system and interconnected forest of hierarchies of component systems. After that we present the extension of the ambient calculus with component system and finally we instantiate our component concept with a formalism based on algebra transformation system and we present an example.

2   The Component Concept

As we mentioned in the introduction, components can be seen as independent units with a specific task or functionality which can be used in as many environments and applications as possible [4]. To achieve this, in a component there must be a clear distinction between its body, where the functionality of the component is described (or implemented) in detail, and its interfaces, describing how the component is related to the outside world.

In [2], a generic component framework is presented. This framework is generic not only concerning the underlying specification formalism used inside the components, but also concerning the concept of transformations in order to model abstraction and refinement between interfaces and body of one component, or between import and export interfaces between different components. The only requirement for a specification formalism to be used in connection to our component framework is to show that the formalism is a transformation framework. More precisely:

Definition 1. *A transformation framework $\mathcal{T}$ over a class of specifications consists of a class of transformations and a subclass of embeddings, which include identical transformations and identical embeddings, both are closed under composition and satisfy the following properties:*

- *For each transformation $t_1 : SPEC_1 \Rightarrow SPEC_3$, and each embedding $i_1 : SPEC_1 \hookrightarrow SPEC_2$ which can be represented as in Figure 1, there is a selected transformation $t_2 : SPEC_3 \Rightarrow SPEC_4$, with embedding $i_2 : SPEC_2 \hookrightarrow SPEC_4$, called the extension of t_1 with respect to i_1. t_2 is also denoted as $E_{i1}(t_1)$.*
- *Horizontal and vertical composition of extension diagrams are required in the usual way.*

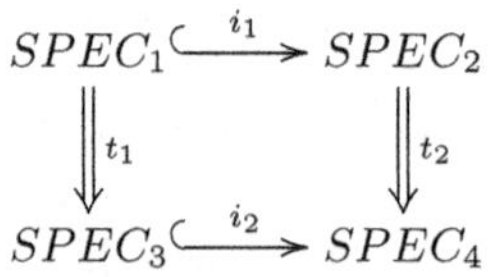

$$SPEC_1 \overset{i_1}{\hookrightarrow} SPEC_2$$
$$\Big\Downarrow t_1 \qquad \Big\Downarrow t_2$$
$$SPEC_3 \overset{i_2}{\hookrightarrow} SPEC_4$$

Fig. 1. Extension Diagram

In this paper, we consider that components may have several import and export interfaces allowing us to connect a given component with several other components. In particular, export interfaces describe services that a component offers to the outside world, while import interfaces specify services used inside a component that are assumed to be defined by other components. Export interfaces are assumed to be different abstractions of the body and each of them is related to the body by a transformation modelling a refinement. Import interfaces are assumed to be related to the body by *independent* embeddings. More precisely:

Definition 2. *A component consists of a body specification with a list of independent import specifications together with the corresponding embeddings and a list of export specifications together with the corresponding transformations into the body specifications. Thus, a component will have this general form*

$$(BOD, < IMP_1, i_1, \ldots, IMP_n, i_n >, < EXP_1, e_1, \ldots, EXP_m, e_m >)$$

A possible graphical representation is in figure 2.

As said above, we assume that the family of embeddings $i_j : I_j \to BOD$ for each connector is **independent**. This means intuitively that the import interfaces

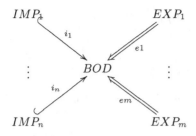

Fig. 2. Diagram of a component

IMP_j of B are pairwise disjoint. Formally, independence of embeddings can be expressed by the following definition:

Definition 3. *A family of embeddings* $i_j : SPEC_j \hookrightarrow SPEC, (j = 1..n)$ *is independent if the following properties are satisfied:*

- *For every family of transformations* $t_j \in Trafo(SPEC_j), (j = 1..n)$, *there exists a selected transformation* $t \in Trafo(SPEC)$ *and selected independent embeddings* i'_j, i'_k *such that the diagram 3 commutes.* t *is called the parallel extension of* $\{t_j\}_{j=1..n}$ *with respect to* $\{i_j\}_{j=1..n}$ *and will be denoted as* $PE_{\{i_j\}_{j=1..n}}(\{t_j\}_{j=1..n})$
- *For any* $SPEC_j, 1 \le j \le n$, *given the extension diagram of figure 4 and any* $SPEC_k, 1 \le k \le n$, *we have that the diagram in figure 5 is a parallel extension diagram, where* i''_k *is the composition of* i_k *and* t''. *Note that, in this case, we are asking that the composition of the embedding* i_k *and the transformation* t'' *should be an embedding*
- *Parallel extension diagrams can be composed vertically.*

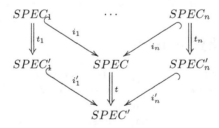

Fig. 3. Parallel Extension

Now, in this paper we provide a new semantics to components which can be seen as a variation on the semantics defined in [2]. In particular, in this paper we consider that the semantics of a component is defined as a class of tuples of transformations (refinements) of the export interfaces that have to satisfy some

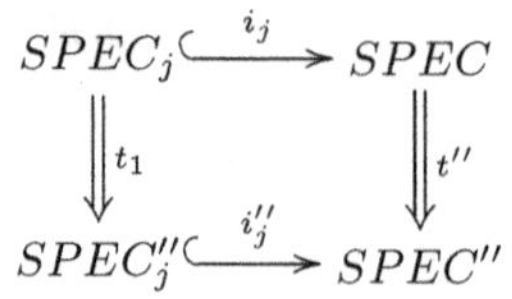

$$SPEC_j \overset{i_j}{\hookrightarrow} SPEC$$

$$\Big\downarrow t_1 \qquad \Big\downarrow t''$$

$$SPEC_j'' \overset{i_j''}{\hookrightarrow} SPEC''$$

Fig. 4. Extension Diagram

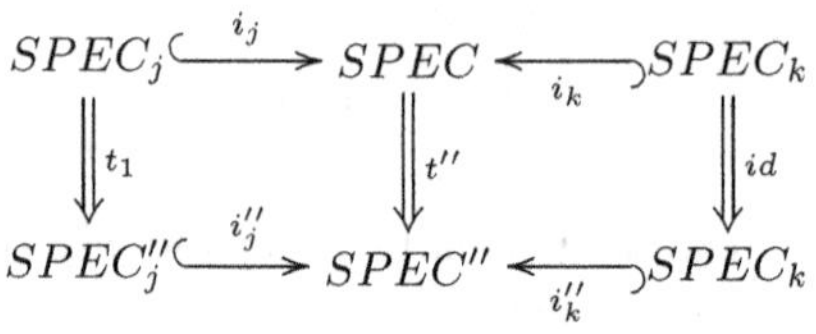

$$SPEC_j \overset{i_j}{\hookrightarrow} SPEC \overset{}{\underset{i_k}{\longleftarrow}} SPEC_k$$

$$\Big\downarrow t_1 \qquad \Big\downarrow t'' \qquad \Big\downarrow id$$

$$SPEC_j'' \overset{i_j''}{\hookrightarrow} SPEC'' \overset{}{\underset{i_k''}{\longleftarrow}} SPEC_k$$

Fig. 5. Extension as parallel extension

constraints. On one hand, these transformations must be defined as the composition of the corresponding export connections e_k and a given transformation t which is a refinement of the body specification. On the other hand, t must be a parallel extension of some family of refinements of the import specifications. More precisely:

Definition 4. *The semantics of a component COMP:*

$$(BOD, < IMP_1, i_1, \ldots, IMP_n, i_n >, < EXP_1, e_1, \ldots, EXP_m, e_m >)$$

is defined as follows:

$$TrafoSem(COMP) = \{< t \circ e_1, \ldots, t \circ e_m > \mid t \in Trafo(BOD) \text{ and}$$
$$\exists t_1 \in Trafo(IMP_1). \ldots . \exists t_n \in Trafo(IMP_n) \quad t = PE_{\{i_j\}_{j=1..n}}(\{t_j\}_{j=1..n})\}$$

where $Trafo(SPEC)$ denotes the class of all transformations defined over $SPEC$.

3   Component Systems

A component system is a set of components together with a set of connections. More precisely, every component j can have m_j import interfaces and p_j export interfaces, and we can connect an import interface of a component j with an export interface of a component k via a transformation c_{jk}. We will assume that every component has a name. More precisely:

Definition 5. *A component system CS consists of a list of components and a list of connections which are denoted as $Comp(CS)$ and $Conn(CS)$ respectively. Each component in $Comp(CS)$ has the form:*

$$C : (BOD^C, < IMP_1^C, i_1, \ldots, IMP_{n_C}^C, i_{n_C} >, < EXP_1^C, e_1, \ldots, EXP_{m_C}^C, e_{m_C} >)$$

where C is the name of the component, and each connection in $Conn(CS)$ is a transformation:

$$conn_{C,C'} : IMP_j^C \Rightarrow EXP_k^{C'}$$

where IMP_j^C is an import interface of the component C and $EXP_k^{C'}$ is an export interface of the component C'.

As in the case of single components, we define the semantics of a component system as a set of tuples of transformations of the export specifications in the system that satisfy a given set of constraints. In particular, given a refinement for each import specification of the given system, we require that the transformation associated to a given export specification should be the composition of the corresponding export connection and the parallel extension of the import refinements. This means that these transformations must belong to the semantics of the components of the system. On the other hand, each connection, $conn_{C,C'} : IMP_j^C \Rightarrow EXP^{C_k}$, induces the additional constraint that the transformation associated to $EXP_k^{C'}$ composed with $conn_{C,C'}$ coincide with the given refinement of IMP_j^C. More precisely:

Definition 6. *Given a component system CS, where $Comp(CS)$ consists of components:*

$$C_j : (BOD^{C_j}, < IMP^{C_j}, i_1^{C_j}, \ldots, IMP_{n_{C_j}}^{C_j}, i_{n_{C_j}}^{C_j} >,$$

$$< EXP_1^{C_j}, e_1^{C_j}, \ldots, EXP_{m_{C_j}}^{C_j}, e_{m_{C_j}}^{C_j} >)$$

with $1 \leq j \leq p$ and $Conn(CS)$ consists of connections:

$$conn_{C_{j_1}, C_{j_2}} : IMP_{k_1}^{C_{j_1}} \Rightarrow EXP_{k_2}^{C_{j_2}})$$

where C_{j_1} and C_{j_2} are components in $Comp(CS)$, the semantics of CS is defined as:

$$TrafoSem(CS) = \{ < t^{C_1} \circ e_1^{C_1}, \ldots, t^{C_1} \circ e_{m_{C_1}}^{C_1}, \ldots, t^{C_p} \circ e_1^{C_p}, \ldots, t^{C_p} \circ e_{m_{C_p}}^{C_p} > \; | $$
$$\forall j (1 \leq j \leq p) \; \forall k (1 \leq k \leq n_{C_j}) \; \exists t_k^{C_j} \in Trafo(IMP_k^{C_j}) \text{ such that}$$
$$(t^{C_j} = PE_{\{i_k^{C_j}\}_{k=1..n_{C_j}}} (\{t_k'^{C_j}\}_{k=1..n_{C_j}}) \text{ and}$$
$$\forall conn_{C_{j_1}, C_{j_2}} \in Conn(CS)(t^{C_{j_2}} \circ e_{k_2}^{C_{j_2}} \circ conn_{C_{j_1}, C_{j_2}} = t_{k_1}'^{C_{j_1}}) \}$$

In the rest of this section, we define some operations on component systems. We present an operation to connect two component system by a single connection, an operation to add a connection to a component system and an operation to make non-deterministically some or all the connections between two component systems.

Definition 7. *If C_1 and C_2 are components in CS_1 and CS_2, respectively, then the operation to connect CS_1 and CS_2:*

$$CS1 \circ_{con:IMP^{C_1} \Rightarrow EXP^{C_2}} CS2$$

defines the following component system:

$$(Comp(CS1) \cup Comp(CS2), Conn(CS1) \cup Conn(CS2) \cup \{con\})$$

Definition 8. *If C_1 and C_2 are components in CS, then the operation to add a connection to CS*

$$add\_conn(con : IMP^{C1} \Rightarrow EXP^{C_2}, CS)$$

defines the following component system:

$$(Comp(CS), Conn(CS) \cup \{con\})$$

The next definition is required for the definition of the next operation:

Definition 9. *Given two specifications $SP1, SP2$, $Trans(SP1, SP2)$ holds if and only if there exist a transformation $t : SP1 \Rightarrow SP2$.*

Definition 10. *The non-deterministic operation to make some or all the connections between two component systems*

$$make\_con(CS1, CS2)$$

with arbitrary set CON such that

$$con : IMP^{C1} \Rightarrow EXP^{C2} \in CON \Leftrightarrow C1 \in Comp(CS1) \land$$
$$C2 \in Comp(CS2) \land Trans(IMP^{C1}, EXP^{C2})\})$$

defines the following component system:

$$(Comp(CS1) \cup Comp(CS2), Conn(CS1) \cup Conn(CS2) \cup CON)$$

As we explained in the introduction, we need this non-deterministic operation because we consider our components as autonomous agents which can decide which interconnections are going to make. Moreover, these interconnections can vary over time. The only restriction in making a connection is that an import interface of a component of the first given system and an export interface of a component of the second given system must be compatible. It may happen that the import interface in the first system can have several compatible export interfaces corresponding to different components of the second system. In this case, the first component system can choose which connection to make and at a later stage disable the first connection made and make a different new connection. One may ask if all these connections will always make sense. In principle, they should. However, it will depend on how much detail we have in the interface specifications. For instance, if the interface specifications just describe signatures, without any semantic descriptions of their behaviour, then, probably, the connection will only be reasonable from a syntactic point of view. But if the interfaces provide enough semantic detail then all the connections will make sense. In the example in section 5, we will have a component system CSSRK which will make a connection with another component system CSSR, and after changing the location by CSSRK, CSSRK will loose the connection with CSSR and it will make a new connection with the component system CSCL1.

In order to extend the component system with the ambient calculus, it is necessary to define an interconnected forest of hierarchies of component systems. Thus, an ambient expression will denote an interconnected forest of hierarchies of component systems. Every component system in the hierarchy can have associated a forest of sub-hierarchies. The interconnections among the component systems of the forest can be made autonomously by the component systems. The interconnections among component systems of the forest which we allow are from a component system to its immediate ancestor component system in the hierarchy. For this, we have to use a similar operation to make_con but now for two separate component systems in a hierarchy of component systems. This operation will be called make_con_uh. Additionally, we allow interconnections among component systems of the same hierarchy. For this, we will use the operation make_con_wh.

The constructor operation for hierarchies (besides the empty hierarchy) is the following:

- **make_hier:** Operation which given a component system and a forest returns a hierarchy of interconnected component systems.

The operations which we will use in the definition of the forest associated to an expression of the ambient calculus are the following (here we just provide an intuitive description of these operations, although it would not have been difficult to define them formally):

- **empty_forest:** Operation which returns the empty interconnected forest.
- **add_forest:** Operation which given a component system and an interconnected forest of hierarchies returns a forest with just one hierarchy with root the given component system and the forest of hierarchies as associated sub-hierarchies.
- **union_forest:** Operations which given two interconnected forest of hierarchies returns the union of the two interconnected forests.

Now we describe the operations make_con_uh and make_con_wh which we have mentioned above to make interconnections among the component systems of a forest:

- **make_con_uh:** Operation which given two hierarchies and an interconnected forest such that the second hierarchy is a sub-hierarchy of the first hierarchy in the given interconnected forest, performs non-deterministically a set of connections from the component system of the root of the sub-hierarchy to the component system of the root of the hierarchy similar to the operation make_con.

- **make_con_wh:** Operation which given two hierarchies and an interconnected forest such that the two hierarchies are sub-hierarchies with the same root in the given interconnected forest, performs non-deterministically a set of connections from the component system of the root of the first sub-hierarchy to the component system of the root of the second sub-hierarchy similar to the operation make_con.

These two last operations are used in the following operation which is used to define the interconnected forest associated to an ambient calculus expression:

- **make_interconnections:** Operation which applies the operations make-_con_uh or make_con_wh appropriately to all possible pairs of component systems which can interconnect.

4   Extension of the Component System with the Ambient Calculus

In this section, we extend the component system with operators of the ambient calculus [1]. We have chosen the ambient calculus because it is a well-established calculus to describe mobility of processes in a hierarchical physical space of computing sites represented by ambients. This hierarchical space can change over time, having the possibility to move an ambient inside another ambient, to move an ambient out of another ambient and to open an ambient. All these moves change the hierarchical space of computing sites.

The main idea of the extension is to define an ambient calculus where component systems may be attached to the name of an ambient. Thus, in our case, the hierarchical space of computing sites is a forest of component systems.

The calculus of components with ambients is defined as follows:

$$P ::= \ P \ \| \ Q \,|\, (m, CS) \,|\, (m, CS)[P] \,|\, !P \,|$$
$$M.P \,|\, \nu n.P \,|$$
$$M ::= \ in\,n \,|\, out\,n \,|\, open\,n \,|\, M.M' \,|$$

As in the definition of the ambient calculus n and m range over names, P and Q over processes and M and M' over capabilities, which are actions to make a move. CS ranges over component systems.

As we can observe, the rest of the operators are very similar to those of the ambient calculus. We have changed the syntax of the parallel operator $\|$ where in [1] is $|$. See also [1] for an informal explanation of the rest of the operators, and the formal definition of the function free names which can be extended to our calculus very easily.

As we mentioned in the previous section, an expression of our calculus denotes an interconnected forest of hierarchies of component systems. We define this semantics in two steps. First, a function called *Forest* defines the forest associated to each (sub)expression. The second step permits the definition of connections between components belonging to different subsystems. In particular, we assume that our components are autonomous agents that can establish connections with other components in a given scope. In this sense, a function called *make_interconnections* is assumed to define, nondeterministically, the interconnections that the component systems can perform autonomously as we described in the previous section.

Definition 11. *Given an expression of the ambient calculus (aexpr), the interconnected forest of hierarchies of component systems associated to this expression is defined as follows:*

$$Intconforest(aexpr) = make\_interconnections(Forest(aexpr))$$

where the function Forest is inductively defined as follows:

$$Forest(P \parallel Q) = union\_forest(Forest(P), Forest(Q))$$
$$Forest((n, CS)) = add\_forest((n, CS), empty\_forest())$$
$$Forest((n, CS)[Q]) = add\_forest((n, CS), Forest(Q)))$$
$$Forest(M.P) = Forest(P)$$
$$Forest(\nu n.P) = Forest(P)$$
$$Forest(!P) = union\_forest(P, Forest(!P))$$

In the following we explain some of the cases of this definition:

- The forest associated to an expression of the form $P \parallel Q$ is the union of the forests of P and Q, and the forest associated to $!P$ is the infinite forest built as the union of infinite occurrences of the forest associated to P.
- To build the forest associated to the restriction operator $\nu n.P$ we should assign a fresh name to any occurrence of n in P. We do not do so because we assume that our ambient expression has no name conflicts because it has been previously renamed appropriately. Otherwise the function Forest would be incorrect.
- The forest associated to the expression M.P is the forest of P because this function denotes the forest associated to P before performing the actions M. After performing the actions M we will have a different forest.

The definition of the structural congruence between expressions of the new calculus and its operational semantics can be defined in a very similar way as in the original presentation in [1]. We just present the operational semantics which is used in the example of the next section. As we can see, this semantics coincides essentially with the original semantics of the ambient calculus:

$$(n, CS)[in\,m.P \parallel Q] \parallel (m, CS')[R] \rightarrow (m, CS')[(n, CS)[P \parallel Q] \parallel R]$$
$$(m, CS)[(n, CS')[out\,m.P \parallel Q] \parallel R]] \rightarrow (n, CS')[P \parallel Q] \parallel (m, CS)[R]$$
$$open\,m.P \parallel (m, CS)[Q] \rightarrow P \parallel Q$$
$$P \rightarrow Q \Rightarrow (\nu n)P \rightarrow (\nu n)Q$$
$$P \rightarrow Q \Rightarrow (n, CS)[P] \rightarrow (n, CS)[Q]$$
$$P \rightarrow Q \Rightarrow P \parallel R \rightarrow Q \parallel R$$
$$P' \equiv P, P \rightarrow Q, Q \equiv Q' \Rightarrow P' \rightarrow Q'$$

5   Example

This example describes a server with a firewall together with two clients trying to obtain a software component. This example is based on an example from [1].

The formalism which we use for the specification of components are algebra transformation systems as in [5]. Thus, specifications denote a class of computations where we have states which are represented by Σ-algebras and computation steps are partial functions from Σ-algebras to Σ-algebras.

We assume predefined the component systems CS and CSPROD which denote the component system of the client to be finished and the component system of the server which the client needs to finish an application, respectively. Additionally, we will use the component systems CSCL1, CSCL1K', CSCL1K", CSSR, CSSRK and CSCL2.

In what follows, we will give some details of the definition of the component system of the server CSSR. This component system will consist of two components: one defining the proper server and another one defining a firewall. The server component has two import algebra transformation systems: one from a client and another one from the firewall. The whole component system is defined as follows:

$$(CSSR, < CSR : (BOD^{csr}, < IMP_1^{csr}, i_1^{csr}, IMP_2^{csr}, i_2^{csr} >, < EXP_1^{csr}, e_1^{csr} >),$$
$$CFW : (BOD^{cfw}, <>, < EXP_1^{cfw}, e_1^{cfw} >) >,$$
$$< c_{csr1cfw1} : IMP_1^{csr} \rightarrow EXP_1^{cfw} >)$$

We will denote the part of the component system of the server which is not a name as CSSRG adding a G at the end of the name of the component system. We will proceed in the same way with the other component systems.

The attributes of the body of the server will include the following:

$$CPU, memory, threads, adm\_domains : integer$$
$$sockets : list[queue[cl\_descr]]$$
$$queue\_sentag : queue[agentid\_plus\_descr]$$
$$queue\_recvag : queue[agentid\_plus\_descr]$$
$$queue\_sentprod : queue[agentid\_plus\_descr]$$

And the attributes imported by the firewall component will include the following:

$$client\_list : list[client\_address]$$
$$socket\_fw : list[queue[cl\_descr]]$$
$$queue\_recv\_server : queue[agentid\_plus\_descr]$$
$$queue\_recv\_client : queue[agentid\_plus\_descr]$$
$$queue\_recv\_prod : queue[agentid\_plus\_descr]$$

Comments:

– CPU, memory, threads and adm_domains are the resources which the server has to assign to incoming mobile agents from clients.

- Sockets in Application Program Interfaces of protocol software for client-server applications are implemented in a similar way than operating systems implement I/O operations to transfer data to or from a file. In both cases the concept of descriptor is used. Because of lack of space we will not describe the main operations of sockets.
- The abstract data types of the attributes use standard parameterized specifications of lists and queues. The queues of agentid_plus_descr are used to enqueue agents to be sent or to be received. The lists of queues are lists of sockets which can have different clients associated to the socket. The list of client addresses are the client addresses from where the firewall accepts connection.
- We also use the specification of client descriptor (cl_descr), client address (client_address) and agent identification and descriptor (agentid_plus_descr) for which we do not give details.

The algebra transformation system of the body of the server will include computation steps to make the following algebra transformations:

- to initialize the values of the server.
- to create a new socket for the server.
- to accept a client to a socket of the server.
- to assign resources to an agent of a client.
- to enqueue an agent in the first queue of the server to send the agent to a client accepted in the socket of the server.
- to dequeue an agent in the first queue of the server and enqueue the agent in the first queue of the firewall.

The two last computation steps can be defined as follows:

- If ag is an agent_plus_descr such that the descriptor of ag belongs to the attribute sockets of the server in the given state algebra, then this computation step enqueues the agent in the attribute queue_sentag of the server of the given state algebra.
- If the attribute queue_sentag of the server of the given state algebra is not empty, this computation rule dequeues an agent of the attribute queue_sentag of the server, and enqueues the obtained agent in the attribute queue_recv_client of the firewall.

The component system CSSRK, which is the component system associated to the agent of the server, is defined with the component CSRK with just an import algebra transformation system. The import algebra transformation system will include an operation to assign resources which will be connected to the equivalent operation of the export algebra transformation system of the server component of the server component system.

The component system CSCL1 is the component system associated to the client and it is defined with the component CCL1 with two export algebra transformation systems. One export algebra transformation system includes the operations of the import algebra transformation system of the server component and the other an operation to assign resources to incoming mobile agents.

The component system CSCL1K' is the component system associated to the agent of the client and it is defined with the component CCL1K' with an import and an export algebra transformation system. The import algebra transformation system will also include an operation to assign resources which initially will be connected to the equivalent operation in the export algebra transformation system of the component of the client component system.

The component system CSCL1K" is defined with the component CCL1K" with an import and an export algebra transformation system.

The definition of component system CSCL2 is not necessary to be specified.

The expression of the first client (CLIENT1) is the following:

$$(CSCL1, CSCL1G)[(CSCL1K', CSCL1K'G)[$$
$$open\, CSSRK.(CSCL1K'', CSCL1K''G)[open\, PROD]] \parallel$$
$$open\, CSCL1K''.CS]$$

The expression of the second client (CLIENT2) is the following:

$$(CSCL2, CSCL2G)[(AGENT, (<>, <>))[out\, CSCL2.in\, CSSR.R] \parallel S]$$

The expression of the server (SERVER) is the following:

$$(\nu CSSR)(CSSR, CSSRG)[$$
$$(CSSRK, CSSRKG)[$$
$$out\, CSSR.in\, CSCL1.in\, CSCL1K'.out\, CSCL1.in\, CSCR] \parallel$$
$$open\, CSCL1K'.(PROD, (<>, <>))$$
$$[in\, CSCL1K''.out\, CSSR.in\, CSCL1.CSPROD]]$$

Initially, we have the two clients and the server in parallel:

$$CLIENT1 \parallel CLIENT2 \parallel SERVER$$

The interconnected forest of hierarchies of component systems has the hierarchies of CLIENT1 (with root the component system CSCL1 and two subhierarchies, one with root CSCL1K' and the application CS), CLIENT2 (with root CSCL2 and subhierarchy AGENT) and SERVER (with root CSSR and two subhierarchies, one with root CSSRK and the product PROD). Some of the interconnections of the forest of hierarchies are the following:

- An interconnection from an import algebra transformation system of the component CSR of CSSR to the export algebra transformation system of the component CCL1 in CSCL1.
- An interconnection from the import algebra transformation system of the component CSRK in CSSRK to an export algebra transformation system of the component CSR in CSSR. The import and export algebra transformation systems contain an operation to assign resources to the agent CSSRK.

Client2 cannot access the server. Client1 can access the server and obtain the software component. In the following, we will see how the protocol works.

In the first sequence of the reductions, a server agent with component system CSSRK, enters client with component system CSCL1.

The result expression is the following:

$$(CSCL1, CSCL1G)[(CSSRK, CSSRKG)[$$
$$in\,CSCL1K'.out\,CSCL1.in\,CSSR]] \parallel$$
$$(CSCL1K', CSCL1K'G)[$$
$$open\,CSSRK.(CSCL1K'', CSCLK''G)[open\,PROD]$$
$$\parallel\, open\,CSCL1K''.CS] \parallel (\nu CSSR)(CSSR, CSSRG)[$$
$$open\,CSCL1K'.(PROD, (<>, <>))$$
$$[in\,CSCL1K''.out\,CSSR.in\,CSCL1.CSPROD]] \parallel$$
$$CLIENT2$$

Now we analyze the dynamic reconfiguration of the interconnected forest of hierarchies. The hierarchy of SERVER has not got the server agent with component system CSSRK anymore, and now it is in the CLIENT1 hierarchy. Additionally, the component system of the server agent CSSRK looses his connection with the global component system of the server CSSR, and when entering the client, it establishes a new connection with the global component system of the client CSCL1. This new connection will allow the agent of the server to gain resources in the client, which are shared with the resources which use the agent client with component system CSCL1K'.

In the following sequence of reductions, the agent with component system CSSRK enters the ambient with component system CSCL1K':

$$(CSCL1, CSCL1G)[(CSCL1K', CSCL1K'G)[$$
$$open\,CSSRK.(CSCL1K'', CSCL1K''G)[open\,PROD] \parallel$$
$$(CSSRK, CSSRKG)[out\,CSCL1.in\,CSSR]] \parallel open\,CSCL1K''.CS] \parallel$$
$$(\nu CSSR)(CSSR, CSSRG))[open\,CSCL1K'.(PROD, (<>, <>))$$
$$[in\,CSCL1K''.out\,CSSR.in\,CSCL1.CSPROD]] CLIENT2$$

Next, the agent with component system CSCL1K' of the client with component system CSCL1 enters the server:

$$(CSCL1, CSCL1G)[open\,CSCL1K''.CS] \parallel$$
$$(\nu\,CSSR)(CSSR, CSSRG)[(CSCL1K', CSCL1K'G)[$$
$$(CSCL1K'', CSCL1K''G)[open\,PROD]] \parallel open\,CSCL1K'.(PROD,$$
$$(<>, <>))[in\,CSCL1K''.out\,CSSR.in\,CSCL1.CSPROD]] \parallel CLIENT2$$

Making this move, the component system CSCL1K' of the agent looses the connection with the component system of the client CSCL1, and when entering the server, it establishes a connection with the global component system CSSR of the server. This connection will allow the agent to gain resources in the server.

Finally, the agent with component system CSCL1K' of the client which is now in the server takes the component and moves it to the client. The final expression is the following:

$$(CSCL1, CSCL1G)[CS \parallel CSPROD] \parallel$$
$$(\nu CSSR)(CSSR, CSSRG)[] \parallel CLIENT2$$

6   Conclusions and Future Work

In this paper we have presented an extension of an existing generic component system to be able to develop distributed and mobile applications. The extension is based on the ambient calculus. In particular, this extension is based on attaching component systems to ambients, in such a way that the typical operations on ambients, such as *in*, *out* or *open*, imply the mobility of the associated components. We would have liked to extend our generic approach to components with an (equally) generic approach to mobility. Nevertheless, we think that this is an important step to provide the basis for a general approach for the development of component-based distributed and mobile applications. In particular, we are now able to describe applications which dynamically reconfigurate an interconnected forest of hierarchies of component systems as we have seen in a simple and comprehensive example.

Acknowledgments: We would like to thank the anonymous referees for their comments.

References

1. L. Cardelli and A. D. Gordon. Mobile ambients. In *In Maurice Nivat, editor, Proc. FOSSACS'98, International Coference on Foundations of Software Science and Computation Structures, volume 1378 of Lecture Notes in Computer Science, pages 140–155. Springer-Verlag*, 1998.
2. Hartmut Ehrig, Fernando Orejas, Benjamin Braatz, Markus Klein, and Martti Piirainen. A generic component framework for system modeling. In *FASE 2002 (LNCS 2306)*, 2002.
3. Ole Jensen and Robin Milner. Bigraphs and mobile processes. Technical report, University of Cambridge, UCAM-CL-TR-57.
4. Stefan Mann, Alexander Borusan, Hartmut Ehrig, Martin Grosse-Rhode, Rainer Mackenthun, Asuman Sunbul, and Herbert Weber. Towards a component concept for continuos software engineering. Technical Report Bericht 55/00, Institut Software-und Shystemtechnik, 2000.
5. Fernando Orejas and Hartmut Ehrig. Components for algebra transformation systems. In *Electronic Notes in Theoretical Computer Science 82 N. 7*, 2003.

Application and Formal Specification of Sorted Term-Position Algebras

Arnd Poetzsch-Heffter and Nicole Rauch

University of Kaiserslautern
{poetzsch, rauch}@informatik.uni-kl.de

Abstract. Sorted term-position algebras are an extension of term algebras. In addition to sorted terms with constructor and selector functions, they provide term positions as algebra elements and functions that relate term positions. This paper describes possible applications of term-position algebras and investigates their formal specification in existing specification frameworks. In particular, it presents an algebraic specification of term-positions in CASL and in a higher-order logic.

1 Introduction

Sorted term algebras are a very helpful and flexible concept for modeling and programming. In particular, they provide the foundation for the datatype declarations in functional programming languages and sorted specification languages (see e.g. [1, 2]). *Term-position algebras*, or *tepos-algebras* for short, are an extension of term algebras. Conceptually, a term position is a node within a given sorted tree. While for a constructor term it only makes sense to ask for its subterms, term positions enable to refer to parent positions and, more generally, to the upper tree context of positions. Formally, a term position p in a constructor term t is the occurrence of a subterm s of t in t. We call s the *term belonging to* p and t the *root term* of p. The tepos-algebra for a given sorted term algebra $\mathcal{A}$ and a sort S of $\mathcal{A}$ is an extension of $\mathcal{A}$ by all positions in constructor terms of sort S.

An important aspect for the practical use of term-position algebras is that they need no further declaration constructs and almost no additional declaration work by the user[1]. They are defined based on the usual language constructs for datatype declaration. In this paper, we investigate the design and the formal specification of the semantics of sorted tepos-algebras. The goal is to use existing specification and verification frameworks for the semantics specification so that their tooling and verification support can be exploited. As specification frameworks, we consider CASL [3–5] and Isabelle/HOL [6]. The contribution of the paper has different aspects: It introduces tepos-algebras as a powerful language concept and their formalization as an interesting specification challenge. In the main parts of the paper, we describe how this challenge can be solved in CASL and Isabelle/HOL and compare the two specifications.

---

[1] By a *user*, we mean a person who writes programs or specifications based on term and tepos-algebras.

J.L. Fiadeiro, P. Mosses, and F. Orejas (Eds.): WADT 2004, LNCS 3423, pp. 201–217, 2005.
© Springer-Verlag Berlin Heidelberg 2005

Overview. The rest of the paper is structured as follows. Section 2 provides an informal introduction to the use of tepos-algebras by a small example. Section 3 explains the design choices underlying the specification of tepos-algebras and formulates the specification challenge. Section 4 presents the specification of tepos-algebras in CASL. Section 5 shows how tepos-algebras can be specified in Isabelle/HOL. Section 6 discusses the approach in relation to other work. Section 7 contains the conclusions.

2   Tepos-Algebras at Work

In this section, we show how tepos-algebras can be used in programming and specification. With this introduction, we pursue four goals:

- The reader should get some intuitive understanding of how tepos-algebras can be applied. According to our experiences[2], working with constructor terms and term positions, that is, with two tree representations at once, is unfamiliar at the beginning, but well accepted after having studied some examples.
- We want to give some idea of how tepos-algebras can be integrated into programming or specification languages.
- To motivate the study of tepos-algebras, we like to demonstrate that they enable new specification techniques. In the example below, we show two such aspects from the area of programming language specification: 1. Simplifying the formulation of context conditions. 2. Avoiding continuation semantics for a language with gotos.
- A subset of the example will later be used to illustrate the formal specification of tepos-algebras.

For illustration purposes, we assume a fictitious programming or specification language TePos with a datatype construct for the declaration of free recursive datatypes with constructors and selectors (such datatype declarations are available in most typed functional programming languages and specification languages).

Datatype Declaration. In TePos, the declaration of the abstract syntax of a small imperative programming language with gotos is as follows:

```
datatype SIMPL is
  Prog = prgm( stm: Stmt )
  Stmt = assg( lhs: Idt , rhs: Expr )
       | sequ( fst: Stmt, scd: Stmt )
       | loop( cnd: Expr, bod: Stmt )
       | goto( tid: Idt )
       | labl( lid: Idt,  stm: Stmt )
  Expr = vare( idt: Idt )
       | cons( val: Int )
       | plus( fst: Expr, scd: Expr )
end
```

---

[2] Most of our experiences were made with students in compiler construction courses, in which we used a tool based on tepos-algebras [7, 8].

This declaration uses the sorts `Idt` for identifiers and `Int` for integer constants, it introduces SIMPL[3] as a name for the declaration, and defines new sorts `Prog` with constructor `prgm` as well as `Stmt` and `Expr` with constructors for the different statement and expression kinds. Besides term sorts and constructors, it provides term selectors like `stm`, `lhs`, and `rhs` that allow to select the subterm of a given term. Selectors are partial functions. For example, the evaluation of `fst( assg("a", cons(8)) )` is not defined, because `fst` is a selector that only works for terms constructed by `sequ`. How partiality is handled in TePos is irrelevant for this paper. We allow overloading of selector names if their domain sorts are different. Otherwise overloading is not allowed.

Tepos-Algebra Declaration. TePos supports a declaration that provides the elements and features of a tepos-algebra. The tepos-algebra is defined as an extension of a datatype (here SIMPL) and one of its sorts (here `Prog`). As a third argument, it takes a string (here `"Pos"`) that is used to name position sorts. Here is the declaration for our example:

```
datatype SIMPLPOS is tepos of SIMPL, Prog, "Pos" end
```

This one-line declaration defines the tepos-algebra with a number of sorts and functions. It defines the sorts `ProgPos`, `StmtPos`, `ExprPos`, `IdtPos`, and `IntPos` of positions in terms of sort `Prog`. For example, an element of sort `StmtPos` represents a subterm occurrence of sort `Stmt` in a term of sort `Prog`. The declaration also defines the overloaded functions

```
term: ProgPos -> Prog        pos:  Prog     -> ProgPos
term: StmtPos -> Stmt        root: StmtPos -> ProgPos
term: ExprPos -> Expr        root: ExprPos -> ProgPos
term: IdtPos  -> Idt         root: IdtPos  -> ProgPos
term: IntPos  -> Int         root: IntPos  -> ProgPos
```

The function `term` yields the term belonging to a position (as defined in Sect. 1); `pos` yields the root position of a term of sort `Prog`; and `root` yields the root position for a given position. Thus, `root` is a first example of a function on positions p that refers to the upper tree context of p.

To reach child positions, that is, positions down the tree, the declaration SIMPLPOS defines selectors for positions. To keep the naming simple, we overload the term selectors. For example, the selector `cnd: Stmt -> Expr` is overloaded by a selector `cnd: StmtPos -> ExprPos`. Both selectors are partial functions, and the position selector is defined for a position p if and only if the term selector is defined for the term belonging to p. Altogether, we get two tree representations linked by the functions `pos` and `term`. Figure 1 illustrates this for a simple term.

By distinguishing between datatype constructors and other functions, Figure 1 also indicates a central aspect of how tepos-algebras are specified. Argument flow of datatype constructors is denoted by solid arrows. For the other

---

[3] Simple Imperative Programming Language.

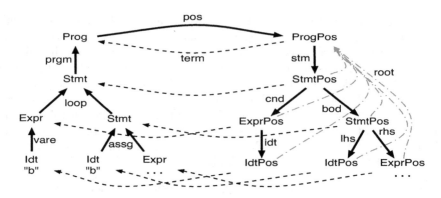

Fig. 1. Illustration of tree representations with terms and positions

functions, we use dashed arrows. Note that the position tree is constructed from the root to the leaves (see Sect. 4).

Tepos-Algebra Extension and Use. A fully-fledged language supporting tepos-algebras would provide further features. In this section, we illustrate and use subsorting on term and position sorts and an extended form of pattern matching. In Section 4, we show how a supersort of all position sorts can be specified and used as a basis for further functions.

Subsorting for free datatypes is naturally defined by the constructors. All terms constructed by a constructor c form one subsort of the range sort of c. We denote the subsorts by the constructor name with a capitalized first letter. For example, `Goto` denotes the goto-statements, that is, the subsort of `Stmt` that contains exactly those terms constructed by `goto`. The corresponding subsorts on positions are denoted by the postfix "Pos" (for example, `GotoPos`). It should be clear that such a subsorting needs no special declarations by the user but can be implicitly provided by the language used.

Based on the declarations `SIMPL` and `SIMPLPOS`, we can define interesting language properties in an elegant declarative way. We start with context conditions. For example, labels must be unique in SIMPL-programs. That is, two different labeled statements $lp1$, $lp2$ in the same program ($\text{root}(lp1) = \text{root}(lp2)$) must have different label identifiers:

$\forall$ `LablPos` $lp1, lp2$:
$\quad lp1 \neq lp2 \,\wedge\, \text{root}(lp1) = \text{root}(lp2) \,\Rightarrow\, \text{term}(\text{lid}(lp1)) \neq \text{term}(\text{lid}(lp2))$

Recall that applying the selector `lid` to a labeled statement position yields an identifier position. To get the identifier at that position, we have to apply the function `term` (cf. Fig. 1). The second context condition states that for each goto statement there must be a corresponding labeled statement:

$\forall$ `GotoPos` gp $\exists$ `LablPos` lp :
$\quad \text{root}(gp) = \text{root}(lp) \,\wedge\, \text{term}(\text{tid}(gp)) = \text{term}(\text{lid}(lp))$

It is worth noting that, without positions, these properties can only be formalized by a nontrivial environment or symboltable mechanism.

For SIMPL-programs satisfying the context conditions, we can define a function target that yields for a goto statement the unique target statement:

```
target:  GotoPos -> LablPos
```

The specification of target is given in the appendix. Here, it is of interest that the user of target need not know about the specification details. The function target links one position of the tree to another one. In particular, we can use it to express an operational semantics for SIMPL without continuations (see [9] for a discussion on continuation semantics). We present such a semantics here as an example to discuss pattern matching on positions. Let State be the sort of mappings from identifiers to integers, eval be a function evaluating an expression in a state, and update be a function that takes a state st, an identifier id, and a value v and yields a "new" state nst such that $nst(i) = st(i)$ for all $i \neq id$ and $nst(id) = v$:

```
State  =  Idt -> Int              update :  State x Idt x Int -> State
eval   :  Exp x State -> Int
```

Based on these notions, the execution of a SIMPL-program p in state st is defined by $\texttt{exec}(\texttt{stm}(\texttt{pos}(p)),st)$ where exec is specified as follows:

```
exec:  StmtPos x State -> State
exec(sp, st) = case sp of
    assg<v,e>      =>  update(st,term(v),eval(term(e),st))
  | sequ<sp1,sp2> =>  exec(sp2, exec(sp1,st))
  | loop<e,bod>    =>  if eval(term(e),st)=0  then st
                          else exec(sp,exec(bod,st))
  | Goto<_>        =>  exec(target(sp),st)
  | Labl<_,sp0>    =>  exec(sp0,st)
```

The case expression is similar to that of functional programming languages. The difference is that matching works on positions. For example, the pattern assg<v,e> matches statement positions of sort AssgPos with child positions v and e. The reason to use a position instead of a term representation of statements is that the execution of goto statements refers to the target statement in the upper context. This can not directly be expressed by constructor terms.

This section should have given some idea of how tepos-algebras can be used in programming and specification. Further examples as well as language and implementation issues are described in [7]. The following sections focus on the challenge of how tepos-algebras can be formally specified.

3   Specification Challenge

On the meta-level, term positions are usually formalized as pairs with the root term as first component and a sequence of natural numbers as second component.

The number sequence describes the selection path from the root position of the term to the subterm position. To illustrate this, let t be the term of Fig. 1:

```
prgm( loop( vare("b"), assg("b",plus(vare("b"),vare("c"))) ) )
```

Using brackets to enclose list elements in the meta-notation, the position at the root of the program is denoted by $(t, [])$, the position of the loop statement by $(t, [1])$, of the assignment by $(t, [1, 2])$ and of the identifier "b" on the left hand side of the assignment by $(t, [1, 2, 1])$. For object-level specifications, this approach has the following four disadvantages: (a) Positions are not sorted; (b) selection by numbers is error-prone; (c) modifications or extensions of the term algebra (e.g. adding a parameter to a constructor) cause subtle modifications of the position handling; (d) the algebraic laws of term positions are hidden. To overcome these disadvantages, tepos-algebras should be formalized within a specification framework in a way that positions are ordinary sorted elements.

The main design problem for tepos-algebras pertains to the sorting/typing discipline for the positions. Essentially, there are four options:

1. All positions of all terms are in one sort.
2. Positions are sorted according to the term sorts they correspond to. That is, there is exactly one position sort for each term sort.
3. In addition to the second option, position sorts are distinguished with respect to the *sort* of the root term. That is, a position sort captures the information about the sort of the root.
4. Position sorts are dependent sorts, depending on the root term.

For the following reasons, we chose the third design option: It is sufficiently fine grained for the applications that we are interested in and that we can imagine so far (see Sect. 2 and [7]). The more coarse grained sorts of the first and second option can be realized within this option by introducing further supersorts. We avoid dependent sorts that are not supported by many specification frameworks. Based on this design decision, the *specification challenge* is as follows:

> Given a sorted free datatype specification with sorts $S_0, ..., S_n$, suitable constructors and selectors, and a sort $S \in \{S_0, ..., S_n\}$, specify the corresponding tepos-algebra with suitable sorts and functions.

Essentially, there exist two approaches to formalize new language concepts or constructs. Either one writes a freestyle mathematical definition, or one uses existing specification languages and frameworks. The first approach provides more flexibility, the second approach allows to inherit the techniques and tools underlying the specification framework. Here, we investigate the second approach. As specification frameworks, we use the algebraic order-sorted specification language CASL and the higher-order many-sorted specification language of Isabelle/HOL. For both frameworks, we specify tepos-algebras by a shallow embedding, that is, we define how a tepos-algebra declaration like that for SIMPLPOS given above is translated into the specification language.

4   Specifying Tepos-Algebras in an Algebraic Framework

The specification of tepos-algebras in CASL is described in two steps. In the first step, we concentrate on the kernel of tepos-algebras containing the position sorts, the selectors on position sorts, and the functions **pos** and **term**. As an introduction, we demonstrate the shallow embedding of the kernel by a representative example (Sect. 4.1). Then, we define it for the general case (Sect. 4.2). In the second step, we explain how extensions of the kernel can be formalized in CASL (Sect. 4.3).

4.1   Introduction to the Tepos-Algebra Specification in CASL

The declaration of a tepos-algebra consists of three parts:

1. a declaration of a free datatype,
2. a declaration of the sort of terms for which the positions should be defined,
3. declarations for the naming of new sorts and functions.

In CASL, the free datatype can be given as a named specification based on some externally declared sorts. As a tiny example, we consider a subset of the abstract syntax of SIMPL (cf. Sect. 2). The extension to SIMPL is straightforward. In CASL syntax, we get the following declaration:

```
spec SIMPLS = sort Idt   then free types
     Prog ::= prgm( stm:? Stmt );
     Stmt ::= assg( lhs:? Idt ; rhs:? Expr )
            | sequ( fst:? Stmt; scd:? Stmt )
            | loop( cnd:? Expr; bod:? Stmt );
     Expr ::= vare( idt:? Idt )
            | plus( fst:? Expr; scd:? Expr )
```

The question mark after the selector names indicates that selectors are partial functions. Note that CASL allows overloading of functions as demonstrated by the selector **fst**. To declare the tepos-algebra for SIMPLS, we could imagine an extension of CASL allowing declarations like:

```
spec SIMPLSPOS = tepos(SIMPLS,Prog,"Pos")
```

The meaning of this declaration is defined by giving a CASL specification for it. The *basic idea* underlying this specification is taken from the meta-level representation of a position as a pair of the root term and a list of natural numbers describing the selection path – recall the example $(t, [1, 2, 1])$ from above. To express the position at the root, we use a constructor **pos**, that is, we write $\mathbf{pos}(t)$ instead of $(t, [])$. The selection of child positions is denoted by unary functions as well. For convenience, we reuse the names of the selectors on the term side for these functions. For example, $(t, [1, 2, 1])$ would be denoted on the object-level as $\mathbf{lhs}(\mathbf{bod}(\mathbf{stm}(\mathbf{pos}(t))))$. This overloading can be handled by CASL if position sorts are different from term sorts and if position sorts corresponding to different term sorts are different as well. Our approach fulfills this requirement; recall our design decision described in the previous section. Following this basic idea leads to two specification problems:

1. How do we specify that different selection paths yield different positions?
2. How do we distinguish "valid" selection paths from "invalid" ones, that is, from paths that do not denote a position in the root term?

The first question has a canonical answer: Use a free type specification in which the selection functions stm, bod, etc. are the constructors of the position sorts. Unfortunately, this leads to a conflict with the second problem, because we get many invalid paths. To overcome this conflict, we can use partial constructors and enforce that they are defined if and only if the path is valid. As a partial function yields "undefined" in a free specification whenever we do not force it explicitly to yield a defined value, we only have to specify in which cases the paths are valid.

A path is valid iff all selection steps are valid. A selection step by selection function sel on a position pp is valid iff the selection by sel is defined on the term belonging to pp. To formalize this, we have to specify a function term that yields for each position the term belonging to it. term can be defined recursively: For the root position of a term p, we have $\text{term}(\text{pos}(p)) = p$. Otherwise, if pp is a position and sel is a selection function for the sort of pp, then $\text{term}(sel(pp)) = sel(\text{term}(pp))$.

The main challenge now is that the specification of the partial constructors and the recursive specification of term are mutually dependent. Thus, in order to implement these ideas in a specification framework, it has to support free specifications of this kind for types with partial constructors and for total recursive functions. CASL meets this challenge. Thus, our specification approach can directly be formulated in CASL. Figure 2 demonstrates this for SIMPLSPOS.

The next subsection provides a complete description of the embedding that we illustrated here by the example.

4.2 Complete Description of the Embedding

In this subsection, we describe how the tepos-algebra for a given datatype declaration is specified in general. Furthermore, we discuss validation issues. Tepos-algebras are declared based on datatype declarations of the following form:

spec DT = sorts $U_1, \ldots, U_p$ then free types

$$S_1 ::= con_1^1 \ (\ sl_{1,1}^1 \quad :? \ T_{1,1}^1 \quad ; \ldots; sl_{1,n(1,1)}^1 \quad :? \ T_{1,n(1,1)}^1 \quad)$$

$$\ldots$$

$$| \ con_{m(1)}^1 \ (\ sl_{m(1),1}^1 :? \ T_{m(1),1}^1 ; \ldots; sl_{m(1),n(1,m(1))}^1 :? \ T_{m(1),n(1,m(1))}^1);$$

$$\ldots$$

$$S_r ::= con_1^r \ (\ sl_{1,1}^r \quad :? \ T_{1,1}^r \quad ; \ldots; sl_{1,n(r,1)}^r \quad :? \ T_{1,n(r,1)}^r \quad)$$

$$\ldots$$

$$| \ con_{m(r)}^r \ (\ sl_{m(r),1}^r :? \ T_{m(r),1}^r ; \ldots; sl_{m(r),n(r,m(r))}^r :? \ T_{m(r),n(r,m(r))}^r)$$

where S_i are different sort names and $T_{j,k}^i$ denote sorts that are either in the defined sorts $\{S_1, \ldots, S_r\}$ or in the used sorts $\{U_1, \ldots, U_p\}$. We assume that the

```
spec SIMPLSPOS = SIMPLS then free {
types
   ProgPos ::= pos( Prog );
   StmtPos ::= stm( ProgPos )?
             | fst( StmtPos )?
             | scd( StmtPos )?
             | bod( StmtPos )?;
   ExprPos ::= rhs( StmtPos )?
             | cnd( StmtPos )?
             | fst( ExprPos )?
             | scd( ExprPos )?;
   IdtPos  ::= lhs( StmtPos )?
             | idt( ExprPos )?;
ops
   term : ProgPos -> Prog;
   term : StmtPos -> Stmt;
   term : ExprPos -> Expr;
   term : IdtPos  -> Idt;

vars p: Prog; pp: ProgPos;
     sp: StmtPos; ep: ExprPos;
. term(pos(p))  = p
. term(stm(pp)) = stm(term(pp))
. term(fst(sp)) = fst(term(sp))
. term(scd(sp)) = scd(term(sp))
. term(bod(sp)) = bod(term(sp))
. term(rhs(sp)) = rhs(term(sp))
. term(cnd(sp)) = cnd(term(sp))
. term(fst(ep)) = fst(term(ep))
. term(scd(ep)) = scd(term(ep))
. term(lhs(sp)) = lhs(term(sp))
. term(idt(ep)) = idt(term(ep))
}
```

Fig. 2. CASL specification for SIMPLSPOS

specification does not use the names pos and term, that all constructor names con_j^i are different, and that selectors are only overloaded if they have different domain sorts, that is, selector names $sl_{j_1,k_1}^{i_1}$ and $sl_{j_2,k_2}^{i_2}$ may only be equal if $i_1 \neq i_2$. To keep the following construction simple, we assume that there is at least one ground term for each used and defined sort. We say that a string π is an *admissible postfix* for a set $\mathcal{T}$ of sort names if no sort name in $\mathcal{T}$ ends with π. For brevity, we will not distinguish between sorts and their names in the following.

The declaration of a tepos-algebra for a datatype declaration DT with defined sorts $\mathcal{S}$ and used sorts $\mathcal{U}$ consists of a sort S in $\mathcal{S}$ and a postfix π admissible for $\mathcal{S} \cup \mathcal{U}$. To formalize the meaning of such a declaration, we need some notions and notations. We say that T is a sort *reachable* from S iff there is a term of sort S with a subterm of sort T. In particular, S is reachable from S (recall that there is a term in S). The set of sorts reachable from S in DT is denoted by $\mathcal{R} = \{R_1, \ldots, R_q\}$. Without loss of generality, we assume that S equals R_1. Furthermore, we need new schematic names:

- The selector names with range type R_i and a domain type in $\mathcal{R}$ are denoted by $slr_1^i, \ldots, slr_{l(i)}^i$. Note that each schematic name slr_j^i denotes the same name as one of the schematic names $sl_{l,m}^k$.
- The index of the domain type of selector slr_j^i is denoted by $dom(i,j)$, that is, the domain type is $R_{dom(i,j)}$.
- R_{ix}^π denotes the sort name obtained from R_{ix} by appending π where ix is a single or double index.

Based on these notations, the tepos-algebra for DT, R_1, and π is defined by the CASL specification shown in Fig. 3.

Validation. As the specification given in Fig. 2 *defines* the meaning of the tepos-algebra for datatype DT, it can only be validated and not verified. Validation

$$
\begin{aligned}
&\text{DT } \textbf{then free } \{ \\
&\quad R_1^\pi ::= \textbf{pos} \quad (R_1 \qquad) \\
&\qquad\quad | \;\; slr_1^1 \;\; (R_{dom(1,1)}^\pi \;\;)? \\
&\qquad\qquad \cdots \\
&\qquad\quad | \;\; slr_{l(1)}^1 (R_{dom(1,l(1))}^\pi)?; \\
&\quad R_2^\pi ::= slr_1^2 \;\; (R_{dom(2,1)}^\pi \;\;)? \\
&\qquad\qquad \cdots \\
&\quad R_q^\pi ::= slr_1^q \;\; (R_{dom(q,1)}^\pi \;\;)? \\
&\qquad\qquad \cdots \\
&\qquad\quad | \;\; slr_{l(q)}^q (R_{dom(q,l(q))}^\pi)?; \\
&\textbf{ops} \\
&\quad \textbf{term} \; : R_1^\pi \rightarrow R_1 \; ; \\
&\qquad\qquad \cdots \\
&\quad \textbf{term} \; : R_q^\pi \rightarrow R_q \; ; \\
&\textbf{vars } \; x : R_1; \; x_1 : R_1^\pi; \ldots; \; x_q : R_q^\pi; \\
&\quad \cdot \; \textbf{term}(\, \textbf{pos}(x) \,) \qquad\qquad\qquad = x \\
&\quad \cdot \; \textbf{term}(\, slr_1^1(x_{dom(1,1)}) \,) \qquad = slr_1^1(\; \textbf{term}(x_{dom(1,1)}) \,) \\
&\qquad\qquad\qquad \cdots \\
&\quad \cdot \; \textbf{term}(\, slr_{l(1)}^1(x_{dom(1,l(1))}) \,) = slr_{l(1)}^1(\, \textbf{term}(x_{dom(1,l(1))}) \,) \\
&\qquad\qquad\qquad \cdots \\
&\quad \cdot \; \textbf{term}(\, slr_1^q(x_{dom(q,1)}) \,) \qquad = slr_1^q(\; \textbf{term}(x_{dom(q,1)}) \,) \\
&\qquad\qquad\qquad \cdots \\
&\quad \cdot \; \textbf{term}(\, slr_{l(q)}^q(x_{dom(q,l(q))}) \,) = slr_{l(q)}^q(\, \textbf{term}(x_{dom(q,l(q))}) \,) \\
&\}
\end{aligned}
$$

Fig. 3. Complete embedding schema for tepos-algebras

here means to check that the specification formalizes our informal understanding and that it has the properties we expect (see [10] for a discussion). An essential property is for example that the extension exists and is unique (up to isomorphism). This holds because we used a free construction based on equational axioms only.

A second important validation property is that the elements in the position sorts represent exactly the positions in the terms of sort R_1. To show this and to illustrate where the CASL semantics comes in, let us assume that $\mathcal{A}$ is a partial algebra satisfying the specification. In the following, we consider all terms to be interpreted in $\mathcal{A}$. We first prove an auxiliary lemma. Then, we come back to the validation property.

Lemma 1. Let t be a term of sort R_1, let $sl_1, \ldots, sl_n$ be some selectors, and let $sl_n(\ldots sl_1(t) \ldots)$ be well-sorted. Then $sl_n(\ldots sl_1(\mathtt{pos}(t)) \ldots)$ is well-sorted and:

1. $sl_n(\ldots sl_1(t) \ldots) = \mathtt{term}(sl_n(\ldots sl_1(\mathtt{pos}(t)) \ldots))$ \qquad (strong equality)

2. $sl_n(\ldots sl_1(t) \ldots)$ is defined $\Leftrightarrow$ $sl_n(\ldots sl_1(\mathtt{pos}(t)) \ldots)$ is defined

Proof of lemma 1: $sl_n(\ldots sl_1(\mathtt{pos}(t)) \ldots)$ is well-sorted according to the construction of the specification. The first property is proved by induction on n. For $n = 0$, we get $\mathtt{term}(\mathtt{pos}(t)) = t$ as a direct consequence of the first axiom. Now, let us assume $sl_n(\ldots sl_1(t) \ldots) = \mathtt{term}(sl_n(\ldots sl_1(\mathtt{pos}(t)) \ldots))$ and let sl be a constructor such that $sl(sl_n(\ldots sl_1(t) \ldots))$ is well-sorted. We derive:

$sl(sl_n(\ldots sl_1(t) \ldots))$
$= (*$ by induction hypothesis $*)$
$sl(\mathtt{term}(sl_n(\ldots sl_1(\mathtt{pos}(t)) \ldots)))$
$= (*$ by the axiom corresponding to sl $*)$
$\mathtt{term}(sl(sl_n(\ldots sl_1(\mathtt{pos}(t)) \ldots)))$

The second property is derived from the first. (1) If $sl_n(\ldots sl_1(t) \ldots)$ is defined, then $\mathtt{term}(sl_n(\ldots sl_1(\mathtt{pos}(t)) \ldots))$ is defined because of the strong equality. Because the interpretation of $\mathtt{term}$ is strict, $sl_n(\ldots sl_1(\mathtt{pos}(t)) \ldots)$ is defined as well. (2) If $sl_n(\ldots sl_1(\mathtt{pos}(t)) \ldots)$ is defined, then $\mathtt{term}(sl_n(\ldots sl_1(\mathtt{pos}(t)) \ldots))$ is defined, because $\mathtt{term}$ is specified as a total function. Because of strong equality, the second property yields that $sl_n(\ldots sl_1(t) \ldots)$ is defined as well. \qquad QED

The second validation property says that the elements of the position sorts represent exactly the valid selection paths for the terms of sort R_1:

Lemma 2. Let MetaPos(R_i) be the set of valid selection paths from a term t of sort R_1 to a subterm of sort R_i and let $R_i^\pi(\mathcal{A})$ denote the carrier set of sort R_i^π in $\mathcal{A}$. Then the following mappings ρ_i, $i \in \{1, \ldots, q\}$, are bijective:

$\rho_i : \mathrm{MetaPos}(R_i) \rightarrow R_i^\pi(\mathcal{A})$
$\rho_i(\ (t, [sl_1, \ldots, sl_n])\ ) =_{def} sl_n(\ldots (sl_1(\mathtt{pos}(t))) \ldots)$

Proof sketch of lemma 2: We have to show that the mappings ρ_i are well-defined, injective, and surjective:

1. Well-defined: The second property of lemma 1 guarantees well-definedness.

2. Injective: It is easy to show that the standard algebra of term positions as introduced informally in Sect. 2 is a model of the specification. In that algebra, different selection paths yield different positions. Since $\mathcal{A}$ is an initial algebra, this property holds for $\mathcal{A}$ as well.

3. Surjective: According to the CASL semantics, the position sorts are generated by the constructors. That is, each element p of a position sort $R_i^\pi(\mathcal{A})$ has a representation of the form $p = sl_k(...(sl_1(\text{pos}(t)))...)$. Consequently, $sl_k(...(sl_1(\text{pos}(t)))...)$ is defined. According to lemma 1, this implies that $sl_k(...sl_1(t)...)$ is defined as well. Thus, we have a preimage for each element of a position sort.   QED

4.3   Extending the Tepos-Algebra Kernel

In Sect. 2, we worked with a tepos-algebra that contained more sorts and functions than the tepos-algebra kernel described above. For example, we used a function `root` and subsorts `GotoPos` and `LablPos`. Such extensions can easily be declared on top of the kernel. In CASL, their specification is straightforward. We show here only how the function `root` and some subsorts can be specified. Other examples would be a supersort for all positions and functions operating on such supersorts (for instance, a function `parent` that yields for each position the parent position). Which of these extensions are included in tepos-algebras is mainly a language design issue and beyond the scope of this paper.

We illustrate the specification of additional functions and subsorts based on the example specification SIMPLS. The function `root` can be recursively defined:

```
vars t : Prog; pp : ProgPos; sp : StmtPos; ep : ExprPos;
·  ¬def root(pos(t))
·  root(stm(pp)) = pp when pos(term(pp)) = pp else root(pp)
·  root(fst(sp)) = root(sp)
·  root(scd(sp)) = root(sp)        · root(fst(ep)) = root(ep)
·  root(bod(sp)) = root(sp)        · root(scd(ep)) = root(ep)
·  root(rhs(sp)) = root(sp)        · root(lhs(sp)) = root(sp)
·  root(cnd(sp)) = root(sp)        · root(idt(ep)) = root(ep)
```

In a handwritten specification, the case for constructor `stm` can be simplified into `root(stm(pp))=pp`, because in the abstract syntax of SIMPL a term of sort `Prog` never occurs as a subterm. However, in general, terms of the root sort can occur as subterms. Thus, a case distinction can be necessary. Finally, we show how subsorts of sorts with multiple constructors can be specified. CASL allows to introduce new subsorts in a convenient way by set comprehension:

```
sort Assg   = { t : Stmt. ∃ id : Idt, e : Expr. t = assg(id, e) }
sort AssgPos = { p : StmtPos. term(p) ∈ Assg }
...
```

All such specifications extending the tepos-algebra kernel can be generated automatically without needing any further declaration support from the user.

5   Specifying Tepos-Algebras in Higher-Order Logic

When we first looked at different specification frameworks to specify the semantics of tepos-algebras, CASL seemed not only appealing because of its support for partial functions and constructors. Appealing was as well the tool HOL-CASL [11] that allows to generate a higher-order theory from a CASL specification. Unfortunately, the current version of HOL-CASL does not support partial constructors in free specifications. Furthermore, HOL-CASL uses a special bottom element to encode partiality into HOL which only supports total functions. In our experiments, it turned out that for our verification goals it is more suitable and elegant to use a different encoding. That is why we developed our own embedding into the Isabelle/HOL framework.

The basic idea of our embedding is as follows. Partiality of a function f is handled by a definedness predicate def_f that yields true for all values on which f is defined. For values x with $\neg\ def\_f(x)$, we specify that $f(x) = arbitrary$ where $arbitrary$ is some arbitrary element of the range of f. (Isabelle/HOL guarantees that sorts are nonempty and uses the Hilbert operator to formalize $arbitrary$.)

A typical application of this technique is the specification of the selectors for datatypes. The standard datatype construct of Isabelle/HOL does not support selectors. Thus, they have to be specified separately. As in Sect. 4, we use the specification SIMPLS to demonstrate the embedding. For example, the selectors stm and lhs are specified as follows:

$$
\begin{aligned}
stm\ (x::Prog) &\equiv case\ x\ of\ prgm\ y \Rightarrow y \\
def\_stm\ x &\equiv case\ x\ of\ prgm\ y \Rightarrow True
\end{aligned}
$$

$$
\begin{aligned}
lhs\ (x::Stmt) &\equiv case\ x\ of\ assg\ (y,\ z) \Rightarrow y \\
&\quad |\ sequ\ (y,\ z) \Rightarrow arbitrary \\
&\quad |\ loop\ (y,\ z) \Rightarrow arbitrary \\
def\_lhs\ x &\equiv case\ x\ of\ assg\ (y,\ z) \Rightarrow True \\
&\quad |\ sequ\ (y,\ z) \Rightarrow False \\
&\quad |\ loop\ (y,\ z) \Rightarrow False
\end{aligned}
$$

Starting from the datatype and selector specification, we specify the position sorts. As Isabelle/HOL does not support partial constructors, we have to do this in several steps:

1. In the first step, we freely-generate sorts that contain more elements than we have positions. We call the sorts $ProgPosU$, $StmtPosU$, $ExprPosU$, and $IdtPosU$ where "U" stands for unrestricted.
2. Then, we define functions corresponding to $term$ on these sorts.
3. Using these functions, we define subsets of the unrestricted position sorts.
4. By lifting the subsets, we define the new sorts $ProgPos$, $StmtPos$, $ExprPos$, and $IdtPos$.

The datatype specification for the unrestricted sorts looks as follows – we append the character "U" to the end of the constructor names because Isabelle/HOL does not allow to overload the names of the selector functions by constructor names (for brevity, we leave out some productions):

datatype *ProgPosU* = *posU Prog*
and *StmtPosU* = *stmU ProgPosU*
     | *fstU StmtPosU*
     | *scdU StmtPosU*
     | *bodU StmtPosU*
and *ExprPosU* = *rhsU StmtPosU*
     | ...

The *term*-function is defined via primitive recursion. Since Isabelle/HOL does not allow overloading of primitive recursive functions, we specify one *term*-function for each sort. For brevity, we only show parts of the specifications and simplify the original Isabelle/HOL source a bit:

primrec
termP (*posU p*) = *p*
termS (*stmU p*) = *stm*(*termP p*)
termS (*fstU p*) = *fst*(*termS p*)

...

Using the *term*-functions, we inductively define the sets of all valid positions. Starting from a valid position, if the application of a selector on the term side is defined, then the application on the position side yields another valid position. These sets are denoted with a postfix "S" (for "set"). Again, we display only a small part of the specification.

ProgPosS :: *ProgPosU set*
inductive *ProgPosS*
(*posU x*) ∈ *ProgPosS*

StmtPosS :: *StmtPosU set*
inductive *StmtPosS*
(*x*::*ProgPosU*) ∈ *ProgPosS* ∧ (*def_stm* (*termP x*)) $\implies$ (*stmU x*) ∈ *StmtPosS*
(*x*::*StmtPosU*) ∈ *StmtPosS* ∧ (*def_fst* (*termS x*)) $\implies$ (*fstU x*) ∈ *StmtPosS*

Isabelle enables to specify types/sorts[4] for such sets provided the sets can be proven to be non-empty. This can always be achieved by specifying a witness, that is an element of the set. Based on this, we can declare the sorts *ProgPos*, *StmtPos*, etc.:

typedef *ProgPos* = *ProgPosS*
typedef *StmtPos* = *StmtPosS*

For these types, Isabelle/HOL automatically provides us with representation and abstraction functions. For example, the representation function *Rep_ProgPos* takes an argument of type *ProgPos* and yields the corresponding element of

---

[4] In Isabelle, sorts are called types.

the underlying set *ProgPosS*, that is, it has range type *ProgPosU*. The abstraction function *Abs_ProgPos* has domain type *ProgPosU*. It maps elements from *ProgPosS* to their abstraction in *ProgPos*. Elements not contained in *ProgPosS* are mapped to *arbitrary*. (Additionally, Isabelle provides a number of lemmas regarding injectivity, inversion, and so on.)

Finally, we have to lift the functions on positions to the new restricted sorts. We demonstrate this here for *fstU*. The corresponding partial function from *StmtPos* to *StmtPos* is denoted by *fstP* ("P" for "partial"). *fstP* and the corresponding definedness predicate *def_fstP* are defined as follows:

$$fstP \ y \equiv Abs\_StmtPos \ (fstU \ (Rep\_StmtPos \ y))$$
$$def\_fstP \ y \equiv (fstU \ (Rep\_StmtPos \ y)) \in StmtPosS$$

Discussion. It is interesting to compare the CASL and the Isabelle/HOL specifications. The CASL specification is much shorter and, what is more important, the underlying idea of the specification technique is directly visible. This is possible because CASL supports partial constructors in a free specification where the partiality depends on an inductively specified total function. Of course, this elegance comes at the price that consistency checking of the specification is more complex. Whereas the Isabelle/HOL theory provides by construction a conservative extension of the datatype specifying the term algebra, extensions in CASL can lead to inconsistent specifications. Considering our work as a specification case study, we learned two lessons:

1. It is helpful to start with a loose specification. We first tried to develop the formalization of tepos-algebras directly in HOL – and almost gave up. Then, we learned about CASL and that its nice, well-integrated features allow for a very concise specification. This was the step when we identified the kernel of tepos-algebras. Finally, we could construct a HOL specification, focussing on design issues simplifying verification.
2. Formalizing partial functions by adding a bottom element to the range and domain types (e.g. by using the type constructor *option*) is not always a good choice. Using the Hilbert-operator and a definedness predicate can lead to more practical specifications, that is, to specifications that simplify the formal verification using interactive tactical provers.

6   Related Work

To our knowledge, this is the first work on formal specification of sorted term positions at the object level. We developed tepos-algebras as a foundation for language specification and implementation tools. Having a rich tree representation enables to use language specification techniques that do not work for free constructor terms. That is why most language specifications with abstract state machines are based on such rich tree represenations (as one example, see [12]).

Depending on the application area, other tree representations and formalization techniques are used. Higher-order abstract syntax (see [13]) is particularly well suited for matching, substitution in terms, and unification. It allows to abstract over parameterized subterm positions. This is very helpful to express name bindings and consistent renamings. So far, we have not looked at how practical it is to specify substitution in a tepos-algebra framework.

Special logics become more and more popular to describe certain kinds and properties of trees or to discover the shape of trees. The logic underlying Mona [14] can for example be used to describe pointer structures as part of a decidable program logic. Similarly, shape analysis uses a logic as a basis for automated analyses for programs with pointers (see e.g. [15]).

7   Conclusions

We demonstrated how tepos-algebras can be used and formally specified. As application area, we looked at language specifications and showed how a continuation semantics can be avoided if the abstract syntax trees of the language are represented by a tepos-algebra. Similarly, complex environments can be avoided by using the position of the declaration to access the declaration information of a program elements.

The main part of the paper explained shallow embeddings of tepos-algebras into CASL and into Isabelle/HOL. Our conclusion is that such frameworks should be used in combination. The powerful CASL language allows to exploit and compare different specification techniques, which is very helpful in the design phase of the specification. On the other hand, Isabelle/HOL provides more automated checks for the specification.[5] Furthermore, it enables to refine the specification towards effective verification applications. Future work in that direction is the development of proof principles and proof strategies for tepos-algebras.

References

1. Milner, R., Tofte, M., Harper, R.: The Definition of Standard ML. MIT Press (1990)
2. Guttag, J.V., Horning, J.J.: Larch: Languages and Tools for Formal Specification. Springer-Verlag (1993)
3. Astesiano, E., Bidoit, M., Kirchner, H., Krieg-Brückner, B., Mosses, P.D., Sannella, D., Tarlecki, A.: CASL: The Common Algebraic Specification Language. Theoretical Computer Science **286** (2002) 153–196
4. Bidoit, M., Mosses, P.D.: CASL User Manual. LNCS 2900. Springer-Verlag (2004)
5. CoFI (The Common Framework Initiative): CASL Reference Manual. LNCS 2960. Springer-Verlag (2004)
6. Nipkow, T., Paulson, L.C., Wenzel, M.: Isabelle/HOL — A Proof Assistant for Higher-Order Logic. LNCS 2283. Springer-Verlag (2002)
7. Poetzsch-Heffter, A.: Prototyping realistic programming languages based on formal specifications. Acta Informatica **34** (1997) 737–772

---

[5] There are as well tools for consistency checking of CASL specifications (see www.informatik.uni-bremen.de/cofi/ccc/). However, we had problems to apply them to specifications with partial constructors.

8. Bauer, B., Höllerer, R.: Übersetzung objektorientierter Programmiersprachen. Springer-Verlag (1998)
9. Slonneger, K.: Executing continuation semantics: A comparison. Software – Practice and Experience **23** (1993) 1379–1397
10. Roggenbach, M., Schröder, L.: Towards trustworthy specifications I: Consistency checks. In Cerioli, M., Reggio, G., eds.: Recent Trends in Algebraic Specification Techniques, 15th International Workshop, WADT 2001. LNCS 2267, Springer-Verlag (2001)
11. Mossakowski, T.: Introduction into HOL-CASL (Version 0.82). Technical report, University of Bremen (2002)
12. Börger, E., Schulte, W.: Programmer Friendly Modular Definition of the Semantics of Java. In Alves-Foss, J., ed.: Formal Syntax and Semantics of Java. LNCS 1523. Springer-Verlag (1998)
13. Pfenning, F., Elliott, C.: Higher-order abstract syntax. In Wise, D.S., ed.: SIGPLAN '88 Conference on Progamming Language Design and Implementation. SIGPLAN Notices 23(7), ACM Press (1988) 199–208
14. Klarlund, N.: Mona & Fido: The logic-automaton connection in practice. In: Computer Science Logic, CSL '97. LNCS 1414, Springer-Verlag (1998) 311–326
15. Sagiv, M., Reps, T., Wilhelm, R.: Solving shape-analysis problems in languages with destructive updating. In: Proceedings of the 23rd ACM SIGPLAN-SIGACT symposium on Principles of programming languages, ACM Press (1996) 16–31

A   Declaration of Function target

The declaration of function **target** shows how functions can be used to represent references from one tree node to another in a declarative way:

```
datatype LablPosNil = lbpos( lbp: LablPos )   | nil  end

labl_lkup: Idt x StmtPos -> LablPosNil
  labl_lkup(id,sm) = case sm of
    sequ<sm1,sm2> => if labl_lkup(id,sm1)!=nil then  labl_lkup(id,sm1)
                                               else  labl_lkup(id,sm2)
  | loop<e,body>  => labl_lkup(id,body)
  | labl<lip,sm0> => if  id = term(lip)  then  sm
                                         else  labl_lkup(id,sm0)
  | _             => nil

target: GotoPos -> LablPos
  target(gp) = lbp(labl_lkup(term(tid(gp)), stm(root(gp))))
```

From Conditional to Unconditional Rewriting

Grigore Roşu

Department of Computer Science,
University of Illinois at Urbana-Champaign
grosu@cs.uiuc.edu

Abstract. An automated technique to translate conditional rewrite rules into unconditional ones is presented, which is suitable to implement, or compile, conditional rewriting on top of much simpler and easier to optimize unconditional rewrite systems. An experiment performed on world's fastest conditional rewriting engines shows that speedups for conditional rewriting of an order of magnitude can already be obtained by applying the presented technique as a front-end transformation.

1 Introduction

Conditional term rewriting is a crucial paradigm in the algebraic specification of abstract data types, since it provides a natural means for executing equational specifications. Many specification languages today, including Maude [4], ELAN [3], OBJ [9], CafeOBJ [6], provide conditional rewriting engines to allow users to execute and reason about specifications. Conditional rewriting also plays a foundational role in functional logic programming [10]. Additionally, there are many researchers, including the author, considering rewriting a powerful programming paradigm by itself, who are often frustrated that conditional rewrite "programs" are significantly slower than unconditional ones doing the same thing.

Conditional rewriting is, however, rather inconvenient to implement directly. To reduce a term, a rewriting engine needs to maintain a *control context* for each conditional rule that is tried. Due to the potential nesting of rewrite rule applications, such a control context may grow arbitrarily. Our technique automatically translates conditional rewrite rules into unconditional rules, by *encoding the necessary control context into data context*. The obtained rules can be then executed on (almost) any unconditional rewriting engine, whose single task is to *match-and-apply* unconditional rules. Such a simplified engine can be seen as a *rewrite virtual machine*, which can be even implemented in hardware for increased efficiency, and our transformation technique can be seen as a compiler.

Experiments performed on two fast rewriting engines show that speedups of an order of magnitude can be obtained right now if one uses our transformation technique as a front-end. However, since these rewrite engines are optimized for conditional rewriting, we expect significant further increases in efficiency if one just focus on the much simpler problem of developing optimized unconditional rewrite engines and use our technique as a front-end. Even though presented as a

J.L. Fiadeiro, P. Mosses, and F. Orejas (Eds.): WADT 2004, LNCS 3423, pp. 218–233, 2005.

translation of conditional rewrite systems into unconditional ones, our technique can easily be adapted and used as a means to implement conditional rewriting also without applying an explicit transformation. We will discuss this elsewhere.

The proofs of Proposition 1 and Theorem 1 can be found in [16].

Related Work. Stimulated by the benefits of transforming conditional term rewrite systems (CTRSs) into equivalent unconditional term rewrite systems (TRSs), there has been much research on this topic. Despite the apparent simplicity of most transformations, they typically work for restricted CTRSs and their correctness, when they are correct, is quite technical and tricky to prove. A large body of literature has been dedicated to transformations preserving only certain properties of CTRSs, e.g., termination and/or confluence. We do not discuss these here; the interested reader is referred, e.g., to Ohlebusch [14].

In this paper we focus on transformations that generate TRSs *computationally equivalent* to CTRSs, i.e., the TRSs can be *transparently* used to reduce terms in the original CTRSs. The first attempt in this category is due to Bergstra and Klop [2], for a restricted class of CTRSs (whose underlying unconditional TRS is left-linear and without superposition); unfortunately, this transformation was shown to be unsound by Dershowitz and Okada [5]. The transformation in Giovannetti and Moiso [8] works only under severe restrictions on the original CTRS: no superposition, simply terminating (enforced by the requirement of a simplification ordering), and non-overlapping of conditions with left-hand-side (lhs) terms. Hintermeier [11] proposes a technique where an "interpreter" for CTRS is defined as a TRS, providing explicit rewrite definitions for matching and applications of rewrite rules. Besides being technically very intricate and practically inefficient, this transformation is proven to be correct only when the original CTRS is confluent and strictly terminating (i.e., decreasing). Our work in this paper was motivated by efforts in *rewriting logic semantics* [12], where rewriting logic is used as a core mechanism to give operational semantics to concurrent programming languages. In this framework, as well as in many others, restrictions such as termination and/or confluence are unacceptably strong. Indeed, in any programming language there are programs which do not terminate, and concurrency leads quickly to non-confluence (e.g., data-races).

Our technique was presented at WADT'04 and was developed independently from that of Viry [17]. However, the two techniques have many similarities[1]. They are both based on decorations of terms, obtained by adding as many auxiliary arguments to each operation f as conditional rules in the original CTRS having f at the top of their lhs. The procedure in [17] encodes the condition of each rule within a special data-structure that occurs as the corresponding auxiliary argument associated to the operation occurring at the top of its lhs. Two unconditional rules are added in the generated TRS for each conditional rule in the original CTRS, one for initializing the special data-structure and the other for continuing the rewriting process when the condition was evaluated. For example, the CTRS (taken from [17]) $\mathcal{R}$ below is transformed into $\mathcal{R}'$:

---

[1] We thank Bernhard Gramlich for making us aware of Viry [17].

(R)
$$\begin{cases} f(g(x)) \to p(x) \text{ if } c(x) \to^* true \\ f(h(x)) \to q(x) \text{ if } d(x) \to^* true \\ c(a) \to true \end{cases}$$

(R')
$$\begin{cases} f(g(x) \mid \bot, z) \to f(g(x) \mid [c(x), (x)], z) \\ f(x \mid [true, (y)], z) \to p(y) \\ f(h(x) \mid z, \bot) \to f(h(x) \mid z, [d(x), (x)]) \\ f(x \mid z, [true, (y)]) \to q(y) \\ c(a) \to true \end{cases}$$

where "$|$" is syntactic sugar for "$,$", separating the normal arguments from the auxiliary ones; "$\bot$" is a special constant whose occurrence states that the corresponding conditional rule has not been tried yet on the current position; a structure $[u, \vec{s}]$ occurring during a rewriting sequence as an auxiliary argument of an operation, means that u is the current reduction status of the corresponding condition that started to be evaluated at some point, and that $\vec{s}$ was the substitution at that point that allowed the lhs of that rule to match. The substitution is needed by the second unconditional rule associated to a conditional rule, to correctly initiate the reduction of the rhs of the original conditional rule.

Despite being proved sound and complete by Viry [17], the procedure above, unfortunately, cannot be used *as is* to interpret any CTRS on top of a TRS. That is because it destroys the confluence of the original CTRS, thus leading to normal forms in the TRS which can be further reduced in the CTRS. Indeed, let us consider the following CTRS $\mathcal{R}$, from Antoy, Brassel and Hanus [1], together with Viry's transformation $\mathcal{R}'$:

(R)
$$\begin{cases} f(g(x)) \to x \text{ if } x \to^* 0 \\ g(g(x)) \to g(x) \end{cases}$$

(R')
$$\begin{cases} f(g(x) \mid \bot) \to f(g(x) \mid [x, (x)]) \\ f(x \mid [0, (y)]) \to y \\ g(g(x)) \to g(x) \end{cases}$$

$\mathcal{R}$ is confluent but $\mathcal{R}'$ is not: $f(g(g(0)) \mid \bot)$ can be reduced to both 0 and $f(g(0) \mid [g(0), (g(0))])$; the latter occurs because the "conditional" rule is first tried and "failed", then the "unconditional" one is applied successfully thus changing the context so that the "conditional" rule becomes conceptually applicable, but it fails to apply since it was already marked as "tried". To solve this problem, Viry [17] proposes a reduction strategy within the generated TRS, called *conditional eagerness*, stating that $t_1, ..., t_n$ must be already in normal form before a "conditional" rule can be applied on a term $f(t_1, ..., t_n \mid \bot, ..., \bot)$. This way, in the example above, $g(g(0))$ is enforced to be first evaluated to $g(0)$ and only then $f(g(0) \mid \bot)$ is applied the "conditional" rule and eventually reduced to 0. However, conditional eagerness does not seem to be trivial to enforce in an unconditional rewriting engine, unless that is internally modified. One simple, but very restrictive, way to ensure conditional eagerness is to enforce innermost rewriting both in the original CTRS and in the resulting TRS.

A different fix to Viry's technique was proposed by Antoy, Brassel and Hanus [1], namely to restrict the input CTRSs to *constructor-based* ones, i.e., ones in which the operations are split into *constructors* and *defined*, and the lhs of each rule is a term of the form $f(t_1, ..., t_n)$, where f is defined and $t_1, ..., t_n$ are all constructor terms. The problematic CTRS above is *not* constructor-based, so Viry's procedure is not guaranteed to work correctly on it. While constructor-baseness is an easy to check and automatic correctness criterion, we believe that it is an unnecessary strong restriction on the input CTRS, which may make the translation useless in many situations of practical interest.

An additional drawback of Viry's transformation is that it increases the number of rewrite rules having the same operator at the root of their lhs, which tends to be a source of matching overhead on many rewrite engines, especially in the context of very large CTRSs[2]. Therefore, we are still left with no satisfactory translation of CTRSs into equivalent TRSs. In this paper we give a practical solution to this problem, which imposes *no restrictions* on the original CTRS, which adds exactly one unconditional rule for each conditional rule in the original CTRS, and which is shown to bring a significant speedup on current conditional rewrite engines if applied as a front-end transformation. Our translation is *almost* ideal, in that it still requires some special support from the underling unconditional rewrite engine: to provide (1) a binary equality operation, denoted $equal?(t, t')$ in this paper, returning *true* iff the normal forms of t and t' are identical, and (2) a conditional $if(b, t, t')$ which is eager in b and lazy in t and t'. However, all rewriting engines that we know provide them [9, 3, 4, 6, 18]. They can also be easily defined if the rewriting engine provides support for simple contextual strategies, which all rewriting engines that we know do.

2 Preliminaries

We recall some basic notions of conditional rewriting, referring the interested reader to [14] for more details. An (unsorted) *signature* Σ is a finite set of operational symbols, each having zero or more arguments. We let $\Sigma_n \subseteq \Sigma$ denote the set of operations of n arguments. The operations of zero arguments in Σ_0 are called *constants*. We assume an infinite set of *variables* $\mathcal{X}$. Given a signature Σ and a set of variables $X \subseteq \mathcal{X}$, we let $T_\Sigma(X)$ denote the algebra of Σ-*terms* over variables in X. A term without variables is called *ground*. A map $\theta : \mathcal{X} \to T_\Sigma(\mathcal{X})$ can be uniquely extended to a morphism of algebras $T_\Sigma(\mathcal{X}) \to T_\Sigma(\mathcal{X})$ replacing each variable in x by a term $\theta(x)$; to keep the notation simple, we let θ also denote this map. A conditional Σ-rewrite rule has the form

$$l \to r \text{ if } u_1 = v_1, \cdots, u_m = v_m,$$

where l, r, u_1, v_1, ..., u_m, v_m are Σ-terms in $T_\Sigma(\mathcal{X})$. The term l is called the *left-hand-side (lhs)*, r is called the *right-hand-side (rhs)*, and $u_1 = v_1, \cdots, u_m = v_m$ is called the *condition* of the rewriting rule above. As usual, we disallow rewriting rules whose lhs is a variable. Further, we assume that the lhs of a rewriting rule contains all the variables that occur in that rule, that is, following the terminology in [13] our rewrite systems are *of type 1*. If $m = 0$, the rewrite rule is called *unconditional* and written $l \to r$. Unless specified differently, by conditional rule we mean a rule with $m \geq 1$. A *conditional (unconditional) Σ-term rewrite system* $\mathcal{R} = (\Sigma, R)$, abbreviated *CTRS (TRS)*, consists of a

---

[2] We have encountered CTRSs of thousands of rules in the context of rewriting logic semantics of programming languages [12].

finite set R of conditional (unconditional) Σ-rewrite rules. Any Σ-rewrite system $\mathcal{R} = (\Sigma, R)$ generates a relation $\rightarrow_{\mathcal{R}}$ on $T_{\Sigma}(\mathcal{X})$, defined recursively as follows. For any $\theta : \mathcal{X} \rightarrow T_{\Sigma}(\mathcal{X})$, $t[\theta(l)] \rightarrow_{\mathcal{R}} t[\theta(r)]$ whenever there exists some s_i such that $\theta(u_i) \rightarrow_{\mathcal{R}}^{\star} s_i$ and $\theta(v_i) \rightarrow_{\mathcal{R}}^{\star} s_i$ for any $1 \leq i \leq m$, where t is a term having one occurrence of a special variable, say $*$, $t[\theta(l)]$ is the term obtained by substituting $*$ with $\theta(l)$ in t, and $\rightarrow_{\mathcal{R}}^{\star}$ is the reflexive and transitive closure of $\rightarrow_{\mathcal{R}}$. Hence, $\alpha \rightarrow_{\mathcal{R}} \beta$ iff α has a subterm matching the lhs of a rule in R via some substitution, s.t. all the terms in each equality in the condition can be iteratively reduced to a common term. Such CTRSs are also called *join* or *standard* [14]. Alternative interpretations of equalities are also possible, and we will discuss transformations of those elsewhere soon. However, as their name suggests, standard conditional rewrite systems are the most common ones and major rewriting engines, e.g., Maude [4] and ELAN [3], support them. These systems perform millions of rewrites per second on standard PCs and are, at our knowledge, the fastest rewriting engines.

Terms which cannot be reduced any further in $\mathcal{R}$ are called *normal forms* for $\mathcal{R}$. Rewriting of a given term may not terminate for two reasons: either the reduction of the condition of a rule does not terminate, or there are some rules that can be applied infinitely often on the given term. On systems like Maude or ELAN, the effect in both situations is the same: the system loops forever unless it crashes running out of memory. Because of this reason, we do not make any distinction between the two causes, and simply call a Σ-rewriting system *terminating* iff it always reduces any Σ-term to a normal form (we let this notion at an intuitive level here, but it can be formalized). Letting $\_; \_$ denote the composition of relations, a relation $\rightarrow$ is *confluent* iff $\leftarrow^{\star}; \rightarrow^{\star} \subseteq \rightarrow^{\star}; \leftarrow^{\star}$.

3   Defining the Basic Infrastructure

We define several operators together with appropriate (unconditional) rules. Most rewriting engines have these basic operators built-in, but here we do not assume any existing operators and therefore define everything needed.

Let *true* and *false* be two constants which are assumed not defined within any given CTRS (otherwise change their name). Let us also assume a fresh binary operator $\wedge$, written in infix associative notation, together with the rules:

$$true \wedge true \rightarrow true, \qquad true \wedge false \rightarrow false,$$
$$false \wedge true \rightarrow false, \qquad false \wedge false \rightarrow false.$$

These will be needed to evaluate conditions that will be translated into corresponding conjunctions of equalities; equalities will be defined shortly.

Let us now consider a special operator $if(\_,\_,\_)$, together with the rules:

$$if(true, x, y) \rightarrow x, \qquad if(false, x, y) \rightarrow y.$$

This operator is assumed *eager in its first argument and lazy in the others*. Most rewrite engines provide it as builtin, so the two rules above are not needed.

We need another special operator, $equal?(\_,\_)$, that reduces its arguments and returns *true* if they are identical and *false* otherwise. One obvious rule to add is $equal?(x,x) \to true$. Moreover, for all $\sigma \in \Sigma_n$ we add

$$equal?(\sigma(x_1,\ldots,x_n),\sigma(y_1,\ldots,y_n)) \to equal?(x_1,y_1) \wedge \cdots \wedge equal?(x_n,y_n), \quad (1)$$

where $x_1,\ldots,x_n,y_1,\ldots,y_n$ are disjoint variables. These rules propagate the equality of two terms having the same operator as root to the equality of their corresponding sub-terms. Note that σ may be a constant in Σ_0, in which case, by convention, $equal?(x_1,y_1) \wedge \cdots \wedge equal?(x_n,y_n)$ is *true*, the unit of $\wedge$. The following rules, one for each pair $\sigma \in \Sigma_n$, $\tau \in \Sigma_m$ of different operations in Σ, state that terms having different operations at root are not equal:

$$equal?(\sigma(x_1,\ldots,x_n),\tau(y_1,\ldots,y_m)) \to false.$$

Note that $equal?$ needs to be *eager in both its arguments*. All rewrite engines we know have such an operator builtin, so these rules are not needed in practice.

For a given signature Σ, let Σ' denote the signature Σ extended with all the auxiliary operations above, and let $\mathcal{I}(\Sigma)$ be the Σ'-rewriting system containing all the rules above. We call $\mathcal{I}(\Sigma)$ the *infrastructure rewriting system of* Σ.

Proposition 1. *Let $\mathcal{R}$ be a Σ-rewrite system, conditional or not. Then*

1. *$\mathcal{I}(\Sigma)$ is a confluent and terminating unconditional Σ'-rewrite system;*
2. *If $u,v \in T_\Sigma(X)$ then $u \; (\to_\mathcal{R}^\star; \leftarrow_\mathcal{R}^\star) \; v$ iff $equal?(u,v) \to_{\mathcal{R} \cup \mathcal{I}(\Sigma)}^\star true$;*
3. *If u,v are ground Σ-terms then a normal form of $equal?(u,v)$ in $\mathcal{R} \cup \mathcal{I}(\Sigma)$ is either true or false;*
4. *$\mathcal{R}$ terminates if and only if $\mathcal{R} \cup \mathcal{I}(\Sigma)$ terminates;*
5. *If $\mathcal{R}$ is confluent and terminates, i.e., it has unique normal forms, then $\mathcal{R} \cup \mathcal{I}(\Sigma)$ is also confluent and terminates.*

By 2., one can replace any equality $u = v$ in the condition of a rule in $\mathcal{R}$ by $equal?(u,v) = true$. Note that the restriction on u and v to be ground is crucial in 3. Suppose, e.g., that u is a variable, say x. Then there is no rule to reduce the term $equal?(x,v)$ to *true* or *false*. Moreover, one does *not* want to add rules of the form $equal?(x,\tau(y_1,\ldots,y_m)) \to false$ to $\mathcal{I}(\Sigma)$ because one would destroy the confluence of $\mathcal{I}(\Sigma)$ and thus the correctness of the definition of $equal?$: indeed, $equal?(\tau(y_1,\ldots,y_m),\tau(y_1,\ldots,y_m))$ would reduce to both *true* and *false* in $\mathcal{I}(\Sigma)$.

4   The Main Transformation

The major reason for which conditional rules are inconvenient to implement in a rewriting engine is because, in order to reduce a term, the rewriting engine needs to maintain a *control context* for each conditional rule that is tried to be applied. By control context we here mean the status of the evaluation of the condition (note that a condition is a set of equalities) plus the right hand term that needs to replace the left hand one in case the condition evaluates to *true*.

Due to the potential nesting of rewrite rule applications, such a control context may grow arbitrarily, meaning that the rewriting engine needs to pay special care to choosing appropriate data-structures to maintain it and to recover the computation in case the evaluation of a condition fails.

Example 1. Let us consider natural numbers built with 0 and successor s, together with the following, on purpose inefficient, conditional rules defining *odd* and *even* operators on natural numbers:

$odd(0) \rightarrow 0,$ $even(0) \rightarrow s(0),$
$odd(s(x)) \rightarrow 0$ if $even(x) = 0,$ $even(s(x)) \rightarrow 0$ if $odd(x) = 0,$
$odd(s(x)) \rightarrow s(0)$ if $even(x) = s(0),$ $even(s(x)) \rightarrow s(0)$ if $odd(x) = s(0).$

In order to check whether a natural number n, i.e., a term consisting of n successor operations applied to 0, is odd, a rewriting engine may need $\mathcal{O}(2^n)$ rewrites in the worst case. Indeed, if $n > 0$ then either the second or the third rule of *odd* can be applied at the first step; however, in order to apply any of those rules one needs to reduce the even of the predecessor of n, twice. Iteratively, the evaluation of each even involves the reduction of two odds, and so on. Moreover, the rewriting engine needs to maintain a control context data-structure, storing the status of the application of each (nested) rule that is being tried in a reduction. It is the information stored in this control context that allows the rewriting engine to backtrack and find an appropriate rewriting sequence.   □

A challenging question motivating the present work is the following: would it be possible to automatically replace conditional rules like the above by unconditional ones, so that a rewriting engine's single job would be to *match-and-apply* rules, without worrying about any control context aspects? A positive answer to this question could potentially lead to a new generation of efficient rewriting engines, which would take advantage of today's increasingly highly parallel computing architectures and would potentially allow optimizations that were not possible for conditional rewriting. In this section we show how a conditional rewrite system $\mathcal{R}$ can be automatically transformed into an unconditional one $\overline{\mathcal{R}}$, which practically preserves all the properties of $\mathcal{R}$. The major idea is, like in the use of continuations (see [15] for a discussion on several independent discoveries of continuations, and [7] for a pragmatic presentation of continuation), to convert the control context into data context. This way, the term to be rewritten is enriched at appropriate positions to contain *all* the information needed to continue its reduction. The rewriting engine does not need to maintain any auxiliary information about the status of the rewriting process: it only needs to find a redex in the term to rewrite and apply a corresponding unconditional rewrite rule, a simple process amenable to high parallelization and optimization.

4.1   An Unsatisfactory Transformation

Once one generates the infrastructure (unconditional) Σ'-rewrite system $\mathcal{I}(\Sigma)$, a simple-minded way to transform a conditional Σ-rewrite system R into an unconditional one is to translate each conditional rule

$$l \rightarrow r \text{ if } u_1 = v_1, \cdots, u_m = v_m$$

into an unconditional rewrite rule

$$l \rightarrow if(equal?(u_1, v_1) \wedge \cdots \wedge equal?(u_m, v_m), r, l).$$

Such a transformation has the desirable property that both the conditional rewrite system and its unconditional variant can "reach", by reduction in zero or more steps starting with a given Σ-term, the same set of Σ-terms. In other words, if a and b are Σ-terms then $a \rightarrow^* b$ in the conditional Σ-rewrite system if and only if $a \rightarrow^* b$ in the unconditional Σ'-rewrite system. Therefore, if reachability analysis is what one is interested in then this simple translation provides an effective method to reduce the problem to unconditional rewrite systems. This rewrite system transformation can be useful in systems like Maude, providing commands of the form "search a =>* b" searching for a sequence of applications of rewrite rules transforming a into b.

However, this translation cannot be used to execute conditional rewriting on top of an unconditional rewriting engine. Indeed, if the conjunction of equalities reduces to *false* then the unconditional rewrite system leads to an infinite rewriting sequence, by keeping applying the rule above. Would it be possible to properly *mark* the term to rewrite whenever a rule is tried and its condition reduces to *false*, so that that rule will not be applied anymore on that position?

4.2   Adding Control Context Arguments

Like in Viry [17], the idea is to add a few auxiliary arguments to some operators to keep the necessary control context information. This way, terms to rewrite will store information about the conditional rules that can be potentially applied on each of their subterms. Let $\mathcal{R} = (\Sigma, E)$ be any Σ rewriting system. For each n and each $\sigma \in \Sigma_n$, let us associate a unique number between 1 and k_σ to each conditional rewrite rule in R whose lhs is rooted in σ, that is, a rule of the form

$$\sigma(t_1, \ldots, t_n) \rightarrow r \text{ if } u_1 = v_1, \cdots, u_m = v_m,$$

with $t_1, \ldots, t_n, r, u_1, v_1, \ldots, u_m, v_m$ terms and $m \geq 1$, where k_σ is the total number of such rules. Note that k_σ is 0 if there is no rule having σ as a root of its lhs, or if all such rules are unconditional.

Let us next define a signature $\overline{\Sigma}$, replacing each $\sigma \in \Sigma_n$ by an operator of $n + k_\sigma$ arguments, $\overline{\sigma} \in \overline{\Sigma}_{n+k_\sigma}$. The additional k_σ arguments are written at the right of the other n arguments, and they can take only two possible values (or constant terms): *true* or *false*. An important step in our transformation technique is to replace all the operations in Σ by corresponding operations in $\overline{\Sigma}$. The intuition for the additional arguments comes from the overall idea of passing the control context (due to conditional rules) into data context: the additional i-th argument of an operation $\overline{\sigma}$ staying at some position in a term to rewrite, tells whether the i-th rule having σ at the root of its lhs is enabled or not at that position; if *true* then it means that the rule can potentially be applied, and if *false* then it means that the rule has been already tried at that position but its condition failed to evaluate to *true*, so there is no need to try it anymore. Let us extend this to Σ-terms, by letting the variables unchanged and replacing each operator σ by $\overline{\sigma}$ with the k_σ additional arguments all *true*. Formally, let $\overline{\cdot} : T_\Sigma(\mathcal{X}) \to T_{\overline{\Sigma}}(\mathcal{X})$ be a map from Σ-terms to $\overline{\Sigma}$-terms defined inductively as

- $\overline{x} = x$ for any variable $x \in \mathcal{X}$, and
- $\overline{\sigma(t_1, \ldots, t_n)} = \overline{\sigma}(\overline{t_1}, \ldots, \overline{t_n}, true, \ldots, true)$ for any $\sigma \in \Sigma_n$ and any terms $t_1, \ldots, t_n \in T_\Sigma(\mathcal{X})$.

Let's define another useful map from Σ-terms to $\overline{\Sigma}$-terms, $\widetilde{\cdot}^X : T_\Sigma(X) \to T_{\overline{\Sigma}}(\mathcal{X})$, but this time indexed by a finite set of variable $X \subseteq \mathcal{X}$, as follows:

- $\widetilde{x}^X = x$ for any variable $x \in X$, and
- $\widetilde{\sigma(t_1, \ldots, t_n)}^X = \overline{\sigma}(\widetilde{t_1}^X, \ldots, \widetilde{t_n}^X, b_1, \ldots, b_{k_\sigma})$ for any $\sigma \in \Sigma_n$ and any terms $t_1, \ldots, t_n \in T_\Sigma(X)$, where $b_1, \ldots, b_{k_\sigma} \in \mathcal{X} - X$ are some arbitrary but fixed different fresh variables that do not occur neither in X nor in $\widetilde{t_1}^X, \ldots, \widetilde{t_n}^X$.

Therefore, $\widetilde{t}^X$ transforms the Σ-term t into a $\overline{\Sigma}$-term, replacing each operation $\sigma \in \Sigma$ by $\overline{\sigma} \in \overline{\Sigma}$ and adding distinct variables for the additional arguments, following some arbitrary but deterministic conventions. Given a Σ-term t in $T_\Sigma(X)$ of the form $\sigma(t_1, \ldots, t_n)$ for some operation $\sigma \in \Sigma_n$, and given a natural number i between 1 and k_σ, then we let $\widetilde{t}^X_{i/true}$ denote the $\overline{\Sigma}$-term $\overline{\sigma}(\widetilde{t_1}^X, \ldots, \widetilde{t_n}^X, b_1, \ldots, b_{i-1}, true, b_{i+1}, \ldots, b_{k_\sigma})$, that replaces b_i in $\widetilde{t}^X$ by *true*. Similarly, $\widetilde{t}^X_{i/false}$ denotes $\overline{\sigma}(\widetilde{t_1}^X, \ldots, \widetilde{t_n}^X, b_1, \ldots, b_{i-1}, false, b_{i+1}, \ldots, b_{k_\sigma})$, that replaces b_i in $\widetilde{t}^X$ by *false*. Thus, $\widetilde{t}^X_{i/true}$ (resp. $\widetilde{t}^X_{i/false}$) contains the additional control context information whether the i-th conditional rule of σ is enabled.

4.3   An Almost Correct Transformation

For a given conditional Σ-rewrite system $\mathcal{R}$, we can now define an unconditional $\overline{\Sigma}'$-rewrite system[3] $\mathcal{R}'$ by adding to $\mathcal{I}(\overline{\Sigma})$ the following unconditional $\overline{\Sigma}'$-rewrite rules. For each conditional ($m \geq 1$) rule $l \to r$ if $u_1 = v_1, \cdots, u_m = v_m$ over variables X in $\mathcal{R}$, say the i-th among the conditional rewrite rules in $\mathcal{R}$ having the root operation of l as a root of their lhs, add to $\mathcal{R}'$ the unconditional rule

$$\widetilde{l}^X_{i/true} \to if(equal?(\overline{u_1}, \overline{v_1}) \wedge \cdots \wedge equal?(\overline{u_m}, \overline{v_m}), \overline{r}, \widetilde{l}^X_{i/false}).$$

For each unconditional rewrite rule $l \to r$ in $\mathcal{R}$ over variables X, add to $\mathcal{R}'$ an unconditional rewriting rule $\widetilde{l}^X \to \overline{r}$.

Therefore, for each conditional rule in $\mathcal{R}$ we add an unconditional one in $\mathcal{R}'$, whose corresponding additional argument of its transformed lhs is *true*. By throwing the control context's ball into matching's court, this intuitively says that such a rule can be applied on a (sub)term only if it is "enabled" in that (sub)term. Its rhs term has a conditional operation at its root, which first evaluates the conjunction of all the equalities of pairs of terms occurring in the condition of the conditional rule; note that these terms are properly transformed into

---

[3] Note that $\overline{(\Sigma')} = (\overline{\Sigma})'$, so we take the liberty to denote this signature $\overline{\Sigma}'$.

Σ-terms enabling all possible rules at any of their positions. If the conjunction evaluates to *true* then the rhs of the conditional rule is returned, also modified to enable all possible rules on it. If the condition reduces to *false* then the only thing to do is to "disable" this current rule. Due to the change of the corresponding argument from *true* to *false*, note that matching will disallow this rule to be applied anymore on that (sub)term. Since unconditional rules are always enabled, they are transformed into unconditional rules ignoring the control context arguments in the lhs and enabling all possible rules on its rhs. Note that $\mathcal{R}'$ does not modify any rule in $\mathcal{R}$ if $\mathcal{R}$ already contains only unconditional rules.

Example 2. Let us apply the translation technique above on the conditional rewriting system for odd/even in Example 1. Since there are two conditional rules whose root of lhs is *odd* and two whose root of lhs is *even*, each of these operators will be enriched with two additional arguments. The new, unconditional rewriting system is then:

$$\overline{odd}(0, b_1, b_2) \;\rightarrow\; 0,$$
$$\overline{odd}(s(x), true, b_2) \;\rightarrow\; if(equal?(\overline{even}(x, true, true), 0),\; 0,\; \overline{odd}(s(x), false, b_2)),$$
$$\overline{odd}(s(x), b_1, true) \;\rightarrow\; if(equal?(\overline{even}(x, true, true), s(0)),\; s(0),\; \overline{odd}(s(x), b_1, false)),$$
$$\overline{even}(0, b_1, b_2) \;\rightarrow\; s(0),$$
$$\overline{even}(s(x), true, b_2) \;\rightarrow\; if(equal?(\overline{odd}(x, true, true), 0),\; 0,\; \overline{even}(s(x), false, b_2)),$$
$$\overline{even}(s(x), b_1, true) \;\rightarrow\; if(equal?(\overline{odd}(x, true, true), s(0)),\; s(0),\; \overline{even}(s(x), b_1, false)).$$

The unconditional rule for $\overline{odd}$ says that 0 is not an odd number, regardless of the control context. The first conditional rule for $\overline{odd}$ has the constant *true* as the first auxiliary argument of its lhs, telling the matching procedure that this rule can be applied only if it was not previously disabled. If the condition of *if* evaluates to *true* then 0 is returned, otherwise the same term as the lhs, except that *true* is replaced by *false*, thus disabling the current conditional rule to avoid getting into non-terminating rewriting. The variable argument b_2 says that it does not matter whether the second conditional rule is enabled or not (but this information will be preserved in case the first conditional rule is disabled). The other conditional equations are similar. If one wants to test whether a number n, i.e., n consecutive applications of successor on 0, is odd, one should reduce the term $\overline{odd}(n)$, i.e., $odd(n, true, true)$, under the unconditional rewrite system. Note that the operations 0 and s are not added auxiliary arguments because they do not occur as a root of a lhs of any conditional rule in the original conditional rewriting system.   □

Unfortunately, the translation above suffers from the same problem as that of Viry [17]: for some CTRSs, the generated TRSs have additional normal forms corresponding to terms which could be further reduced in the original CTRS.

Example 3. Consider the problematic CTRS from Section 1:

$$f(g(x)) \rightarrow x \text{ if } x = 0,$$
$$g(g(x)) \rightarrow g(x),$$

whose corresponding TRS, according to the transformation above, is:

$$f(g(x), true) \rightarrow if(equal?(x, 0),\ x,\ f(g(x), false)),$$
$$g(g(x)) \rightarrow g(x).$$

Then note that even if $f(g(g(0)))$ admits a unique normal form in the original CTRS, $f(g(g(0)), true)$ admits two normal forms in its corresponding TRS:

$$f(g(g(0)), true) \rightarrow f(g(0), true) \rightarrow if(equal?(0, 0),\ 0,\ f(g(0), false)) \rightarrow^* 0,$$
$$f(g(g(0)), true) \rightarrow if(equal?(g(0), 0),\ g(0),\ f(g(g(0)), false)) \rightarrow^*$$
$$\rightarrow^* f(g(g(0)), false) \rightarrow f(g(0), false).$$

The latter cannot be further reduced with the rules in the TRS.   □

The problem here, like in Viry's transformation [17], is that a successful application of a rewrite rule may enable some application of a conditional rule that has already been tried before, but at that time failed to apply. One unsatisfactory way to fix this problem is, like in [17], to enforce conditional eagerness on the generated TRS; another, even more unsatisfactory, is to reduce the applicability of the transformation to only innermost, or eager, CTRSs. We next show how to fix this problem in general.

4.4   The Correct Transformation

To fix the problem in the previous subsection, we need a mechanism to "inform" the term to reduce, after each successful application of a rewrite rule, that some "conditional" rules that have been tried before and failed may succeed now. More precisely, we need to traverse the term along the path from the current position (where the successful rule was applied) to its root, and make all the auxiliary arguments of the operations on this path *true*. This can be accomplished, for example, by considering a new (unary) operator, say {_}, stating that the enclosed term has just been modified, together with appropriate rewrite rules to propagate this information upwards, updating the "applicability bits": for each $\sigma \in \Sigma_n$ and each $1 \leq i \leq n$, consider a rule

$$\sigma(x_1, ..., x_{i-1}, \{x_i\}, x_{i+1}, ..., x_n, b_1, ..., b_{k_\sigma}) \rightarrow$$
$$\rightarrow \{\sigma(x_1, ..., x_{i-1}, x_i, x_{i+1}, ..., x_n, true, ..., true)\}.$$

The applicability information of an operation can be updated from several of its subterms; to keep this operation idempotent, we also consider the rule

$$\{\{x\}\} \rightarrow \{x\}.$$

Formally, for a given conditional Σ-rewrite system $\mathcal{R}$, we let $\overline{\Sigma}'_{\{\}}$ define the signature $\overline{\Sigma}'$ in the previous subsection extended with the unary operator $\{\_\}$ above, and we let $\overline{\mathcal{R}}$ be the unconditional $\overline{\Sigma}'_{\{\}}$-rewrite system extending $\mathcal{I}(\Sigma)$ with the operator $\{\_\}$ together with its unconditional rewrite rules above, as well as with the following rules. For each conditional ($m \geq 1$) rule $l \rightarrow r$ if $u_1 = v_1, \cdots, u_m = v_m$ over variables X in $\mathcal{R}$, say the i-th among the conditional rewrite rules in $\mathcal{R}$ having the root operation of l as a root of their lhs, add to $\overline{\mathcal{R}}$:

$$\widetilde{l}^X_{i/true} \to if(equal?(\{\overline{u_1}\}, \{\overline{v_1}\}) \land \cdots \land equal?(\{\overline{u_m}\}, \{\overline{v_m}\}), \{\overline{r}\}, \widetilde{l}^X_{i/false}).$$

For each unconditional rewrite rule $l \to r$ in $\mathcal{R}$ over variables X, add to $\overline{\mathcal{R}}$ an unconditional rewriting rule $\widetilde{l}^X \to \{\overline{r}\}$.

Before we formalize the exact relationship between CTRSs and their unconditional variants, let us define another map of terms, this time from $\overline{\Sigma}$-terms to Σ-terms. Let $\widehat{\cdot}: T_{\overline{\Sigma}}(\mathcal{X}) \to T_\Sigma(\mathcal{X})$ be the map defined inductively as

- $\widehat{x} = x$ for any variable $x \in \mathcal{X}$, and
- $\overline{\sigma}(t'_1, \ldots, t'_n, s_1, \ldots, s_{k_\sigma}) = \sigma(\widehat{t'_1}, \ldots \widehat{t'_n})$ for any operator $\sigma \in \Sigma_n$ and any terms $t'_1, \ldots, t'_n, s_1, \ldots, s_{k_\sigma} \in T_{\overline{\Sigma}'}(\mathcal{X})$.

Therefore, $\widehat{t'}$ forgets all the auxiliary arguments of each operation occurring in t'. Note in particular that $\widehat{\widetilde{t}} = \widehat{\overline{t}} = t$ for any $t \in T_\Sigma(\mathcal{X})$.

Example 4. Let us consider the problematic CTRS in example 3. Its corresponding TRS generated as above contains the following rules:

$$\{\{x\}\} \to \{x\}$$
$$f(\{x\}, b) \to \{f(x, true)\}$$
$$g(\{x\}) \to \{g(x)\}$$

$$f(g(x), true) \to if(equal?(\{x\}, \{0\}), \{x\}, f(g(x), false)),$$
$$g(g(x)) \to \{g(x)\}.$$

Then the term $f(g(g(0)), true)$ admits the normal form $\{0\}$:

$$f(g(g(0)), true) \to if(equal?(\{g(0)\}, \{0\}), \{g(0)\}, f(g(g(0)), false)) \to^*$$
$$\to^* f(g(g(0)), false) \to f(\{g(0)\}, false) \to$$
$$\to \{f(g(0), true)\} \to \{0\}.$$

Note that the normal form $\{0\}$ is possible exactly because the information that a subterm has been rewritten is transmitted upwards via the operator $\{_-\}$ and its associated rules. The obtained TRS is not confluent, because the term above also admits the normal form 0, but this time the (at most two) normal forms that a term can have (t and/or $\{t\}$) are very closely related and one can easily infer the desired normal form in the original CTRS. In order to have a unique normal form in the TRS, we will actually enclose the original term into curly brackets before we reduce it, as the theorem below suggests.

Theorem 1. *If $\mathcal{R}$ is a conditional Σ-rewriting system then*

1. *For any ground Σ-terms α and β, $\alpha \to^*_\mathcal{R} \beta$ if and only if there is some ground $\overline{\Sigma}$-term γ such that $\widehat{\gamma} = \beta$ and $\{\overline{\alpha}\} \to^*_{\overline{\mathcal{R}}} \{\gamma\}$;*
2. *$\mathcal{R}$ terminates on a ground Σ-term α if and only if $\overline{\mathcal{R}}$ terminates on $\{\overline{\alpha}\}$;*
3. *$\overline{\mathcal{R}}$ terminates if and only if it terminates on all terms $\{\overline{\alpha}\}$ with α a Σ-term;*
4. *If γ in 1. is a normal form (in $\overline{\mathcal{R}}$) then β is also a normal form (in $\mathcal{R}$);*
5. *If $\mathcal{R}$ terminates then $\mathcal{R}$ is ground confluent iff $\overline{\mathcal{R}}$ is ground confluent.*

5   Putting Them all Together

We can now present the main result of this paper, namely a technique providing a rewriting engine that accepts conditional rewrite rules, obtained by appropriately wrapping a simpler rewriting engine that only accepts unconditional rules.

> **Input:** a conditional Σ-rewrite system $\mathcal{R}$ and a Σ-term α over variables X to be reduced with $\mathcal{R}$.
> **Step 1:** Add the variables in X as fresh constants into Σ, so that α becomes a ground Σ-term;
> **Step 2:** Generate $\overline{\Sigma}$ like in Subsection 4.2, by adding to each operator $\sigma \in \Sigma$ as many auxiliary arguments as conditional rules of lhs rooted in σ are in $\mathcal{R}$;
> **Step 3:** Generate the infrastructure unconditional $\overline{\Sigma}'$-rewrite system $\mathcal{I}(\overline{\Sigma})$ by adding to $\overline{\Sigma}$ the operators *equal?* and *if* $(\_, \_, \_)$ as well as their corresponding rules described in Section 3;
> **Step 4:** Generate the unconditional $\overline{\Sigma}'$-rewrite system $\overline{\mathcal{R}}$ by adding to $\mathcal{I}(\overline{\Sigma})$ the operation $\{\_\}$ and its rules, as well as the unconditional $\overline{\Sigma}'$-rewrite rules associated to the rules in $\mathcal{R}$ as shown in Subsection 4.4;
> **Step 5:** Reduce the term $\{\overline{\alpha}\}$ to a normal form in $\overline{\mathcal{R}}$, say $\{\gamma\}$, using any engine for unconditional rewriting;
> **Step 6:** Return the Σ-term $\widehat{\gamma}$.

We claim that the steps above, by applying a series of simple and totally automatic syntactic transformations to the input conditional rewrite system and term to rewrite, yield a rewriting engine that accepts conditional rewrite systems.

Step 1 shows a usual way to reduce terms with variables: interpret the variables as constants. However, in our framework it is quite important to add these variables explicitly as constants early in the reduction process. This is because the equality operator will need to consider these constants as distinct operations, so that it can add appropriate rules, including ones of the form "*equal?* $(a, \sigma(x_1, \ldots, x_n)) \to false$" for any such constant a and operation $\sigma \in \Sigma_n$.

Step 2 modifies the signature Σ into $\overline{\Sigma}$, by analyzing the rules in $\mathcal{R}$ and adding an appropriate number of arguments to operations in Σ. Since the constants added to Σ at Step 1 are fresh, no rule in $\mathcal{R}$ has them as lhs terms, so these constants will not be changed in $\overline{\Sigma}$. Note, however, that other constants in Σ may get translated into operations with several arguments.

Step 3 adds the auxiliary equality and conditional operators that are needed to translate the conditional rules into unconditional ones. Note, again, that the constants added at Step 1 will increase the number of rules for *equal?*.

Step 4 generates the unconditional rewrite system $\overline{\mathcal{R}}$, by adding exactly one unconditional rule per conditional or unconditional rule in $\mathcal{R}$. Once $\overline{\mathcal{R}}$ is available, any engine for unconditional rewriting can be used to reduce $\overline{\alpha}$ under $\overline{\mathcal{R}}$, as done in Step 5. Note that the normal form γ of $\overline{\alpha}$ in $\overline{\mathcal{R}}$, if it exists, is a $\overline{\Sigma}$-term. Therefore, since the hat function is defined as $\widehat{\ } : T_{\overline{\Sigma}}(\mathcal{X}) \to T_{\Sigma}(\mathcal{X})$, the term $\widehat{\gamma}$ returned at Step 6 is indeed a Σ-term, as the result of reducing α under $\mathcal{R}$ is expected to be.

	Maude		Elan	
n	Conditional	Unconditional	Conditional	Unconditional
11	0.007	0.003	0.604	0.061
13	0.027	0.009	2.309	0.244
15	0.142	0.037	8.922	0.959
17	0.459	0.154	34.2	3.8
21	7.9	2.6	548.1	62.9
23	29.2	10.4	-	-
25	117.2	49.3	-	-
31	7431	2489	-	-

Fig. 1. Times in seconds to reduce $odd(n)$ using Maude and ELAN, using both the conditional rewrite system in Example 1 and its unconditional variant in Example 2

Theorem 2. *The algorithm above, taking a conditional Σ-rewrite system $\mathcal{R}$ and a Σ-term α as input, terminates iff $\mathcal{R}$ terminates on α. If the algorithm terminates and outputs a Σ-term β, then β is a normal form of α in $\mathcal{R}$.*

Proof. The algorithm terminates if and only if its Step 5 terminates, because all the other steps are nothing but simple syntactic translations over a finite signature and number of rewrite rules. Step 5 terminates if and only if $\overline{\mathcal{R}}$ terminates on $\{\overline{\alpha}\}$. By 2 in Theorem 1, this is equivalent to saying that $\mathcal{R}$ terminates on α. Now suppose that the algorithm terminates and that it outputs a Σ-term β. By Steps 5 and 6 and the discussion preceding this theorem, this happens if and only if there is some $\overline{\Sigma}$-term γ such that $\{\overline{\alpha}\} \rightarrow^*_{\overline{\mathcal{R}}} \{\gamma\}$ and $\widehat{\gamma} = \beta$, which, by 1 in Theorem 1, is equivalent to $\alpha \rightarrow^*_{\mathcal{R}} \beta$. Since γ is a normal form in $\overline{\mathcal{R}}$, it follows by 4 in Theorem 1 that β is a normal form in $\mathcal{R}$.

6   Preliminary Experiments

As mentioned previously in the paper, our original purpose for translating conditional rewrite systems into unconditional ones was to ease the process of implementing rewriting engines, at the same time aiming at highly efficient implementations of rewriting based on a fast and simple *rewriting virtual machine*. To test the computational equivalence between conditional rewriting systems and their corresponding unconditional variants, we have applied the translation presented in Section 5 manually for the conditional rewrite system in Example 1, and performed some experiments on a 2.4GHz PC machine using two major rewriting engines: Maude [4] and Elan [3]. The results of these experiments are listed in Figure 1, in seconds; the "-" should be read "core dump".

It was an unexpected surprise to see that the unconditional variant was much faster than its corresponding conditional rewrite system: almost 3 times faster for Maude and 10 times for ELAN. Since the implementation details of these rewrite engines are not well documented, we do not know the exact reasons for which conditional rewriting is so slow in these systems in comparison to unconditional rewriting. However, the preliminary results in Figure 1 tell us that maintaining the control context required to backtrack through conditional rewrite rules is a non-trivial matter, and that the translation technique proposed in this paper can be perhaps implemented in these systems to bring immediate benefits to conditional rewriting.

7   Conclusion and Future Work

An automatic technique to transform a conditional term rewriting system into an unconditional one was presented, which preserves all the major properties. The technique consists of adding some key auxiliary arguments to certain operations, which are used to maintain the control context as data context.

Only first steps towards a generic transformation procedure for general conditional rewrite systems have been made here. We have not considered rewriting modulo axioms, such as associativity, commutativity and identity, but these will be considered soon. Also, the current technique can be improved: not all the auxiliary arguments added to operations seem to be necessary. Can we statically reduce their number?

Future work will also investigate how well analysis techniques for unconditional rewriting systems translate into corresponding analysis techniques for conditional ones via our transformation. One complication here is that some of the operations that we introduce require to be evaluated eagerly in some arguments. This is also a drawback of our technique if one wants to use it for theorem proving purposes. Can one find a transformation which imposes no evaluation strategies?

References

1. S. Antoy, B. Brassel, and M. Hanus. Conditional narrowing without conditions. In *5th ACM SIGPLAN international conference on Principles and practice of declarative programming (PPDP'03)*, pages 20–31. ACM Press, 2003.
2. J. Bergstra and J. Klop. Conditional rewrite rules: Confluence and termination. *Journal of Computer and System Sciences*, 32(3):323–362, 1986.
3. P. Borovansky, H. Cirstea, H. Dubois, C. Kirchner, H. Kirchner, P. Moreau, C. Ringeissen, and M. Vittek. *ELAN: User Manual*, 2000. Loria, Nancy, France.
4. M. Clavel, F. Durán, S. Eker, P. Lincoln, N. Martí-Oliet, J. Meseguer, and C. Talcott. *Maude 2.0 Manual*, 2003. http://maude.cs.uiuc.edu/manual.
5. N. Dershowitz and M. Okada. A rationale for conditional equational programming. *Theoretical Computer Science*, 75:111–138, 1990.
6. R. Diaconescu and K. Futatsugi. *CafeOBJ Report: The Language, Proof Techniques, and Methodologies for Object-Oriented Algebraic Specification*. World Scientific, 1998. AMAST Series in Computing, volume 6.
7. D. P. Friedman, C. T. Haynes, and M. Wand. *Essentials of programming languages*. MIT Press, 1992.
8. E. Giovannetti and C. Moiso. Notes on the elimination of conditions. In *1st International Workshop on Conditional Term Rewriting Systems (CTRS'87)*, volume 308 of *LNCS*, pages 91–97. Springer, 1987.

9. J. Goguen, T. Winkler, J. Meseguer, K. Futatsugi, and J.-P. Jouannaud. Introducing OBJ. In *Software Engineering with OBJ: algebraic specification in action*, pages 3–167. Kluwer, 2000.
10. M. Hanus. The integration of functions into logic programming: From theory to practice. *The Journal of Logic Programming*, 19 & 20:583–628, 1994.
11. C. Hintermeier. How to transform canonical decreasing ctrss into equivalent canonical trss. In *4th International Workshop on Conditional and Typed Rewriting Systems (CTRS'94)*, volume 968 of *LNCS*, pages 186–205, 1994.
12. J. Meseguer and G. Roşu. Rewriting logic semantics: From language specifications to formal analysis tools. In *2nd International Joint Conference on Automated Reasoning (IJCAR'04)*, LNCS, to appear, 2004.
13. A. Middeldorp and E. Hamoen. Completeness results for basic narrowing. *Journal of Applicable Algebra in Eng., Communication and Computing*, 5:313–353, 1994.
14. E. Ohlebusch. *Advanced Topics in Term Rewriting*. Springer, 2002.
15. J. C. Reynolds. The discoveries of continuations. *LISP and Symbolic Computation*, 6(3–4):233–247, 1993.
16. G. Roşu. From conditional to unconditional rewriting. Technical Report UIUCDCS-R-2004-2471, University of Illinois at Urbana-Champaign, Aug. 2004.
17. P. Viry. Elimination of conditions. *Journal of Symbolic Computation*, 28:381–401, Sept. 1999.
18. E. Visser. Stratego: A language for program transformation based on rewriting strategies. System description of Stratego 0.5. In *Rewriting Techniques and Applications (RTA'01)*, volume 2051 of *LNCS*, pages 357–361. Springer, May 2001.

Type Class Polymorphism in an Institutional Framework

Lutz Schröder, Till Mossakowski, and Christoph Lüth

BISS, Department of Computer Science,
University of Bremen

Abstract. Higher-order logic with shallow type class polymorphism is widely used as a specification formalism. Its polymorphic entities (types, operators, axioms) can easily be equipped with a 'naive' semantics defined in terms of collections of instances. However, this semantics has the unpleasant property that while model reduction preserves satisfaction of sentences, model expansion generally does not. In other words, unless further measures are taken, type class polymorphism fails to constitute a proper institution, being only a so-called rps preinstitution; this is unfortunate, as it means that one cannot use institution-independent or heterogeneous structuring languages, proof calculi, and tools with it.

Here, we suggest to remedy this problem by modifying the notion of model to include information also about its potential future extensions. Our construction works at a high level of generality in the sense that it provides, for any preinstitution, an institution in which the original preinstitution can be represented. The semantics of polymorphism used in the specification language HASCASL makes use of this result. In fact, HASCASL's polymorphism is a special case of a general notion of polymorphism in institutions introduced here, and our construction leads to the right notion of semantic consequence when applied to this generic polymorphism. The appropriateness of the construction for other frameworks that share the same problem depends on methodological questions to be decided case by case. In particular, it turns out that our method is apparently unsuitable for observational logics, while it works well with abstract state machine formalisms such as state-based CASL.

1 Introduction

The idea that a *logic* is something that comes with signatures, models, sentences and a satisfaction relation is formalized in the notion of *institution* as introduced in [15]. In practice, this concept is exploited to support genericity and heterogeneity in specification frameworks. For example, the semantics and proof calculus for structured and architectural specifications in CASL [27] is generic over institutions, and heterogeneous CASL [25, 26] uses a graph of institutions for heterogeneous specification. The central condition governing the behaviour of institutions is the *satisfaction condition*, stating that satisfaction of sentences is preserved under both model expansion and reduction.

J.L. Fiadeiro, P. Mosses, and F. Orejas (Eds.): WADT 2004, LNCS 3423, pp. 234–251, 2005.
© Springer-Verlag Berlin Heidelberg 2005

Type class polymorphism has been used in programming languages like Haskell [31], as well as in the higher-order logic of Isabelle [38]. It is one of the central features of the recently developed specification language HASCASL [35, 36]. Little attention has been paid in the literature to the question whether type class polymorphism can be formalized as an institution, the main problem here being that with the 'naive' semantics, the satisfaction condition fails in the sense that satisfaction of polymorphic axioms is preserved only by model reduction, not by model expansion, because expanded models may have more types. Thus, the naive semantics defines only a so-called rps preinstitution [32] rather than an institution.

The work of [28] is an initial attempt to define an institution for polymorphism but imposes severe restrictions on signature morphisms by simply ruling out the introduction of new types. For the case of polymorphism without type classes, one solution is to parametrize the notion of model by a fixed universe of types [7, 19]; this solution, however, does not seem to be suitable for type class polymorphism.

The main goal of the present work is to provide a semantics that avoids both problems, i.e. caters for type classes and works with the usual structured specification style where the signature is built up successively. In particular, we wish to avoid restrictions on *signature morphisms*; instead, we argue that the failure of the satisfaction condition points to a flaw in the notion of *model*. The key idea is to notice that polymorphic axioms are intended as statements about all types including those yet to be declared, and that therefore models should take into account future extensions. Starting from this observation, we obtain a general procedure that transforms a preinstitution into an institution, the so-called institution of *extended models*. This construction is employed in the semantics of HASCASL. It turns out that the notions of semantic consequence and model-expansive extension engendered by the construction agree with intuitive expectations, at least in sufficiently rich logics such as the logic of HASCASL.

More generally, HASCASL's treatment of polymorphic sentences can be subsumed under a definition of polymorphic formulae in institutions introduced here. Such generic polymorphic frameworks are perfect candidates for the extended model construction, and indeed it turns out that the notion of semantic consequence in the institution of extended models over a generic polymorphic framework is simpler and more natural than the original notion.

There are several other known examples of logical frameworks where the satisfaction condition fails unless restrictions are imposed. E.g. in observational logics, signature morphisms are usually not allowed to introduce new observers [5, 16], precisely in order to rescue the satisfaction condition. Moreover, in the (non-)institution of SB-CASL [3, 4], the satisfaction condition fails for signature morphisms that introduce additional state components [3]. We discuss both these examples from a methodological perspective; it turns out that our construction cannot be recommended for the observational case, since it suppresses coinduction, while the semantics obtained for SB-CASL arguably provides the 'right' notion of semantic consequence.

The material is organized as follows. Sections 2 and 3 provide preliminary material concerning the institution-theoretic background and type class polymorphism in HASCASL. The failure of the satisfaction condition in the various settings mentioned above is treated in detail in Section 4. Section 5 defines polymorphic formulae over an institution. The construction of an institution from a given preinstitution is introduced in Section 6, and applied to generic polymorphic frameworks in Section 7. The issue of model-expansive extensions (also referred to as model-theoretically conservative extensions or, e.g. in the semantics of CASL, just as conservative extensions) is discussed in Section 8. Section 9 provides some observations on how the generic mechanism instantiates in frameworks other than type class polymorphism.

2   Institutions

A specification formalism is usually based on some notion of signature, model, sentence and satisfaction. These are the ingredients of the notion of *institution* as introduced by Goguen and Burstall [15]. Contrary to Barwise's notion of abstract model theory [2], the theory of institutions does not assume that signatures are algebraic signatures; indeed, nothing at all is said about signatures except that they form a class and that there are *signature morphisms*, which can be composed in some way. This amounts to stating that signatures form a *category*.

There is also nothing special assumed about the form of the *sentences* and *models*. Given a signature Σ, the Σ-sentences form just a set, while the Σ-models form a category (taking into account that there may be *model morphisms*). Signature morphisms lead to *translations* of sentences and of models (thus, the assignments of sentences and of models to signatures are functors). There is a contravariance between the sentence and model translations: sentences are translated *along* signature morphisms, while models are translated *against* signature morphisms.

Following [15], this is formalized as follows.

Definition 1. An *institution* $I = (\mathbf{Sign}^I, \mathbf{Sen}^I, \mathbf{Mod}^I, \models^I)$ consists of

- a category $\mathbf{Sign}^I$ of *signatures*;
- a functor $\mathbf{Sen}^I : \mathbf{Sign}^I \to \mathbf{Set}$ giving, for each signature Σ, the set of *sentences* $\mathbf{Sen}^I(\Sigma)$, and for each signature morphism $\sigma : \Sigma \to \Sigma'$, the *sentence translation map* $\mathbf{Sen}^I(\sigma) : \mathbf{Sen}^I(\Sigma) \to \mathbf{Sen}^I(\Sigma')$, where $\mathbf{Sen}^I(\sigma)(\varphi)$ is often written as $\sigma\varphi$;
- a functor $\mathbf{Mod}^I : (\mathbf{Sign}^I)^{op} \to \mathbf{CAT}$ (where $\mathbf{CAT}$ denotes the quasicategory of categories and functors [1]) giving, for each signature Σ, the category of *models* $\mathbf{Mod}^I(\Sigma)$, and for each signature morphism $\sigma : \Sigma \to \Sigma'$, the *reduct functor* $\mathbf{Mod}^I(\sigma) : \mathbf{Mod}^I(\Sigma') \to \mathbf{Mod}^I(\Sigma)$, where $\mathbf{Mod}^I(\sigma)(M')$, the σ-reduct of M', is often written as $M'|_\sigma$; and
- a satisfaction relation $\models^I_\Sigma \subseteq |\mathbf{Mod}^I(\Sigma)| \times \mathbf{Sen}^I(\Sigma)$ for each $\Sigma \in \mathbf{Sign}^I$,

such that for each $\sigma : \Sigma \to \Sigma'$ in $\mathbf{Sign}^I$, the *satisfaction condition*

$$M' \models^I_{\Sigma'} \sigma\varphi \Leftrightarrow M'|_\sigma \models^I_\Sigma \varphi$$

holds for all $M' \in \mathbf{Mod}^I(\Sigma')$ and all $\varphi \in \mathbf{Sen}^I(\Sigma)$.

The notion of institutions owes much of its importance to the fact that several languages for modularizing specifications are generic over an underlying institution [11, 12, 13, 18, 27, 33]. Furthermore, institutions form the basis of *heterogeneous* frameworks such as heterogeneous CASL [25, 26]. Such frameworks require a means of interrelating institutions, i.e. some notion of morphism between institutions. There are various such notions in the literature; one of the most important ones are institution comorphisms, which essentially express that fact that one institution is *encoded* into another.

Definition 2. Given institutions I and J, an *institution comorphism* [17] (also called a *plain map of institutions* [21]) $\mu = (\Phi, \alpha, \beta) : I \to J$ consists of

- a functor $\Phi : \mathbf{Sign}^I \to \mathbf{Sign}^J$,
- a natural transformation $\alpha : \mathbf{Sen}^I \to \mathbf{Sen}^J \circ \Phi$,
- a natural transformation $\beta : \mathbf{Mod}^J \circ \Phi^{op} \to \mathbf{Mod}^I$

such that the following *satisfaction condition* is satisfied for all $\Sigma \in \mathbf{Sign}^I$, $M' \in \mathbf{Mod}^J(\Phi(\Sigma))$ and $\varphi \in \mathbf{Sen}^I(\Sigma)$:

$$M' \models^J_{\Phi(\Sigma)} \alpha_\Sigma\varphi \Leftrightarrow \beta_\Sigma M' \models^I_\Sigma \varphi.$$

Example 3. Equational logic and first-order logic can be formalized as institutions [15], and the obvious inclusion is a comorphism.

3   Polymorphism in HASCASL

HASCASL is a wide-spectrum language which provides a common framework for algebraic specification and functional programming, oriented in particular towards Haskell. This is achieved by extending the algebraic specification language CASL [6] with higher-order functions in the style of Moggi's partial λ-calculus [23], type constructors, type classes, and constructor classes (for details, see [35, 36]); general recursion is specified on top of this in the style of HOLCF. The semantics of a HASCASL specification is the class of its (set-theoretic) *intensional Henkin models*: function types are interpreted by sets which need not contain all set-theoretic functions, and two functions that yield the same value on every input need not be equal.

The main point of interest for the purposes of this paper is the semantics of HASCASL's type class oriented shallow polymorphism. A type class in HASCASL (for the sake of simplicity, we omit constructor classes here) gives rise to a subset of the *syntactical* set of types, where types are generated from basic types and type constructors, the latter either user-declared or, like function types, built-in. The set of types associated to a class is determined by the explicitly declared instances of the class. Instances may be monomorphic or polymorphic. E.g., the specification

class $Ord < Eq$
vars $a : Type, b : Eq$
types $Nat : Ord$;
       $List\ a$;
       $List\ b : Eq$

declares a class Eq with a subclass Ord, a unary type constructor $List$, and a
type Nat (without *defining* any of these items); moreover, Nat is declared to be
of class Ord, hence also of class Eq, and $List$ is declared to produce types of class
Eq when applied to arguments of class Eq. In the signature determined by the
above declarations, the classes Eq and Ord coincide, both consisting precisely
of the types of the form $List\ (List\ (\dots List\ Nat))$. When further instances are
declared later on in the specification process, the two classes will in general be
different.

Axioms and operators may be polymorphic over classes. E.g., we can write
(continuing the above specification)

var $c : Ord$
op $\_ \leq \_ : Pred(c \times c)$
var $x, y, z : c$
   • $x \leq x$
   • $(x \leq y \wedge y \leq z) \Rightarrow x \leq z$

This means that $\_ \leq \_$ is a polymorphic predicate over class Ord satisfying
reflexivity and transitivity. Operators and axioms may be explicitly tied to a
class by means of a bracket notation, thus making up the *interface* of the class
which generates proof obligations (which, like the proof obligations associated
to CASL's semantic annotations, lie outside the scope of the semantics proper)
for later instantiations.

In general, polymorphic types, operators, and axioms are semantically coded
out by collections of instances. That is, the effect of a polymorphic type is essentially just its contribution to the syntactic type universe; a polymorphic operator
is interpreted as a family of operators, one for each instantiation of its type arguments; and a polymorphic axiom is understood as a collection of axioms, indexed
over all types in the classes named in the quantifiers. This constitutes the *first
level* of the semantics of polymorphism used in HASCASL; as will be explained
in detail in the next section, one does not obtain an institution at this level. This
deficiency is repaired at the *second level* of the semantics; this second level and
the general construction behind it are the subject of this paper. The semantics
of polymorphic formulas at the first level will moreover be identified as a special
case of a general definition of polymorphism in institutions in Section 5.

4   Failures of the Satisfaction Condition

There are various features in modern specification languages that tend to
cause the satisfaction condition (cf. Section 2) to fail; besides polymorphism as
discussed in the previous section, this includes observational satisfaction and

dynamic equations between programs in states-as-algebras frameworks such as SB-CASL [4]. Briefly, the reasons for the failures are as follows:

- **Parametric polymorphism:** if a signature morphism σ introduces additional types, then the translation of a polymorphic axiom φ may fail in a model M although φ holds in the reduct of M along σ, namely if φ holds for the 'old' types, but not for the newly introduced ones.
- **Observational equality:** if a signature morphism σ introduces additional observers, then observational equalities that hold in the reduct of a model M under σ may fail in M, since the new observers may detect previously unobservable differences.
- **dynamic equations:** if a signature morphism σ introduces additional state components (i.e. dynamic functions, predicates, or sorts), then dynamic equations $p = q$ between stateful program expressions [4] that hold in the reduct $M|_\sigma$ of a model M may fail to hold in M, since the interpretations of p and q may differ on the new state components [3].

In all these cases, only one direction of the satisfaction condition holds, so that logics with these features constitute proper rps preinstitutions; we explicitly repeat the definition [32]:

Definition 4. A *preinstitution* consists of a signature category equipped with model and sentence functors and a satisfaction relation in the same sense as an institution (cf. Section 2); these data are not, however, required to obey the satisfaction condition. A preinstitution is called an *rps preinstitution* ('reducts preserve satisfaction') if

$$M \models \sigma\varphi \quad \text{implies} \quad M|_\sigma \models \varphi$$

for all M, σ, φ, and an *eps preinstitution* ('extensions preserve satisfaction') if the reverse implication holds.

Let PI_1, PI_2 be preinstitutions. A *preinstitution comorphism* [24] $\mu : PI_1 \to PI_2$ consists of the same data (Φ, α, β) as an institution comorphism (in particular, sentence translation is covariant and model translation is contravariant), without however being required to obey the satisfaction condition as in Definition 2. A preinstitution comorphism μ is called *rps* if

$$M \models \alpha\varphi \quad \text{implies} \quad \beta M \models \varphi,$$

and *weakly eps* if a model M satisfies $\alpha\varphi$ whenever $\beta K \models \sigma\varphi$ for all K, σ such that $K|_{\Phi\sigma} = M$.

Thus, an institution is a preinstitution that is simultaneously rps and eps, and a preinstitution comorphism between two institutions is an institution comorphism iff it is rps and weakly eps.

The typical remedy used hitherto to obtain institutions in the presence of the mentioned features is to restrict signature morphisms to cases where the full satisfaction condition holds. We discuss this point in more detail in Section 9; here, we just note that this is not an acceptable solution for the case of polymorphism: one has to require that signature morphisms do not introduce additional types, a restriction that effectively prevents the use of structured specifications. We emphasize that this problem is *not* solved by treating quantified types as first-class types (higher rank polymorphism), even if one manages to work around the obstacle that the latter is inconsistent with higher order logic [10]: e.g., the restriction that signature morphisms be surjective on types is imposed also in [28], where it is needed in order to ensure preservation of coherent families of domains in a semantics of higher rank polymorphism in the style of Reynolds. In other words, ensuring coherence of polymorphic operators model-theoretically is not a feasible option.

For plain shallow polymorphism without type classes, a further alternative is to interpret the range of quantification over type variables in a fixed universe of types, i.e. some collection of sets closed under a number of constructions, rather than in the syntactical universe of declared types. This is the approach taken e.g. in [7, 19]; it is not apparently suitable for HASCASL and similar frameworks for two reasons:

- in connection with a Henkin style semantics of function types, it is unclear what closure of the type universe under function types means;
- the type universe does not give an indication of what the interpretation of type classes should be, in particular since type classes on the one hand can be entirely loose and on the other hand are meant to contain only explicitly declared instances rather than, say, all structures matching the interface.

Independently of these specific issues, a further general disadvantage of the universe approach is that the choice of a universe unduly influences semantic consequence — the particularities of the chosen universe may induce unintended semantic consequences in a rather unpredictable way, thus introducing an unnecessary degree of incompleteness of deduction. The solution chosen in the semantics of HASCASL is therefore to add a second level to the model semantics according to the general construction described below.

5   Generic Polymorphism

We now introduce a general notion of syntactic polymorphism in an institution which covers HASCASL's type class polymorphism as a special case. This construction provides a wide range of examples of rps preinstitutions. We will return to this example in Section 7, where we will show that the notion of semantic consequence between polymorphic formulae induced by our generic construction of institutions from preinstitutions is not only in accordance with intuitive expectations, but also greatly simplifies the original notion.

Our construction of polymorphic formulae is similar in spirit to the *open formulae* introduced in [37]: given a signature Σ_1, an open Σ_1-formula is just a sentence ϕ in some extension Σ_2 of Σ_1, and a Σ_1-model M satisfies such a formula if all its expansions to Σ_2 satisfy ϕ. In typical algebraic settings, this produces exactly the right kind of first or higher order quantification if Σ_2 introduces only additional constants or function symbols, respectively; essentially, the new symbols then play the role of universally quantified variables. However, the given notion of satisfaction is rather too strong if Σ_2 introduces additional types; since new sorts and function symbols involving new sorts (including instances of polymorphic operators for new sorts) can be interpreted with arbitrary malevolence in extensions of M, most open formulae involving such a Σ_2 will in fact be unsatisfiable.

Thus, we need a relaxed notion of satisfaction in order to arrive at the right notion of universal quantification over types. The idea is to require satisfaction of ϕ as above not for *all* extensions of M, but only for *extensions by syntactic definition*, i.e. the new signature items in Σ_2 have to be interpreted in terms of the base signature Σ_1. Of course, the involved notion of interpretation will have to be sufficiently general. E.g., we will want to interpret function symbols by terms, type constants by composite types etc. — in other words, we will need to use derived signature morphisms. All this is formalized as follows.

Definition 5. An *institution with signature variables* is an institution I with a distinguished object-full subcategory **Var** of the signature category **Sign** (i.e. **Var** need not be full in **Sign**, but contains all objects of **Sign**) whose morphisms are called *signature variables*. Signature variables are assumed to be *pushout-stable*, i.e. pushouts of signature variables along **Sign**-morphisms exist and are signature variables. (Morphisms in **Sign** should be thought of as derived signature morphisms.)

In I, a *polymorphic formula* $\forall \sigma.\, \phi$ over a signature Σ_1 consists of a signature variable $\sigma : \Sigma_1 \to \Sigma_2$ and a Σ_2-sentence ϕ. A Σ_1-model M *satisfies* $\forall \sigma.\, \phi$ if

$$M \models \tau\phi \qquad \text{for all } \tau \text{ in } \mathbf{Sign} \text{ such that } \tau \circ \sigma = id.$$

A sentence $\tau\phi$ as above is called an *instance* of $\forall \sigma.\, \phi$. The *translation* $\rho(\forall \sigma.\, \phi)$ of $\forall \sigma.\, \phi$ along a signature morphism $\rho : \Sigma_1 \to \Sigma_3$ is defined to be $\forall \bar\sigma.\, \bar\rho\phi$, where

$$
\begin{array}{ccc}
\Sigma_2 & \xrightarrow{\ \bar\rho\ } & \bullet \\[4pt]
{\scriptstyle \sigma}\uparrow & & \uparrow{\scriptstyle \bar\sigma} \\[4pt]
\Sigma_1 & \xrightarrow[\ \rho\ ]{} & \Sigma_3
\end{array}
$$

is a pushout; note that $\bar\sigma$ is indeed a signature variable. (This definition determines the translation only up to isomorphism; for similar reasons as given in Remark 5.1. of [37], this is not actually a problem.)

The *polymorphic preinstitution* Poly(I) *over* I is given as follows: the notions of signature, model, and model reduction are inherited from I; Σ-sentences are polymorphic formulae over Σ; satisfaction and sentence translation are as above.

The sentences of I can be coded in Poly(I): a Σ-sentence ϕ in I is equivalent to the polymorphic formula $\forall id_\Sigma.\, \phi$, where id_Σ is indeed a signature variable thanks to object-fullness of **Var** in **Sign**.

Example 6. An example of the syntactic polymorphism described above is HASCASL's type class polymorphism. The base institution is essentially as in the first level of the HASCASL semantics, except that polymorphic sentences are excluded, so that we actually obtain an institution rather than an rps preinstitution. (Note that the base institution does have polymorphic types and operators. In particular, signature morphisms can translate polymorphic operators only as a whole, not instance by instance.) Signature morphisms are, as announced above, *derived signature morphisms* which map

- operator constants to terms;
- type constructors to λ-expressions which denote composite type constructors possibly containing subtype formation; and
- classes to subsets of the syntactic type universe.

A signature variable in this institution is an injective *plain* HASCASL signature morphism (which maps types to types, operators to operators etc. as usual) which is bijective on all syntactic entities except types. (This illustrates the necessity of the restricted cocompleteness requirement for institutions with signature variables: pushouts of derived signature morphisms in general fail to exist, while pushouts of derived signature morphisms along signature variables do exist; this phenomenon is typical of derived signature morphisms in general.) Then, polymorphic formulae and their satisfaction as defined above coincide with the corresponding notions in HASCASL as explained in Section 3. E.g., if $\sigma : \Sigma_1 \hookrightarrow \Sigma_2$ extends Σ_1 by a single new type constant a, then the polymorphic formula $\forall \sigma. \phi$ is equivalent to the polymorphic HASCASL sentence $\forall a : Type. \phi$: the left inverses τ of σ correspond to the possible instantiations of the type variable a in Σ_1. Note that the interpretation of instances of polymorphic operators involving a is forced by the interpretation of a, since, as emphasized above, signature morphisms map polymorphic operators as single entities.

By the above example and Section 4, it is clear that the polymorphic preinstitution $\text{Poly}(I)$ will in general fail to be an institution. However, we have

Theorem 7. *The polymorphic preinstitution* $\text{Poly}(I)$ *is an rps preinstitution.*

6   A Generic Institutionalization

We now describe a general process that transforms preinstitutions into institutions. We begin with a heuristic observation regarding the intended meaning of polymorphic definitions. Consider the specification

> **spec** COMPOSITION =
>   **vars** $a, b, c : Type$
>   **op** $comp : (b \to c) \to (a \to b) \to a \to c$
>   **vars** $f : b \to c; \; g : a \to b$
> $\bullet \; comp \; f \; g = \lambda x : a \bullet f \, (g \; x)$

where for the sake of the argument we abuse HASCASL as a notation for the simply typed λ-calculus with shallow polymorphism in much the same sense as described in Section 3, the only real point of this being the assumption that, unlike in actual HASCASL, there is no unit type. On the first level of the semantics as described in Section 3, COMPOSITION is model-theoretically entirely vacuous, since the syntactic set of types is empty and hence the polymorphic axiom is trivially satisfied in 'all' models of the signature (there is in fact only one model, since the signature is effectively empty). This is clearly not the intention of COMPOSITION. Indeed this specification is necessarily meant as a building block for other specifications that import the polymorphic operator and its definition, which then induce instances according to the ambient signature. In other words, the real purpose of COMPOSITION is apparently to say something about the interpretation of *comp* at *all* types, even those not yet declared. Thus, a model of the specification should contain information not only about the interpretation of the presently declared signature, but also about all 'future' extensions of this interpretation. This is the motivation for the following definitions:

Definition 8. Let *PI* be a preinstitution. An *extended model* of a signature Σ_1 is a pair (N, σ), where $\sigma : \Sigma_1 \to \Sigma_2$ is a signature morphism and N is a Σ_2-model in *PI*. The *reduct* $(N, \sigma)|_\tau$ of (N, σ) along a signature morphism τ is $(N, \sigma \circ \tau)$. The extended model (N, σ) *satisfies* a sentence φ if

$$N \models \sigma\varphi$$

in *PI*.

We record explicitly

Theorem and Definition 9. *The extended models, together with the original notions of signature and sentence from PI, form an institution, called the* institution of extended models *and denoted* $\mathrm{Ext}(PI)$.

Proof. Functoriality of reduction is easy to see. To check the satisfaction condition, let $\tau : \Sigma_1 \to \Sigma_2$ be a signature morphism, let φ be a Σ_1-sentence, and let (N, σ) be an extended Σ_2-model. Then $(N, \sigma) \models \tau\varphi$ in $\mathrm{Ext}(PI)$ iff $N \models \sigma\tau\varphi$ in *PI* iff $(N, \sigma)|_\tau = (N, \sigma \circ \tau)$ satisfies φ in $\mathrm{Ext}(PI)$. $\square$

The semantic consequence relation in $\mathrm{Ext}(PI)$ is precisely as expected:

Proposition 10. *A Σ_1-sentence ψ is a semantic consequence of a set Φ of Σ_1-sentences in* $\mathrm{Ext}(PI)$ *iff*

$$\sigma\Phi \models \sigma\psi$$

in PI for each signature morphism $\sigma : \Sigma_1 \to \Sigma_2$.

Proof. 'If': trivial.

'Only if': let $\sigma : \Sigma_1 \to \Sigma_2$ be a signature morphism, and let N be a Σ_2-model such that $N \models \sigma\Phi$ in *PI*. Then the extended model (N, σ) satisfies Φ and hence also ψ, i.e. we have $N \models \sigma\psi$. $\square$

That is, a formula is a semantic consequence of a specification $Sp = (\Sigma, \Phi)$ (where Φ is a set of Σ-sentences) iff this is the case, in PI, in all extensions of Sp.

Example 11. In the example specification COMPOSITION from Section 4, all formulae are semantic consequences on the first level, i.e. in PI, since all formulae are vacuously true. This pathology disappears in $\mathrm{Ext}(PI)$, where semantic consequences of the specification are only those formulae that follow from the definition of composition independently of how many types are introduced, such as e.g. associativity of composition. Thus, the notion of semantic consequence at the second level, unlike the one at the first level, conforms to intuitive expectations. We will make this more precise in Section 7.

One can give a concise description of extensions in $\mathrm{Ext}(PI)$:

Lemma 12. *The extensions of an extended model* (N, τ) *along a signature morphism* σ *are precisely the extended models* (N, ρ) *where* $\tau = \rho \circ \sigma$.

We can represent PI in $\mathrm{Ext}(PI)$ by a preinstitution comorphism (cf. Definition 4)

$$\eta : PI \to \mathrm{Ext}(PI)$$

which is the identity on signatures and sentences, and takes every extended model to its base model.

Proposition 13. *The comorphism* η *is weakly eps. Moreover,* η *is rps if* PI *is rps.*

Remark 14. Interestingly, the concept of extended model is close to the *very abstract* or *hyper-loose semantics* as introduced in [9, 30], where models may interpret more symbols than just the ones named in their signature. This is used e.g. in the semantics of RSL [14].

There are two crucial differences here. The first is of motivational nature: the purpose of very abstract semantics is to ensure that refinement is model class inclusion; there is no intended connection with repairing the satisfaction condition, and in fact, the construction described in [9] is explicitly intended as a construction *on institutions* (one of the example applications given in [9, 30] is to the institution of many-sorted first order logic). Note that, when applied to institutions, the very abstract semantics is equivalent to the original semantics in terms of the engendered semantic consequence relation.

Secondly, at a more technical level, the phrase 'models may interpret additional symbols' means that very abstract semantics limits the notion of model to extended models with injective signature morphisms; the main technical content of [9] is to solve the difficulties caused by this restriction w.r.t. model reduction. For the purposes pursued here, the restriction to injective extensions is not only unnecessary, but would indeed invalidate our main result; i.e. for models of polymorphism modeled along the construction of [9], the satisfaction condition would still fail.

Taking PI as the first level of the HASCASL semantics (cf. Section 3), we define the *second level* of the semantics [36] to be given by $\mathrm{Ext}(PI)$.

7   Semantic Consequence for Generic Polymorphism

We now investigate the implications of the extended model construction explained in Section 6 in relation to the generic polymorphism introduced in Section 5 — recall that generic polymorphism in general leads only to an rps preinstitution. For the remainder of this section, let I be an institution with signature variables, and let $\mathrm{Poly}(I)$ denote the polymorphic preinstitution over I as defined in Section 5.

Let $\forall\sigma.\,\phi$ and $\forall\rho.\,\psi$ be polymorphic formulae over a signature Σ_1. It is easy to check that $\forall\rho.\,\psi$ is a semantic consequence of $\forall\sigma.\,\phi$ in $\mathrm{Poly}(I)$ iff

$$\{\tau\phi \mid \tau \circ \sigma = id\} \models \pi\psi$$

in I for each signature morphism π such that $\pi \circ \rho = id$. This is rather unpleasant, since it means we have to prove a possibly infinite number of semantic consequences, one for each instance $\pi\psi$ of $\forall\rho.\,\psi$ in Σ_1. Fortunately, the (stronger) notion of semantic consequence in the institution $\mathrm{Ext}(\mathrm{Poly}(I))$ is much more tractable:

Theorem 15. *In* $\mathrm{Ext}(\mathrm{Poly}(I))$, $\forall\rho.\,\psi$ *is a semantic consequence of* $\forall\sigma.\,\phi$ *iff*

$$\rho(\forall\sigma.\,\phi) \models \psi$$

in $\mathrm{Poly}(I)$ *(or, since ψ enjoys eps, equivalently in* $\mathrm{Ext}(\mathrm{Poly}(I))$)

(Recall that $\rho(\forall\sigma.\,\phi) = \forall\bar\sigma.\,\bar\rho\phi$, where $(\bar\rho, \bar\sigma)$ is the pushout of (σ, ρ)). The above condition can be equivalently rephrased as the semantic consequence

$$\{\lambda\phi \mid \lambda \circ \sigma = \rho\} \models \psi \qquad\qquad (*)$$

in I. Thus, unlike proofs of semantic consequence in $\mathrm{Poly}(I)$ as described above, proofs in $\mathrm{Ext}(\mathrm{Poly}(I))$ are actually feasible, since we have to prove only a single generic instance of the goal, rather than all instances that exist in the base signature due to pure syntactic happenstance. Moreover,

> *any sound and complete deduction system for I induces a sound and complete deduction system for* $\mathrm{Ext}(\mathrm{Poly}(I))$,

while for $\mathrm{Poly}(I)$, one will in general only obtain a sound but not complete deduction system.

The formulation of semantic consequence given in the theorem is exactly what one would intuitively expect: we fix the additional syntactic material quantified over by ρ and prove ψ only for this fixed instance; in the proof, we are allowed to make use of all instances of ϕ, including instances involving the new syntactic material. Proofs of polymorphic formulas e.g. in Isabelle [29] work in precisely this way, which we have now provided with a semantic foundation.

Proof (Theorem 15). 'Only If': by Proposition 10, we have $\rho(\forall\sigma.\,\phi) \models \rho(\forall\rho.\,\psi)$, and ψ is an instance of $\rho(\forall\rho.\,\psi)$. The latter follows from the universal property of the pushout of ρ with itself.

'If': let Σ_1 be the base signature of $\forall\sigma.\,\phi$ and $\forall\rho.\,\psi$, let $\kappa : \Sigma_1 \to \Sigma_2$ be a signature morphism, and let

$$
\begin{array}{ccc}
\bullet \xrightarrow{\ \bar\kappa_\sigma\ } \bullet & & \bullet \xrightarrow{\ \bar\kappa_\rho\ } \bullet \\
\sigma\uparrow\quad\quad\uparrow\bar\sigma & \text{and} & \rho\uparrow\quad\quad\uparrow\bar\rho \\
\Sigma_1 \xrightarrow[\kappa]{} \Sigma_2 & & \Sigma_1 \xrightarrow[\kappa]{} \Sigma_2
\end{array}
$$

be the associated pushout diagrams. Then $\kappa(\forall\sigma.\,\phi) = \forall\bar\sigma.\,\bar\kappa_\sigma\phi$ and $\kappa(\forall\rho.\,\psi) = \forall\bar\rho.\,\bar\kappa_\rho\psi$. By Proposition 10, we thus have to prove

$$
\forall\bar\sigma.\,\bar\kappa_\sigma\phi \models \forall\bar\rho.\,\bar\kappa_\rho\psi
$$

in $\mathrm{Poly}(I)$, i.e. given a model M such that $M \models \forall\bar\sigma.\,\bar\kappa_\sigma\phi$ and τ such that $\tau\bar\rho = id$, we have to show $M \models \tau\bar\kappa_\rho\psi$ in I. Since semantic consequence in I is stable under translation, this reduces by $(*)$ above to showing $M \models \tau\bar\kappa_\rho\lambda\phi$ for all λ such that $\lambda \circ \sigma = \rho$. For such a λ, we have $\tau\bar\kappa_\rho\lambda\sigma = \tau\bar\kappa_\rho\rho = \tau\bar\rho\kappa = \kappa$, so that the pushout property yields ν such that $\nu\bar\sigma = id$ and $\nu\bar\kappa_\sigma = \tau\bar\kappa_\rho\lambda$. Then M satisfies the instance $\nu\bar\kappa_\sigma\phi$ of $\forall\bar\sigma.\,\bar\kappa_\sigma\phi$; but $\nu\bar\kappa_\sigma\phi = \tau\bar\kappa_\rho\lambda\phi$. $\square$

8   Model-Theoretic Conservativity

While the semantic consequence relation engendered by the extended model construction is without further ado precisely the 'right' one, the issue of model expansion, i.e. of conservativity in the model-theoretic sense as used e.g. in CASL, is somewhat more subtle. We recall a few definitions:

Definition 16. A *theory* in a (pre-)institution is a pair $Sp = (\Sigma, \Phi)$ consisting of a signature Σ and a set Φ of Σ-sentences. A *model* of Sp is a Σ-model M such that $M \models \Phi$. A theory is *consistent* if it has a model. A signature morphism $\sigma : \Sigma_1 \to \Sigma_2$ is a *theory morphism* $(\Sigma_1, \Phi_1) \to (\Sigma_2, \Phi_2)$ if

$$
\Phi_2 \models \sigma\Phi_1.
$$

A theory morphism $\sigma : Sp_1 \to Sp_2$ is *model-theoretically conservative* or *model-expansive* if every model M of Sp_1 has an Sp_2-extension, i.e. a model N of Sp_2 such that $N|_\sigma = M$.

Notice that by Proposition 10 and Example 11, the notion of theory morphism in $\mathrm{Ext}(PI)$ is in general properly stronger than in PI.

Proposition 17. *A theory is consistent in an rps preinstitution PI iff it is consistent in $\mathrm{Ext}(PI)$.*

Typical extensions that would be expected to be model-expansive e.g. in HASCASL are (recursive) function definitions, loose declarations of new signature elements, and declarations of free datatypes. An apparent obstacle to model-expansivity of such extensions at the second level of the semantics is Part (i) of the following observation:

Proposition 18. *Let PI be an rps preinstitution, and let $\sigma : (\Sigma_1, \Phi_1) \to (\Sigma_2, \Phi_2)$ be a theory morphism in* $\text{Ext}(PI)$. *Then the following holds:*

(i) If σ is model-expansive in $\text{Ext}(PI)$ *and (Σ_1, Φ_1) is consistent, then σ is a section as a signature morphism; i.e. there exists a signature morphism $\tau : \Sigma_2 \to \Sigma_1$ such that $\tau \circ \sigma = id$.*

(ii) If σ is a section as a theory morphism in $\text{Ext}(PI)$, *i.e. there exists a theory morphism $\tau : (\Sigma_2, \Phi_2) \to (\Sigma_1, \Phi_1)$ such that $\tau \circ \sigma = id$, then σ is model-expansive.*

Proof. *(i):* By assumption and the rps condition, (Σ_1, Φ_1) has a model (M, id) in $\text{Ext}(PI)$. By Lemma 12, existence of an extension of this model along σ implies that σ is a section.

(ii): Straightforward. □

When plain signature morphisms are used, which typically map type constants to type constants, operators to operators etc., then the necessary condition above is clearly too restrictive; essentially, the only model-expansive extensions one obtains are those that define symbols by other symbols already present. The solution to this is to use *derived* signature morphisms instead, which typically are allowed to map, say, type constants to composite types, operators to terms, and the like; by the sufficient condition (ii) above, one then obtains as model-expansive extensions all declarations and definitions which can be implemented by some composite object in the present theory.

In the case of HASCASL, the notion of derived signature morphism required here is the one already given in Example 6. Thanks to the richness of HASCASL specifications, the model-expansive extensions are indeed the expected ones under this definition; this includes

- equational definitions
- well-founded recursive definitions of functions into types that admit a unique description operator [34]
- general recursive definitions over cpo's
- inductive datatype definitions, provided that the base theory already contains the natural numbers (this is a categorical result inherited from topos theory [22])
- class declarations.

In general, it depends on the expressive power of signatures and theories in the preinstitution at hand whether or not using derived signature morphisms leads to a satisfactory notion of model-expansivity. It should however be noted that there is usually quite some latitude in the definition of derived signature morphism; many forms of extensions can be made model-expansive by just giving a more liberal definition of what a derived signature morphism can do.

9   Application to Other Frameworks

We now briefly discuss the effects of the extended model construction in other frameworks where the satisfaction condition may fail, to wit, in observational and state-based frameworks as described in Section 4. Of course, the construction will always work in principle; however, the question remains whether or not the ensuing semantic modifications are methodologically desirable, and what the actual benefits are. Here, we will concentrate on two issues:

A) Is the notion of semantic consequence engendered by the extended model construction the expected one? I.e., in view of Proposition 10, is semantic consequence intended to be independent of the surrounding signature?
B) Is the alternative solution of restricting signature morphisms acceptable?

We have seen that, in the case of type class polymorphism, the answer is 'yes' to Question A) and 'no' to Question B): semantic consequences that hold only due to the particular nature of the presently declared types can be regarded as unwanted side effects, and limiting signature morphisms to be surjective on types is not an option.

The situation is different with observational satisfaction. It is precisely the point of having distinguished observable operations or sorts that these govern the notion of observational equality, and moreover that the given set of observers determines a proof principle for observational equality, namely coinduction. This proof principle is lost when extended models are considered (in a setting with unrestricted signature morphisms): since deduction then has to work within arbitrary signature extensions that may introduce any number of additional observers, the notion of semantic consequence for extended models is just semantic consequence in standard equational logic. This is clearly not the desired effect, so that the notion of extended model cannot in fact be considered suitable for observational specification. It is thus lucky that, given this negative answer to Question A), the answer to Question B) is affirmative: it is common practice to restrict signature morphisms of observational specifications in such a way that extensions never introduce new observers [5, 16]. This forces a specification style where all observers are introduced in one go at the beginning, being regarded as constituting the requirements on the system, and the non-observable part, i.e. the implementation, is added later; indeed, this specification style is explicitly advocated e.g. in [20].

Finally, let us have a look at the specification of stateful systems in the states-as-algebras paradigm as used e.g. in the specification language SB-CASL [4]. The problem here, as pointed out in Section 4, are so-called *dynamic equations* between program-like expressions called *transition terms* in SB-CASL (besides these dynamic equations, SB-CASL also features pre- and postconditions, which are however unproblematic w.r.t. the satisfaction condition). The purpose of dynamic equations lies both in the (possibly recursive) definition of procedures and in their loose equational specification e.g. as inverses of other procedures (a very simple example of this is given in [4]). As indicated in Section 4, dynamic equations may break in model expansions when signature morphisms introduce additional state components.

The methodology of state-based specification in this sense is not as yet well developed, so that we feel entitled to pitch in our own bit of philosophy, as follows. Concerning Question B) above, it seems undesirable to have a development paradigm where the specification process starts with defining the entire state space in full detail and only then allows the formulation of requirements for programs that work on this state space; to the contrary, one would normally wish to start with the requirements, mentioning only the parts of the state space relevant for input and output, and then work out the detailed design of the state space. As to Question A), it appears for rather the same reason that semantic consequences that hold only due to an insufficiently detailed description of the state space should be regarded as spurious, so that the notion of semantic consequence induced by the extended model construction is indeed an improvement over the original one. As an extreme example, consider a specification that introduces some procedure names, but no dynamic signature components at all (presumably with the intention to specify these in later extensions), i.e. induces a trivial state space; in SB-CASL, such a specification might look as follows:

> **spec** SP =
>> **proc** p, q
>> **pre** $p : True$
>> **pre** $q : True$

(the two preconditions express that p and q terminate). Then, unless extended models are used, any two terminating programs (transition terms) would be equal, i.e. their equality is a semantic consequence of the precondition expressing their termination; in particular, the above specification implies the dynamic equation $p = q$. We argue that this sort of semantic consequence is actually a pathology, which is eliminated by our extended model construction.

10   Conclusion

Starting from the problem that type class polymorphism does not enjoy the satisfaction condition of institutions, but only the *reduction preserves satisfaction* (rps) half, we suggest a general construction of institutions from preinstitutions. The construction is based on the idea that a model of a specification should contain information not only about the interpretation of the presently declared signature, but also about all 'future' extensions of this interpretation. Consequently, the *extended models* of a signature are defined to consist of a signature extension and a model of the extended signature. The arising notion of semantic consequence is the expected one, namely, semantic consequence in all signature extensions in the original preinstitution. Moreover, in sufficiently rich logics such as the HASCASL logic, one also obtains the expected model-expansive extensions.

The semantics of polymorphism used in HASCASL makes use of this result, so that HASCASL does indeed fit into the institution-independent framework of CASL. We have also investigated the use of our construction in other frameworks where the satisfaction condition fails for unrestricted signature morphisms, the result being that the implications of our constructions are methodologically undesirable in the case of observational satisfaction, but beneficial in the case of dynamic equations in a states-as-algebras framework. The suitability of our approach for security formalisms, which also exhibit the phenomenon that security assertions tend to be unstable under refinement [8], is under investigation.

A particularly pleasing point is that HASCASL's polymorphic sentences can be subsumed under a general definition of polymorphic formulae over institutions; the extended model construction, when applied to such generic polymorphic frameworks, leads to a very natural notion of semantic consequence which agrees with proof principles used e.g. in Isabelle [29]. In this sense, our method provides a semantic basis for existing proof methods.

Acknowledgements

The authors wish to thank Hubert Baumeister for valuable information about SB-CASL and Tom Maibaum for useful hints and discussions.

References

[1] J. Adámek, H. Herrlich, and G. E. Strecker, *Abstract and concrete categories*, Wiley Interscience, 1990.

[2] J. Barwise, *Axioms for abstract model theory*, Ann. Math. Logic **7** (1974), 221–265.

[3] H. Baumeister, *An institution for SB-CASL*, talk presented at the 15th International Workshop on Algebraic Development Techniques, Genova, 2001.

[4] H. Baumeister and A. Zamulin, *State-based extension of* CASL, Integrated Formal Methods, LNCS, vol. 1945, Springer, 2000, pp. 3–24.

[5] M. Bidoit and R. Hennicker, *On the integration of observability and reachability concepts*, Foundations of Software Science and Computation Structures, LNCS, vol. 2303, Springer, 2002, pp. 21–36.

[6] M. Bidoit and P. D. Mosses, CASL *user manual*, LNCS, vol. 2900, Springer, 2004.

[7] T. Borzyszkowski, *Higher-order logic and theorem proving for structured specifications*, Recent Trends in Algebraic Development Techniques, (WADT 99), LNCS, vol. 1827, Springer, 2000, pp. 401–418.

[8] A. Bossi, R. Focardi, C. Piazza, and S. Rossi, *Refinement operators and information flow security*, Software Engineering and Formal Methods, IEEE Computer Society Press, 2003, pp. 44–53.

[9] M. Cerioli and G. Reggio, *Very abstract specifications: a formalism independent approach*, Math. Struct. Comput. Sci. **8** (1998), 17–66.

[10] T. Coquand, *An analysis of Girard's paradox*, Logic in Computer Science, IEEE, 1986, pp. 227–236.

[11] R. Diaconescu, J. Goguen, and P. Stefaneas, *Logical support for modularisation*, Workshop on Logical Frameworks, Programming Research Group, Oxford University, 1991.

[12] F. Durán and J. Meseguer, *Structured theories and institutions*, Category Theory and Computer Science, ENTCS, vol. 29, 1999.

[13] H. Ehrig and B. Mahr, *Fundamentals of algebraic specification 2*, Springer, 1990.

[14] C. George, P. Haff, K. Havelund, A. E. Haxthausen, R. Milne, C. Bendix Nielson, S. Prehn, and K. R. Wagner, *The Raise Specification Language*, Prentice Hall, 1992.

[15] J. Goguen and R. Burstall, *Institutions: Abstract model theory for specification and programming*, J. ACM **39** (1992), 95–146.

[16] J. Goguen and G. Malcolm, *A hidden agenda*, Theoret. Comput. Sci. **245** (2000), 55–101.

[17] J. Goguen and G. Rosu, *Institution morphisms*, Formal aspects of computing **13** (2002), 274–307.

[18] J. Goguen and W. Tracz, *An implementation-oriented semantics for module composition*, Foundations of Component-Based Systems, Cambridge, 2000, pp. 231–263.

[19] R. Kubiak, A. Borzyszkowski, and S. Sokolowski, *A set-theoretic model for a typed polymorphic lambda calculus — a contribution to MetaSoft*, VDM: The Way Ahead, LNCS, vol. 328, Springer, 1988, pp. 267–298.

[20] A. Kurz, *Logics for coalgebras and applications to computer science*, Ph.D. thesis, Universität München, 2000.

[21] J. Meseguer, *General logics*, Logic Colloquium 87, North Holland, 1989, pp. 275–329.

[22] I. Moerdijk and E. Palmgren, *Wellfounded trees in categories*, Ann. Pure Appl. Logic **104** (2000), 189–218.

[23] E. Moggi, *Categories of partial morphisms and the λ_p-calculus*, Category Theory and Computer Programming, LNCS, vol. 240, Springer, 1986, pp. 242–251.

[24] T. Mossakowski, *Representations, hierarchies and graphs of institutions*, Ph.D. thesis, Universität Bremen, 1996, also: Logos-Verlag, 2001.

[25] _____, *Foundations of heterogeneous specification*, Recent Trends in Algebraic Development Techniques (WADT 02), LNCS, vol. 2755, Springer, 2003, pp. 359–375.

[26] _____, HETCASL - *heterogeneous specification. Language summary*, 2004, http://www.tzi.de/cofi/hetcasl.

[27] P. D. Mosses (ed.), CASL *reference manual*, LNCS, vol. 2960, Springer, 2004.

[28] M. Nielsen and U. Pletat, *Polymorphism in an institutional framework*, Tech. report, Technical University of Denmark, 1986.

[29] T. Nipkow, L. C. Paulson, and M. Wenzel, *Isabelle/HOL — a proof assistant for higher-order logic*, LNCS, vol. 2283, Springer, 2002.

[30] P. Pepper, *Transforming algebraic specifications – lessons learnt from an example*, Constructing Programs from Specifications, Elsevier, 1991, pp. 1–27.

[31] S. Peyton-Jones (ed.), *Haskell 98 language and libraries — the revised report*, Cambridge, 2003, also: J. Funct. Programming **13** (2003).

[32] A. Salibra and G. Scollo, *A soft stairway to institutions*, Recent Trends in Data Type Specification (WADT 91), LNCS, vol. 655, Springer, 1993, pp. 310–329.

[33] D. Sannella and A. Tarlecki, *Specifications in an arbitrary institution*, Information and Computation **76** (1988), 165–210.

[34] L. Schröder, *The* HASCASL *prologue: categorical syntax and semantics of the partial λ-calculus*, available as http://www.informatik.uni-bremen.de/~lschrode/hascasl/plam.ps

[35] L. Schröder and T. Mossakowski, HASCASL: *Towards integrated specification and development of functional programs*, Algebraic Methodology And Software Technology, LNCS, vol. 2422, Springer, 2002, pp. 99–116.

[36] L. Schröder, T. Mossakowski, and C. Maeder, HASCASL – *Integrated functional specification and programming. Language summary*, available under http://www.informatik.uni-bremen.de/agbkb/forschung/formal_methods/CoFI/HasCASL

[37] A. Tarlecki, *Quasi-varieties in abstract algebraic institutions*, J. Comput. System Sci. **33** (1986), 333–360.

[38] M. Wenzel, *Type classes and overloading in higher-order logic*, Theorem Proving in Higher Order Logics, LNCS, vol. 1275, Springer, 1997, pp. 307–322.

Architectural Specifications for Reactive Systems⋆

Artur Zawłocki

Warsaw University, Institute of Informatics,
Banacha 2, 02-097 Warszawa, Poland
zawlocki@mimuw.edu.pl

Abstract. This paper concerns the problem of building reactive systems
in a modular way.

Several institutions have been proposed for the specification of reactive systems throughout the last fifteen years. Based on the institutions,
formalisms for the incremental construction of system specifications have
been developed. Related problem of modular construction of system implementations has received less attention. This paper is the first attempt
to use *architectural specifications* of CASL for that purpose. The semantics of the architectural specifications is based on the underlying institution. We argue that none of the institutions defined so far for reactive
systems is appropriate as a basis for architectural specifications, and
therefore we propose another one, better suited for this task.

We also show how to express synchronisation of reactive systems using
implementation-building operations of CASL architectural specifications.

1 Introduction

This paper concerns the problem of building reactive systems in a modular fashion. Reactive systems can be formalised in various ways. A simple and intuitive
model adopted in this paper is a *labelled transition system* (cf e.g. [NW95–Sec.
2]), enriched with an assignment of *data structures* to system states.

To avoid confusion we reserve the term *system* to mean a labelled transition
system and will refer to reactive systems as *reactive components*, or *components*
for short.

For specifying reactive components we intend to use CTL⋆ logic ([GHR95]),
expressive enough to describe *safety*, *liveness* and *fairness* properties.

1.1 Structuring Specifications

Institutions ([GB92]) provide semantics for the specification-building operations,
as defined e.g. in [ST84], independent of the underlying logic. Operations for

---

⋆ This research was sponsored by the EC 5th Framework project AGILE: Architectures
for Mobility (IST-2001-32747).

J.L. Fiadeiro, P. Mosses, and F. Orejas (Eds.): WADT 2004, LNCS 3423, pp. 252–269, 2005.

composing complicated specifications out of the smaller ones have been incorporated into several specification languages, such as CASL ([CoFI04]) or CAFEOBJ ([DF98]). They can be used for constructing specifications of reactive components in a modular fashion, provided that an appropriate institution is defined, and in the past fifteen years several suitable institutions have been proposed.

Dynamic algebras of Costa and Reggio ([CR97]) and related *entity algebras* of Reggio ([Reg90]) model reactive components as structures of many-sorted partial first-order logic, with distinguished sorts for the states and the labels of a transition system. Transitions are represented as ternary predicates. Static data and dynamic transitions are thus treated uniformly in a familiar algebraic framework, in particular, both states and transition labels can be manipulated as first-class values. Logic introduced for specifying systems is an extension of first-order logic with CTL*-like temporal modalities. Dynamic algebra approach is used in [ACR99] to propose CASL-LTL extension to CASL.

An alternative extension of the algebraic approach towards modelling reactive systems are *D-oids* of Astesiano and Zucca ([AZ95]). A D-oid is a set of *static structures* modelling configurations of a component (e.g. structures of first-order logic) and a set of *dynamic operations*, associating with each structure, and a tuple of arguments from its carrier, a resulting structure together with a partial mapping between the carriers of the two structures. The mappings make possible tracing the identity of an element of the carrier throughout the sequence of dynamic operations and specifying operations that create, merge or delete elements. Related to D-oids is the SB-CASL extension of CASL by Baumeister and Zamulin ([BZ00]), where transitions between states are specified operationally in terms of state updates. However, the formalism of [BZ00] does not constitute an institution.

At the other end of the spectrum there are less algebraic and more process-oriented approaches, where system transitions are labelled with *action symbols* and component behaviour is described in terms of sequences (or trees, if branching structure is taken into account) of actions. Fiadeiro and Maibaum define in [FM92] an institution for linear temporal logic. They use theories as units for building specifications. Colimits in the category of theories are used to construct large specifications from smaller ones — an idea that is further explored in COMMUNITY specification formalism ([FLW03]).

In [CSS98] Sernadas et al use a generic institution for a class of modal logics as a basis for an object specification formalism. Transition systems are used as models, but an occurrence of an action is represented as an atomic proposition holding in a state rather than as a transition labelled with an action symbol. Similar institution is defined for CTL* in [AF95] — again, the transitions are not labelled.

Perhaps the closest to our work is the institution of Cengarle ([Cen98]). There, a model is a labelled transition system with data structures in system states and transitions labelled with action symbols, while the logic is a variant of CTL*. The institution is shown to have the amalgamation property and structuring of theories is also considered.

1.2   Structuring Models

In contrast with structuring the specifications, much less work has been done concerning structuring the models.

Using a formalism like the one of [FM92] one may be tempted to identify a structure of a reactive component with the structure of its specification. However Fiadeiro and Maibaum remark ([FM92–pp. 8]):

> It is the specification activity we want to make modular (...) a specifier will not actually manipulate models but theories.

In this paper we are interested precisely in "manipulating models".

The specification formalism of COMMUNITY ([FLW03]) may be viewed as a further development of ideas from [FM92]. A COMMUNITY *design* is a specification of a reactive component, a *design morphism* identifies a component specified by its source as a part of a component specified by its target. Specifications of complex systems are given by colimits in the category of designs, the structure of a system is thus reflected in the structure of its specification. However, design morphisms are not specification morphisms in the institutional sense — a clear distinction is made in COMMUNITY between *superposition* and *refinement* of designs.

There are nevertheless purely institutional tools developed for structuring the models, such as *architectural specifications* ([BST99]) of CASL. Although CASL is an algebraic specification language, its architectural specifications are given institution-independent semantics and thus can be also used to structure reactive components. However, there have been no attempts yet to apply them in this domain.

Before discussing architectural specifications we first define transition systems formally. Fixing the semantic domain for reactive system specifications will not restrict severely the choice of an institution whereas it will provide the reader with more intuition.

2   Transition Systems

There are two kinds of atomic observations one can make regarding a reactive component — one can notice when a component performs certain *action* and examine current values of component *attributes*. The set of action symbols and the signature of state attributes constitute the signature of a reactive system.

Definition 1 (System signature). *A system signature is a pair* $\Theta = \langle \Gamma, \Sigma \rangle$, *where*

- Γ *is a set of* action symbols,
- Σ *is a many-sorted first-order* data signature.

We could distinguish a subset of *rigid symbols* in Σ to represent state-independent data, such as arithmetical operations, but we prefer to keep the presentation simple.

Definition 2 (Θ-system). *Let $\Theta = \langle \Gamma, \Sigma \rangle$ be a system signature. A Θ-system is a triple $S = \langle W, D, T \rangle$, where*

- *W is a nonempty set of* system states,
- *$D : W \to \mathbf{Str}(\Sigma)$ is a mapping assigning a* data structure *to each state, such that for every $v, w \in W$, $|D(v)| = |D(w)|$,*
- *$T \subseteq W \times \Gamma \times W$ is a set of* transitions.

The common carrier of data structures of S is denoted by $Univ(S)$, and the class of all Θ-systems is denoted by $Sys(\Theta)$.

As usual, we write $v \xrightarrow{g} w \in T$ whenever $\langle v, g, w \rangle \in T$. We also write $v \xrightarrow{\Delta} w \in T$, for $\Delta \subseteq \Gamma$, if $v \xrightarrow{g} w \in T$ for some $g \in \Delta$. We say that v *has no Δ-successor in S*, written $v \xrightarrow{\Delta}\!\!\!\!\!/\;$ if there is no w such that $v \xrightarrow{\Delta} w \in T$. $v \nrightarrow$ stands for $v \xrightarrow{\Gamma}\!\!\!\!\!/\;$.

The identity of system states is not relevant when one is only interested in data contained in states and transitions between them. Therefore we write $\langle W, D, T \rangle = \langle W', D', T' \rangle$ whenever there is a bijection $i : W \to W'$ such that $D = i; D'$ and $T' = \left\{ i(v) \xrightarrow{g} i(w) \mid v \xrightarrow{g} w \in T \right\}$.

3   Modular System Description

Having fixed the semantic domain we describe the syntax and the semantics of a simplified version of architectural specifications (see [CoFI04–Chapt. 5] for precise definitions) using a small example of a network service.

The service repeatedly performs a sequence of three actions: accepts a request, processes it and delivers a response. Moreover, the requests can be classified into several types (we will distinguish just two of them in the specification) and there is also a possibility that the service will forward a request to some other external network service.

We need three intermediate specifications: *BasicService*, describing the basic accept-process-respond cycle, *MultiService*, distinguishing between two request types, and *FwdService*, specifying the forwarding mechanism.

> **arch spec** *Service* **is**
> > **units** S : *BasicService*
> > > F : *BasicService* $\xrightarrow{\vartheta_1}$ *MultiService*
> > > G : *BasicService* $\xrightarrow{\vartheta_2}$ *FwdService*
> > **result** $F(S)$ **and** $G(S)$

The above architectural specification states that the implementation of *Service* is built with three *units*. S is a *basic unit* that implements *BasicService*. F is a *generic unit* implementing an operation that when given an implementation of *BasicService* produces an implementation of *MultiService*, and G is another generic unit producing an implementation of *FwdService*.

The semantics of architectural specification is institution-based: S denotes a *model* of *BasicService*, F and G denote functions on models and $F(S)$ **and** $G(S)$ denotes the *amalgamation* (see [Tar99]) of models.

A signature Θ_{MS} of the result of F is identified as an extension of a signature Θ_{BS} of its argument by a morphism $\vartheta_1 : \Theta_{BS} \to \Theta_{MS}$.[1] The function denoted by F is required to be *persistent*, i.e. $F(A)|_{\vartheta_1} = A$ must hold for any model A of Θ_{BS}. In the CASL terminology $F(A)$ is an *expansion of A along ϑ_1*.

Specifications of individual units will be just sets of sentences in an institution of our choice. Desirable properties of a candidate institution are the cocompleteness of the category of signatures — making composing the specifications possible — and the amalgamation property, ensuring that the amalgamation of units is well-defined. However when one wants to build models incrementally, using generic unit application and unit amalgamation, it is even more important that each model have many non-trivial expansions along a signature morphism.

In most of the institutions considered in Sec. 1.1, the reduct is defined in such a way, that all expansions of a system have the same sets of states and transitions (modulo relabelling) as the original system.[2] With such an institution adopted for the semantics of the example specification, all the models involved will be forced to share a common set of states and transitions. All than can be done by applying generic units and amalgamating models would be a relabelling of transitions and expanding the data structures in states.

What we would like to do instead, is to expand a system by enlarging its state space and by adding new transitions. To be able to do that, we must propose a new institution as a basis for architectural specifications.

4   Specification Logic

By $f : A \rightharpoonup B$ we denote a partial mapping f from A to B. We write $f(a)\downarrow$ if f is defined for $a \in A$, $f(a)\uparrow$ otherwise. $dom(f)$ denotes the set $\{a \in A \mid f(a)\downarrow\}$. The identity $f(a) = f(b)$ implies that both $f(a)$ and $f(b)$ are defined.

4.1   Prefixes and Runs

Formulas of the specification logic will be tested against *runs* — sequences of consecutive transitions performed by a system and of states encountered. Runs may be either finite or infinite but are required to be maximal.

Definition 3 (Prefix, run). *Let* $S = \langle W, D, T \rangle$ *be a* Θ-*system, where* $\Theta = \langle \Gamma, \Sigma \rangle$. *A prefix in* S *is a pair* $\rho = \langle \rho_s, \rho_a \rangle$, *where* $\rho_s : \mathbb{N} \rightharpoonup W$ *and* $\rho_a : \mathbb{N} \rightharpoonup \Gamma$ *are such that*

---

[1] We drop the CASL requirement that the argument signature must be included in the result signature.

[2] One exception are dynamic algebras, where a model corresponds to a set of transition systems — each for every sort of states. It is possible to expand a model by adding entirely new transition system, but still not by enlarging an existing one.

(i) *either ρ_a is total or exists $l \in \mathbb{N}$ such that, for all $n \in \mathbb{N}$, $\rho_a(n){\downarrow}$ iff $n < l$; in the former case ρ is infinite, in the latter it has length l; we write $len(\rho) = \omega$ or $len(\rho) = l$, respectively,*

(ii) $\rho_s(0){\downarrow}$ *and, for all $n > 0$, $\rho_s(n){\downarrow}$ iff $\rho_a(n-1){\downarrow}$,*

(iii) *for all $n \in \mathbb{N}$ such that $\rho_a(n){\downarrow}$, $\rho_s(n) \xrightarrow{\rho_a(n)} \rho_s(n+1) \in T$,*

ρ *is a Δ-prefix if $\overrightarrow{\rho_a}(\mathbb{N}) \subseteq \Delta$. ρ is Δ-maximal if it is a Δ-prefix and is either infinite or $\rho_s(len(\rho)) \overset{\Delta}{\nrightarrow}$.*

ρ *is a run in S (or S-run) if it is a Γ-maximal prefix. ρ' is a prefix of ρ if $len(\rho') \leq len(\rho)$, $\rho'_s(n) = \rho_s(n)$ for all $n \leq len(\rho')$ and $\rho'_a(n) = \rho_a(n)$ for all $n < len(\rho')$.*

Runs(S) denotes the set of all S-runs.

For a prefix ρ such that $len(\rho) \geq n$ we define its *n-th suffix* $\rho^{(n)}$ in a straightforward way:

$$\rho_s^{(n)}(k) = \rho_s(n+k) \text{ if } \rho_s(n+k){\downarrow}, \text{ otherwise } \rho_s^{(n)}(k){\uparrow}$$

$$\rho_a^{(n)}(k) = \rho_a(n+k) \text{ if } \rho_a(n+k){\downarrow}, \text{ otherwise } \rho_a^{(n)}(k){\uparrow}$$

Prefixes ρ and ρ' are *co-initial* if $\rho_s(0) = \rho'_s(0)$.

4.2   Formulas

The logical part of the institution is a rather obvious extension of the standard CTL* ([GHR95]) to deal with first-order data structures in system states and transition labels (usually, CTL* is interpreted over propositional Kripke frames with a single transition relation).

$$\begin{aligned}
\phi \longrightarrow\ & t_1 = t_2 \ \mid\ P(t_1, \ldots, t_k) \ \mid\ \mathit{ff} \ \mid\ \phi_1 \to \phi_2 \ \mid\ \forall x_s \phi_1 \\
& \boldsymbol{X}_\Delta \phi_1 \ \mid\ \phi_1 \, \boldsymbol{U} \, \phi_2 \ \mid\ \boldsymbol{A} \, \phi_1 \\
& \text{where } P \in \Sigma_k, t_i \in T_\Sigma(\mathcal{X}), x_s \in \mathcal{X}_s \text{ and } \Delta \subseteq \Gamma
\end{aligned}$$

Fig. 1. Syntax of Θ-formulas

The syntax of formulas over Θ, or Θ-formulas, is shown on Fig. 1. We assume a countable many-sorted set $\mathcal{X}$ of individual variables. A formula is either atomic or is built with standard propositional connectives, a quantification over first-order variables and temporal modalities.

Instead of a single $\boldsymbol{X}$ ("next") modality we use a family of modalities indexed with sets of actions. $\boldsymbol{U}$ ("until") and $\boldsymbol{A}$ ("for all runs") are standard CTL* modalities.

Besides usual propositional connectives we also introduce some other abbreviations, shown on Fig. 2, including temporal modalities $\boldsymbol{F}$ ("finally"), $\boldsymbol{G}$ ("globally") and $\boldsymbol{W}$ ("weak until").

Forms(Θ) denotes the set of all Θ-formulas.

$$\begin{array}{ll}
\Delta! \equiv X_\Delta\, tt & E\,\phi_1 \equiv \neg A\, \neg\phi_1 \\
[\Delta]\phi_1 \equiv A\,(\Delta! \Rightarrow X\,\phi_1) & \langle\Delta\rangle\phi_1 \equiv \neg[\Delta]\neg\phi_1 \\
F\,\phi_1 \equiv tt\,U\,\phi_1 & G\,\phi_1 \equiv \neg F\,\neg\phi_1 \\
\phi_1\,W\,\phi_2 \equiv G\,\phi_1 \vee \phi_1\,U\,\phi_2
\end{array}$$

Fig. 2. Abbreviations

4.3   Formula Satisfaction

Θ-formula satisfaction is defined for the runs of a Θ-system in a standard way.

Definition 4 (Formula satisfaction). *For a Θ-system $S = \langle W, D, T \rangle$ we define the relation $\models_S \subseteq Runs(S) \times Univ(S)^{\mathcal{X}} \times Forms(\Theta)$ inductively:*

- $\rho, \xi \models_S P(t_1, \ldots, t_k)$ *iff* $D(\rho_s(0)), \xi \models_\Sigma P(t_1, \ldots, t_k)$,
- $\rho, \xi \models_S t_1 = t_2$ *iff* $D(\rho_s(0)), \xi \models_\Sigma t_1 = t_2$,
- $\rho, \xi \not\models_S f\!f$,
- $\rho, \xi \models_S \phi_1 \to \phi_2$ *iff* $\rho, \xi \models_S \phi_1$ *implies* $\rho, \xi \models_S \phi_2$,
- $\rho, \xi \models_S \forall x_s\phi_1$ *iff* $\rho, \xi[a/x_s] \models_S \phi_1$ *for all* $a \in |Univ(S)|_s$,
- $\rho, \xi \models_S X_\Delta\,\phi_1$ *iff* $len(\rho) > 0$, $\rho_a(0) \in \Delta$ *and* $\rho^{(1)}, \xi \models_S \phi_1$,
- $\rho, \xi \models_S \phi_1\,U\,\phi_2$ *iff* *exists* $n \leq len(\rho)$ *such that* $\rho^{(n)}, \xi \models_S \phi_2$ *and, for all* $k < n$, $\rho^{(k)}, \xi \models_S \phi_1$,
- $\rho, \xi \models_S A\,\phi_1$ *iff* $\rho', \xi \models_S \phi_1$ *for every* $\rho' \in Runs(S)$ *co-initial with* ρ.

We write $\rho \models_S \phi$ if $\rho, \xi \models_S \phi$ for all $\xi \in Univ(S)^{\mathcal{X}}$.

In the above definition $\models_\Sigma$ is a satisfaction relation of many-sorted first-order logic.

For a set $\Delta \subseteq \Gamma$ we define the formula

$$MaxPrefix_\Delta = \Delta!\,W\,A\,\neg\Delta!$$

satisfied precisely by those runs which have a maximal Δ-prefix.

4.4   Sentences

We now extend the satisfaction relation from single runs to sets of runs. To that end we define Θ-*sentence* to be a Θ-formula with an operator indicating whether it must hold for all runs of a set or for at least one of them.

Definition 5 (Sentence). *For a signature Θ the set $\mathbf{Sen}(\Theta)$ of Θ-sentences is defined as follows:*

$$\mathbf{Sen}(\Theta) = \{\square\,\phi \mid \phi \in Forms(\Theta)\} \cup \{\Diamond\,\phi \mid \phi \in Forms(\Theta)\}$$

Definition 6 (Sentence satisfaction). *For a signature Θ the satisfaction relation $\models_\Theta \subseteq Sys(\Theta) \times \mathbf{Sen}(\Theta)$ is defined as follows:*

- $S \models_\Theta \square\,\phi$ *iff for all* $\rho \in Runs(S)$, $\rho \models_S \phi$,
- $S \models_\Theta \Diamond\,\phi$ *iff there exists* $\rho \in Runs(S)$ *such that* $\rho \models_S \phi$.

4.5   An Example Specification

For the sake of example we list below several axioms of the specification *BasicService*.

The signature Θ_{BS} of *BasicService* contains action symbols *req* (for *request*), *proc* (for *process*) and *resp* (for *response*) representing actions performed by a service, sorts representing request and response data and constants *req_data*, *resp_data* and *null* (strictly speaking we need two *null* constants of different sorts).

(1) $\Diamond\ req\_data = null \wedge resp\_data \neq null$
(2) $\Box\ (req\_data = null \wedge resp\_data = null) \rightarrow \langle req \rangle tt$
(3) $\Box\ [req](req\_data \neq null \wedge \neg req!\ U\ resp!)$

Axiom (1) state the existence of a state satisfying certain requirements. Axiom (2) gives the sufficient conditions for an action *req* to be enabled. The last axiom describes the effects of *req* and ensures service responsiveness — after an occurrence of *req* a *resp* must eventually follow and no *req* is allowed to occur before that point.

To present other specifications involved in *Service* we need to introduce signature morphisms and associated model and sentence functors.

5   The Institution $\mathbb{SYS}$

5.1   Signature Morphisms

A signature morphism $\vartheta : \Theta \rightarrow \Theta'$ describes an extension of a system signature. Morphisms may rename or add elements to data signature, remove or add action symbols and split existing action symbols into several variants.

Definition 7 (Signature morphism). *Let* $\Theta = \langle \Gamma, \Sigma \rangle$ *and* $\Theta' = \langle \Gamma', \Sigma' \rangle$ *be signatures. A signature morphism* $\vartheta : \Theta \rightarrow \Theta'$ *is a pair* $\langle \gamma, \sigma \rangle$*, where*

- $\gamma : \Gamma' \rightharpoonup \Gamma$ *is a partial mapping,*
- $\sigma : \Sigma \rightarrow \Sigma'$ *is a first-order many-sorted signature morphism.*

If $\langle \gamma, \sigma \rangle : \Theta \rightarrow \Theta'$ is a signature morphism, an action symbol $g' \in \Gamma'$ is called a *variant of* $g \in \Gamma$ if $\gamma(g') = g$. g' is a *new action symbol* if $\gamma(g')\uparrow$.

The idea of an action part of a morphism being a contravariant partial mapping is adopted from COMMUNITY.

Signatures and their morphisms with obvious identities and morphism composition constitute the category **Sign**. Being a product of two cocomplete categories it is also cocomplete.

5.2   The Reduct Operation

Let $\vartheta : \Theta \rightarrow \Theta'$ be a signature morphism and let S be a Θ-model. As argued in Sec. 3, due to the requirement that parametric units may only produce expansions of their arguments, it is important that the class of Θ'-systems that are reduced to S include many non-trivial expansions of S.

In our institution, the expansions can be built by:

- adding new data components, which leads to splitting each state of the original system into several states differing in the values of new state components; the effects of transitions can then be modelled with more detail by describing how the new components are affected;
- splitting actions into variants which can then be specified and modelled (partially) independently;
- introducing new actions which may add new runs to the original system.

The reduct operation is defined in two steps. First, a notion of *prereduct* is introduced. A prereduct of a Θ'-system along $\vartheta : \Theta \to \Theta'$ is a Θ-system for which satisfaction condition holds. A proper reduct is obtained by taking a quotient of a prereduct by a suitable equivalence relation.

Definition 8 (Prereduct). *Let* $S = \langle W, D, T \rangle$ *be a* Θ'-*system and let* $\vartheta = \langle \gamma, \sigma \rangle : \Theta \to \Theta'$ *be a signature morphism. The* prereduct *of* S *along* ϑ *is a* Θ-*system* $S|_{\vartheta}^{\star} = \langle W, D', T' \rangle$, *where*

- $D'(w) = D(w)|_{\sigma}$, *for every* $w \in W$,
- $T' = \{ v \xrightarrow{\gamma(g)} w \mid v \xrightarrow{g} w \in T \text{ and } \gamma(g){\downarrow} \}$

The prereduct of a Θ'-system is obtained by reducing data structures along the data component of a signature morphism, relabelling the transitions according to the action component and removing all the transitions labelled with actions that are new in Θ'. The set of states is not affected, therefore all expansions of a model with respect to the prereduct operation share the common set of states, which clearly is not satisfactory for our purposes.

To define the proper reduct we recall the standard notion of bisimulation.

Definition 9 (Bisimulation). *Let* $S_1 = \langle W_1, D_1, T_1 \rangle$ *and* $S_2 = \langle W_2, D_2, T_2 \rangle$ *be systems. A relation* $\sim \subseteq W_1 \times W_2$ *is a* bisimulation *between* S_1 *and* S_2 *if, for all* $\langle v_1, v_2 \rangle \in \sim$, *the following conditions hold:*

(1) $D_1(v_1) = D_2(v_2)$,
(2) if $v_1 \xrightarrow{g} w_1 \in T_1$, *for some* w_1, g, *then there exists* w_2 *such that* $v_2 \xrightarrow{g} w_2 \in T_2$ *and* $w_1 \sim w_2$.
(3) if $v_2 \xrightarrow{g} w_2 \in T_2$, *for some* w_2, g, *then there exists* w_1 *such that* $v_1 \xrightarrow{g} w_1 \in T_1$ *and* $w_1 \sim w_2$.

We say that S_1 and S_2 are *bisimilar*, written $S_1 \approx S_2$, if there exists a *total* bisimulation $\sim$ between S_1 and S_2, i.e. a bisimulation such that for every $v_1 \in W_1$ there exists $v_2 \in W_2$ such that $v_1 \sim v_2$ and for every $w_2 \in W_2$ there exists $w_1 \in W_1$ such that $w_1 \sim w_2$.

For any Θ-system S, the largest bisimulation on S (the sum of all bisimulations between S and itself) is an equivalence relation.

Definition 10 (Reduct). *Let* $S = \langle W, D, T \rangle$ *be a* Θ'-*system and let* $\vartheta = \langle \gamma, \sigma \rangle : \Theta \to \Theta'$ *be a signature morphism. Denote by* $\approx$ *the largest bisimulation on* $S|_{\vartheta}^{\star}$. *The* reduct *of* S *along* ϑ *is a* Θ-*system* $S|_{\vartheta} = \langle W', D', T' \rangle$, *where*

- $W' = W/\approx$
- $D'([v]_\approx) = D(v)|_\sigma$,
- $T' = \{[v]_\approx \xrightarrow{\gamma(g)} [w]_\approx \mid v \xrightarrow{g} w' \in T \text{ for some } w' \in [w]_\approx \text{ and } \gamma(g)\downarrow\}$.

5.3   Formula Translation

The components of a signature morphism $\vartheta = \langle \gamma, \sigma \rangle : \Theta \to \Theta'$ determine the translation $\bar{\sigma} : T_\Sigma(\mathcal{X}) \to T_{\Sigma'}(\mathcal{X})$ of terms and the translation $\overrightarrow{\gamma}^{-1} : \mathcal{P}(\Gamma) \to \mathcal{P}(\Gamma')$ of sets of action symbols. These two mappings induce the translation function $\bar{\vartheta} : Forms(\Theta) \to Forms(\Theta')$.

Clearly, simple renaming of action and operations in terms and formulas is not enough for the satisfaction condition to hold. The expanded system may contain runs that are removed by the reduct operation. The key idea here is to translate Θ-formulas in such a way, that they concern only those runs of the expanded Θ'-system that *extend* runs of the reduct system.

Definition 11 (Run extension). *Let S be a Θ'-system and let $\vartheta : \Theta \to \Theta'$ be a signature morphism. A run $\rho' \in Runs(S)$ extends $\rho \in Runs(S|_\vartheta^\star)$ if, for all $k \le len(\rho)$, $\rho_s(k) = \rho'_s(k)$ and, for all $k < len(\rho)$, $\gamma(\rho'_a(k)) = \rho_a(k)$.*

Definition 12 (Formula translation). *For a signature morphism $\vartheta : \Theta \to \Theta'$ the formula translation function $\bar{\vartheta} : Forms(\Theta) \to Forms(\Theta')$ is defined inductively as follows:*

$$
\begin{aligned}
\bar{\vartheta}(t_1 = t_2) &= \bar{\sigma}(t_1) = \bar{\sigma}(t_2) & \bar{\vartheta}(P(t_1, \ldots, t_n)) &= P(\bar{\sigma}(t_1), \ldots, \bar{\sigma}(t_2)) \\
\bar{\vartheta}(\mathit{ff}) &= \mathit{ff} & \bar{\vartheta}(\phi_1 \to \phi_2) &= \bar{\vartheta}(\phi_1) \to \bar{\vartheta}(\phi_2) \\
\bar{\vartheta}(\forall x_s \phi_1) &= \forall x_{\sigma(s)} \bar{\vartheta}(\phi_1) & \bar{\vartheta}(\boldsymbol{X}_\Delta \phi_1) &= \boldsymbol{X}_{\overrightarrow{\gamma}^{-1}(\Delta)} \bar{\vartheta}(\phi_1) \\
\bar{\vartheta}(\phi_1 \, \boldsymbol{U} \, \phi_2) &= \bar{\vartheta}(\phi_1) \, \boldsymbol{U}_{dom(\gamma)} \, \bar{\vartheta}(\phi_2) & \bar{\vartheta}(\boldsymbol{A} \, \phi_1) &= \boldsymbol{A}_{dom(\gamma)} \, \bar{\vartheta}(\phi_1)
\end{aligned}
$$

where

$$
\begin{aligned}
\phi_1 \, \boldsymbol{U}_\Delta \, \phi_2 &= (\Delta! \wedge \phi_1) \, \boldsymbol{U} \, \phi_2 \\
\boldsymbol{A}_\Delta \, \phi_1 &= \boldsymbol{A} \, (MaxPrefix_\Delta \to \phi_1)
\end{aligned}
$$

Intuitively, ρ satisfies $\phi_1 \, \boldsymbol{U}_\Delta \, \phi_2$ if a Δ-prefix of ρ satisfies $\phi_1 \, \boldsymbol{U} \, \phi_2$, and ρ satisfies $\boldsymbol{A}_\Delta \, \phi_1$ if ϕ_1 is satisfied by all runs co-initial with ρ which start with a maximal Δ-prefix (cf the definition of $MaxPrefix_\Delta$ in Sec. 4.3).

5.4   Satisfaction Condition

The satisfaction condition for the institution is verified first for prereducts, then it is shown that the prereduct $S|_\vartheta^\star$ is logically equivalent to the reduct $S|_\vartheta$.

Lemma 1. *Let $\vartheta : \Theta \to \Theta'$, $S \in Sys(\Theta')$, $\phi \in Forms(\Theta)$ and $\rho \in Runs(S|_\vartheta^\star)$. Denote by $Ext(\rho)$ the set of those S-runs that extend ρ. The following equivalence holds:*

$$
\rho \models_{S|_\vartheta^\star} \phi \text{ iff, for every } \rho' \in Ext(\rho), \; \rho' \models_S \bar{\vartheta}(\phi)
$$

Definition 13 (Sentence translation). *For a signature morphism* $\vartheta : \Theta \to \Theta'$ *the* sentence translation function $\mathbf{Sen}(\vartheta) : \mathbf{Sen}(\Theta) \to \mathbf{Sen}(\Theta')$ *is defined as follows:*

$$\mathbf{Sen}(\vartheta)(\square\, \phi) = \square \; MaxPrefix_{dom(\gamma)} \to \bar{\vartheta}(\phi)$$
$$\mathbf{Sen}(\vartheta)(\lozenge\, \phi) = \lozenge \; \bar{\vartheta}(\phi) \wedge MaxPrefix_{dom(\gamma)}$$

The following lemma is a consequence of Lemma 1.

Lemma 2. *Let* $\vartheta : \Theta \to \Theta'$, $S \in Sys(\Theta')$ *and* $\psi \in \mathbf{Sen}(\Theta)$. *The following equivalence holds:*

$$S|_{\vartheta}^{\star} \models_{\Theta} \psi \quad \text{iff} \quad S \models_{\Theta'} \mathbf{Sen}(\vartheta)(\psi)$$

To establish the satisfaction condition for reducts it remains to show that bisimilar models satisfy the same sentences.

Lemma 3. *Let* $S_1, S_2 \in Sys(\Theta)$. *If* $S_1 \approx S_2$ *then, for all* $\psi \in \mathbf{Sen}(\Theta)$, $S_1 \models_{\Theta} \psi$ *iff* $S_2 \models_{\Theta} \psi$.

Proposition 4 (Satisfaction Condition). *Let* $\vartheta : \Theta \to \Theta'$, $S \in Sys(\Theta')$ *and* $\psi \in \mathbf{Sen}(\Theta)$. *The following equivalence holds:*

$$S|_{\vartheta} \models_{\Theta} \psi \quad \text{iff} \quad S \models_{\Theta'} \mathbf{Sen}(\vartheta)(\psi)$$

The satisfaction condition is a consequence of Lemmas 2, 3 and the fact that for any Θ'-system S and any signature morphism $\vartheta : \Theta \to \Theta'$, the prereduct $S|_{\vartheta}^{\star}$ is bisimilar to the reduct $S|_{\vartheta}$.

The definition below summarizes this section.

Definition 14 (Institution SYS). *The institution SYS is a tuple* $\langle \mathbf{Sign}, \mathbf{Mod}, \mathbf{Sen}, \models \rangle$, *where*

- *The category of signatures* **Sign** *is defined in Sec. 5.1.*
- *The model functor* **Mod** : **Sign**^{op} → **Cat** *is defined as follows:*
  - **Mod**(Θ) *is a discrete category in which objects are fully abstract* Θ-*systems, i.e. those, for which the largest bisimulation is the identity,*
  - *for* $\vartheta : \Theta \to \Theta'$ *the reduct functor* **Mod**(ϑ) : **Mod**(Θ') → **Mod**(Θ) *is defined in Def. 10.*
- *The sentence functor* **Sen** : **Sign** → **Set** *is defined in Def. 5 and Def. 13.*
- *The satisfaction relation* $\models$ *is defined in Def. 6.*

We consider only fully abstract systems as Θ-models, since otherwise a reduct along id_{Θ} would not be an identity functor. From now on all systems are assumed to be fully abstract.

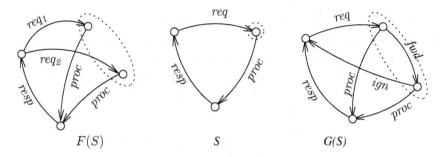

Fig. 3. An example model

5.5   Example Specification Again

Assume, that the signature Θ_{MS} of *MultiService* contains four action symbols: req_1, req_2, *proc* and *resp*, and the signature morphism $\vartheta_1 : \Theta_{BS} \to \Theta_{MS}$ maps both req_1 and req_2 to *req* in the signature Θ_{BS} of *BasicService* (cf. Sec. 4.5), *proc* to *proc* and *resp* to *resp*.

The signature Θ_{FS} of *FwdService* contains two new action symbols besides those from Θ_{BS}: *fwd* (for *forward*) and *ign* (for *ignore*). That last action represents the possibility that the external — and thus unreliable — service to which a request was forwarded did not process it correctly for some reason.

A possible implementation of the unit S and the results of applying parametric units to it are shown on Fig. 3. $F(S)$ adds more detail to S by splitting a state and transitions: to the target state of the *req*-transition in S, correspond two states in $F(S)$, each representing an entry point for a different type of service. A single *req*-transition of S is split into two variants, representing two different types of service requests.

$G(S)$ adds entirely new behaviour to S: new transitions, and thus new runs, appear. The *fwd* transition represents forwarding the request, both the source and the target of this transition correspond to a single state of S. Adding the new *ign* transition introduces the possibility that the request is not served (i.e. no *proc*,*resp* sequence occurs after *req*).

Let ψ be the sentence stated as axiom (3) in the specification *BasicService*. The satisfaction condition ensures that, since ψ holds in S and both $F(S)$ and $G(S)$ are expansions of S, translations of ψ must also be satisfied by those two models.

Sen$(\vartheta_1)(\psi)$ is equivalent to the sentence

$$\Box \, [req\_1, req\_2](req\_data \neq null \land \neg\{req\_1, req\_2\}! \, \boldsymbol{U} \, resp!)$$

and **Sen**$(\vartheta_2)(\psi)$ to

$$\Box \, MaxPrefix_{\{req, proc, resp\}} \to [req](req\_data \neq null \land \{proc, resp\}! \, \boldsymbol{U} \, resp!)$$

Note, that the property expressed by ψ is no longer guaranteed to hold for all runs of a Θ_{FS}-system, but only for those that start with a maximal $\{req, proc, resp\}$-prefix.

6   System Amalgamation

The denotation of $F(S)$ **and** $G(S)$ in the example specification is the amalgamation of models denoted by $F(S)$ and $G(S)$.

Let $\Theta_1 \xleftarrow{\vartheta_1} \Theta \xrightarrow{\vartheta_2} \Theta_2$ be a diagram in **Sign**. Its pushout is a signature $\Theta' = \langle \Gamma', \Sigma' \rangle$ together with morphisms $\upsilon_1 = \langle \delta_1, \tau_1 \rangle : \Theta_1 \to \Theta'$ and $\upsilon_2 = \langle \delta_2, \tau_2 \rangle : \Theta_2 \to \Theta'$, defined as follows:

$$\Gamma' = \{\langle g_1, g_2 \rangle \in \Gamma_1 \times \Gamma_2 \mid \gamma_1(g_1) = \gamma_2(g_2) \text{ or } \gamma_1(g_1)\uparrow, \gamma_2(g_2)\uparrow\}$$
$$+ \{g_1 \in \Gamma_1 \mid \gamma_1(g_1)\uparrow\} + \{g_2 \in \Gamma_2 \mid \gamma_2(g_2)\uparrow\}$$

$$\delta_i(\langle g_1, g_2 \rangle) = g_i, \quad \delta_i(g_i) = g_i, \quad \delta_i(g_{2-i})\uparrow, \quad \text{for } i = 1, 2$$

and $\langle \Sigma', \tau_1, \tau_2 \rangle$ is a pushout of the diagram $\Sigma_1 \xleftarrow{\sigma_1} \Sigma \xrightarrow{\sigma_2} \Sigma_2$ in the category of many-sorted first-order signatures.

Lemma 5 (Weak Amalgamation). *Let* $\langle \Theta', \upsilon_1 : \Theta_1 \to \Theta', \upsilon_2 : \Theta_2 \to \Theta' \rangle$ *be a pushout of the diagram*

$$\Theta_1 \xleftarrow{\vartheta_1} \Theta \xrightarrow{\vartheta_2} \Theta_2$$

Let $S_1 \in \mathbf{Mod}(\Theta_1)$ *and* $S_2 \in \mathbf{Mod}(\Theta_2)$ *be such that* $S_1|_{\vartheta_1} = S_2|_{\vartheta_2}$. *There exists* $S' \in \mathbf{Mod}(\Theta')$ *such that* $S'|_{\upsilon_1} = S_1$ *and* $S'|_{\upsilon_2} = S_2$.

Proof. The amalgamated model, denoted by $S_1 \times_{\vartheta_2}^{\vartheta_1} S_2$, is constructed as follows: let $\approx_1$ and $\approx_2$ denote largest bisimulations on $S_1|_{\vartheta_1}^\star$ and $S_2|_{\vartheta_2}^\star$, respectively. Let us recall from Sec. 2 that the identity $S_1|_{\vartheta_1} = S_2|_{\vartheta_2}$ means that there exists a suitable bijection $i : W_1/\approx_1 \to W_2/\approx_2$.

$S_1 \times_{\vartheta_2}^{\vartheta_1} S_2 = \langle W', D', T' \rangle$ is given by:

- $W' = \{\langle v_1, v_2 \rangle \in W_1 \times W_2 \mid i([v_1]_{\approx_1}) = [v_2]_{\approx_2}\}$
- $D'(\langle v_1, v_2 \rangle) = DS_1(v_1) \times_{\sigma_2}^{\sigma_1} D_2(v_2)$ (the amalgamation of first-order structures), and

$$T' = \left\{ \langle v_1, v_2 \rangle \xrightarrow{\langle g_1, g_2 \rangle} \langle w_1, w_2 \rangle \mid v_1 \xrightarrow{g_1} w_1 \in T_1 \text{ and } v_2 \xrightarrow{g_2} w_2 \in T_2 \right\}$$

$$\cup \left\{ \langle v_1, v_2 \rangle \xrightarrow{g_1} \langle w_1, v_2 \rangle \mid v_1 \xrightarrow{g_1} w_1 \in T_1 \right\} \cup \left\{ \langle v_1, v_2 \rangle \xrightarrow{g_2} \langle v_1, w_2 \rangle \mid v_2 \xrightarrow{g_2} w_2 \in T_2 \right\}$$

Verifying that $S_1 \times_{\vartheta_2}^{\vartheta_1} S_2$ is a Θ'-model and that $(S_1 \times_{\vartheta_2}^{\vartheta_1} S_2)|_{\upsilon_1} = S_1$ and $(S_1 \times_{\vartheta_2}^{\vartheta_1} S_2)|_{\upsilon_2} = S_2$ is straightforward.

The example from Fig. 4 shows that for an arbitrary pair S_1, S_2 of systems extending the common one there may not exist a unique system that extends them both. The signatures of S, S_1 and S_2 are $\langle \emptyset, \Sigma \rangle$, $\langle \{f\}, \Sigma \rangle$ and $\langle \{g\}, \Sigma \rangle$, respectively, for some data signature Σ. The pushout signature is $\langle \{f, g, \langle f, g \rangle\}, \Sigma \rangle$. Four other systems that extend both S_1 and S_2 can be obtained by removing some transitions from $S_1 \times_{\vartheta_2}^{\vartheta_1} S_2$. The construction given in the above lemma always yields the largest (w.r.t. the number of transitions) such system.

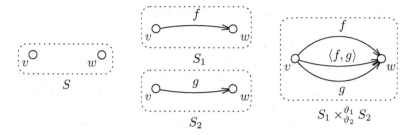

Fig. 4. An example of amalgamation

Amalgamation Lemma provides the semantics for the $F(S)$ **and** $G(S)$ construct in the example. From the persistence of F and G it follows that $F(S)|_{\vartheta_1} = S = G(S)|_{\vartheta_2}$ and thus, by Lemma 5, we conclude the existence of a system $F(S) \times_{\vartheta_2}^{\vartheta_1} F(G)$ which is denoted by $F(S)$ **and** $G(S)$.

7   Describing System Synchronisation

It is important to realise that a reduct of a system should not be, in general, viewed as its subsystem. Intuitively, to consider C a subcomponent of C', one should be able to retrieve from every configuration of C' a corresponding configuration of C. Whenever C' changes its configuration by performing an action, C may have to change configuration accordingly, by performing a corresponding "subaction". In other words, C should *simulate* C'.

The notion of a subcomponent can be expressed on the level of transition systems by defining a notion of a system morphism, that identifies its target as a subsystem of its source.

Definition 15 (Category of systems). *The category of systems* **Sys** *has pairs* $\langle \Theta, S \rangle$, *where* $S \in \mathbf{Mod}(\Theta)$, *as objects. A morphism* $\mu : \langle \Theta', S' \rangle \to \langle \Theta, S \rangle$ *is a pair* $\langle \vartheta, \pi \rangle$, *where*

- $\vartheta = \langle \gamma, \sigma \rangle : \Theta \to \Theta'$ *is a signature morphism,*
- $\pi : S' \to S$ *is a function such that*
  - *(i)* $D(\pi(v')) = D'(v')|_\sigma$,
  - *(ii) if* $v' \xrightarrow{g'} w' \in T'$ *and* $\gamma(g')\!\downarrow$ *then* $\pi(v') \xrightarrow{\gamma(g')} \pi(w') \in T$
  - *(iii) if* $v' \xrightarrow{g'} w' \in T'$ *and* $\gamma(g')\!\uparrow$ *then* $\pi(v') = \pi(w')$.

We will also use the term *system* for objects of **Sys**.

The above definition is very similar to the definition of the category of systems in [NW95–Sec. 1].

Consider again systems depicted on Fig. 4. $\langle \Theta_1, S_1 \rangle$ is not a subsystem of $\langle \Theta', S_1 \times_{\vartheta_2}^{\vartheta_1} S_2 \rangle$, since S_1 cannot simulate the g-transition of $S_1 \times_{\vartheta_2}^{\vartheta_1} S_2$. Similarly, S in not a subsystem of S_1, since it cannot simulate the f-transition. This example shows that systems being amalgamated do not have to share a common subsystem.

The parallel composition of reactive components C_1 and C_2 can be defined as "the smallest" component having both C_1 and C_2 as subcomponents. On the level of systems, a parallel composition of $S_1 \in \mathbf{Mod}(\Theta_1)$ with $S_2 \in \mathbf{Mod}(\Theta_2)$ would be then the categorical product $\langle \Theta_1, S_1 \rangle \times \langle \Theta_2, S_2 \rangle$ (cf [NW95–Sec. 2]). As the following result shows, the product coincides with a special form of the amalgamation, where both amalgamated systems extend the trivial one.

Fact 6. *Denote by Σ_0 the empty many-sorted first-order signature and by A_0 — the unique (empty) Σ_0-structure. $\Theta_0 = \langle \emptyset, \Sigma_0 \rangle$ is the initial system signature and $S_0 = \langle \{\star\}, \{\star \mapsto A_0\}, \emptyset \rangle$ is the unique Θ_0-system.*

Let $S_1 \in \mathbf{Mod}(\Theta_1)$, $S_2 \in \mathbf{Mod}(\Theta_2)$ and let $\vartheta_1 : \Theta_0 \to \Theta_1$ and $\vartheta_2 : \Theta_0 \to \Theta_2$. The product $\langle \Theta_1, S_1 \rangle \times \langle \Theta_2, S_2 \rangle$ is $\langle \Theta_1 + \Theta_2, S \rangle$, where $S = S_1 \times_{\vartheta_2}^{\vartheta_1} S_2$ is the amalgamation of S_1 and S_2.

The signature of the product $\langle \Theta_1, S_1 \rangle \times \langle \Theta_2, S_2 \rangle$, where $\Theta_1 = \langle \Gamma_1, \Sigma_1 \rangle$, $\Theta_2 = \langle \Gamma_2, \Sigma_2 \rangle$, is given by

$$\Theta_1 + \Theta_2 = \langle (\Gamma_1 \times \Gamma_2) + \Gamma_1 + \Gamma_2, \Sigma_1 + \Sigma_2 \rangle$$

(see the definition of the pushout signature in Sec. 6).

Systems $\langle \Theta_1, S_1 \rangle$ and $\langle \Theta_2, S_2 \rangle$ can be *synchronised* by requiring that certain actions of Θ_1 and Θ_2 may only occur simultaneously. This can be done by providing a *synchronisation set* Γ, with mappings $\gamma_1 : \Gamma_1 \rightharpoonup \Gamma$ and $\gamma_2 : \Gamma_2 \rightharpoonup \Gamma$, where Γ_1 and Γ_2 are action components of Θ_1 and Θ_2, respectively. The signature of the synchronised system is given by a pushout $\Theta' = \langle \Gamma', \Sigma_1 + \Sigma_2 \rangle$ of $\Theta_1 \xleftarrow{\langle \gamma_1, \iota_{\Sigma_1} \rangle} \langle \Gamma, \Sigma_0 \rangle \xrightarrow{\langle \gamma_2, \iota_{\Sigma_2} \rangle} \Theta_2$, where ι_Σ denotes the unique signature morphism $\iota_\Sigma : \Sigma_0 \to \Sigma$.

Since Γ' is a subset $(\Gamma_1 \times \Gamma_2) + \Gamma_1 + \Gamma_2$ with "unsynchronised" pairs of actions removed, the synchronisation of $\langle \Theta_1, S_1 \rangle$ and $\langle \Theta_2, S_2 \rangle$ via γ_1, γ_2 can be then defined as the system obtained by removing from the product $\langle \Theta_1, S_1 \rangle \times \langle \Theta_2, S_2 \rangle$ transitions labelled with symbols not in Γ'. Similar approach is taken in [NW95].

Such a restriction of $\langle \Theta_1, S_1 \rangle \times \langle \Theta_2, S_2 \rangle$ can be expressed using the reduct operation, which motivates the following definition.

Definition 16 (System synchronisation). *Let $S_1 \in \mathbf{Mod}(\Theta_1)$, $S_2 \in \mathbf{Mod}(\Theta_2)$, $\Theta_1 = \langle \Gamma_1, \Sigma_1 \rangle$, $\Theta_2 = \langle \Gamma_2, \Sigma_2 \rangle$. Let Γ be a set and let $\gamma_1 : \Gamma_1 \rightharpoonup \Gamma$, $\gamma_2 : \Gamma_2 \rightharpoonup \Gamma$ be partial mappings.*

A synchronisation of S_1 and S_2 via γ_1, γ_2, denoted $S_1 \|_{\gamma_2}^{\gamma_1} S_2$, is a Θ'-system $(S_1 \times_{\upsilon_2}^{\upsilon_1} S_2)|_\iota$, where

- *$\langle \Theta', \vartheta_1 : \Theta_1 \to \Theta', \vartheta_2 : \Theta_1 \to \Theta' \rangle$ is a pushout of $\Theta_1 \xleftarrow{\langle \gamma_1, \iota_{\Sigma_1} \rangle} \langle \Gamma, \Sigma_0 \rangle \xrightarrow{\langle \gamma_2, \iota_{\Sigma_2} \rangle} \Theta_2$, $\Theta' = \langle \Gamma', \Sigma_1 + \Sigma_2 \rangle$,*
- *$\upsilon_1 : \Theta_0 \to \Theta_1$ and $\upsilon_2 : \Theta_0 \to \Theta_2$ are unique morphisms from the initial signature Θ_0,*
- *$\iota = \langle \delta, id_{\Sigma_1 + \Sigma_2} \rangle : \Theta' \to \Theta_1 + \Theta_2$ is such that $\delta(g) = g$ if $g \in \Gamma'$ and $\delta(g)\!\uparrow$ otherwise.*

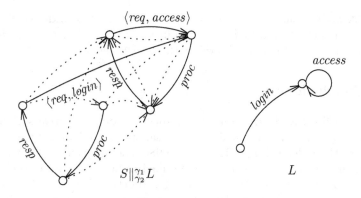

Fig. 5. An example of synchronisation

The morphism $\iota : \Theta' \to \Theta_1 + \Theta_2$ in the above definition exists thanks to the fact that the source $\langle \Gamma, \Sigma_0 \rangle$ of the pushout diagram of signatures has an empty data signature. The synchronisation we consider is therefore less general than a mechanism used in COMMUNITY, where two components may also share some part of data structure besides actions.

7.1   An Architectural Specification of Synchronisation

The synchronisation of the model S of *BasicService* with the model L of simple access-control system is shown on Fig. 5. The synchronisation set is $\Gamma = \{\star\}$, action symbols of S and L are $\Gamma_1 = \{req, proc, resp\}$ and $\Gamma_2 = \{login, access\}$ respectively. Mappings $\gamma_1 : \Gamma_1 \rightharpoonup \Gamma$ and $\gamma_2 : \Gamma_2 \rightharpoonup \Gamma$ are given by

$$\gamma_1(req) = \star = \gamma_2(login) = \gamma_2(access)$$
$$\gamma_1(proc)\!\uparrow, \gamma_1(resp)\!\uparrow$$

The synchronised system has a signature with an action set

$$\Gamma' = \{\langle req, login \rangle, \langle req, access \rangle, proc, resp\}$$

and is a restriction of the product of S and L. Dotted lines on Fig. 5 represent transitions removed from the product.

The synchronisation could be described by a simple architectural specification using a provisional syntax

> **arch spec** *AuthorisingService* **is**
> **units** $S : BasicService$
> $L : AccessControl$
> **result sync** *req* **in** S **with** *login, access* **in** L

where the term **sync** *req* **in** S **with** *login, access* **in** L is a syntactic sugar for

$$(S \text{ and } L) \text{ hide } \langle proc, login \rangle, \langle proc, access \rangle, \langle resp, login \rangle, \langle resp, access \rangle$$

In CASL architectural specifications the semantics of U **hide** *symbols* is the reduct of model denoted by U along an inclusion morphism $\iota : \Theta \to \Theta_U$, where Θ_U is a signature of U and Θ is Θ_U with *symbols* removed.

8   Conclusions

We have defined an institution for the specification of reactive systems. With regard to the syntax of the logic and the satisfaction relation the institution is similar to those defined in [CR97] or [Cen98]. The uniqueness of our institution — compared to institutions briefly described in Sec. 1 — lies in the definition of the reduct operation and the translation function. They are defined so that an expansion of a system can have larger state space and more transitions than the original system.

In Sec. 7 we have briefly discussed the difference between system amalgamation and synchronisation. On the one hand, it is possible to amalgamate systems that do not have a common subsystem. On the other hand, it is possible to synchronise arbitrary systems that do not have a common reduct. Nevertheless, action synchronisation can be expressed in terms of system amalgamation and reduction. Therefore, covering system synchronisation in architectural specifications does not seem to require adding new architectural operations. However, synchronisation by sharing common state, in the style of COMMUNITY, is not supported yet and will require further work.

The category of systems was defined in Sec. 7 only to provide motivation for the definition of system synchronisation. It may be worthwile to investigate more thoroughly the interplay between various constructions in the category of systems and the reduct operation.

The future work will also include incorporating action refinement in the institutional framework. Ideally, the reduct operation would be defined in such a way, that in an expansion of a system a single transition could be replaced by a sequence, or rather a subsystem of "sub-transitions". This will probably require replacing transition systems with a non-interleaving model of concurrent components, such as asynchronous transition systems ([NW95–Sec. 10]).

References

[AF95]   M. Arrais and J.L. Fiadeiro: Unifying Theories in Different Institutions. *Proc. 11th. Workshop on Abstract Data Types (ADT'95)*, LNCS 1130, pp. 81–101, Springer, 1996.

[AZ95]   E. Astesiano and E. Zucca: D-oids: a Model for Dynamic Data-Types. *Mathematical Structures in Computer Science*, 5(2), pp. 257–282, 1995.

[BZ00]   H. Baumeister and A. Zamulin: State-Based Extension of CASL. *Proc. 2nd. Intl. Conf. Integrated Formal Methods (IFM'00)*, LNCS 1945, pp. 3–24, Springer, 2000.

[BST99]   M. Bidoit, D. Sannella and A. Tarlecki: Architectural specifications in CASL. *Proc. 7th Int. Conf. Algebraic Methodology and Software Technology (AMAST'98)*, LNCS 1548, pp. 341–357, Springer, 1999.

[Cen98]   M.V. Cengarle: *The Temporal Logic Institution*, Technical Report 9805, Ludwig-Maximilians-Universität München, 1998.

[CoFI04]   The Common Framework Initiative (CoFI): CASL *Reference Manual*, LNCS 2960 (IFIP Series), Springer, 2004.

[CR97]   G. Costa and G. Reggio: Specification of Abstract Dynamic Data Types: A Temporal Logic Approach. *Theoretical Computer Science*, 173(2), pp. 513–554, 1997.

[DF98]   D. Diaconescu and K. Futatsugi: *CafeOBJ Report: The Language, Proof Techniques, and Methodologies for Object-Oriented Algebraic Specification*, vol. 6 of AMAST series in Computing, World Scientific, 1998.

[FLW03]   J.L. Fiadeiro, A. Lòpes and M. Wermelinger: A Mathematical Semantics for Architectural Connectors. *Generic Programming*, LNCS 2793, pp. 190–234, Springer, 2003.

[FM92]   J.L. Fiadeiro and T. Maibaum: Temporal Theories as Modularisation Units for Concurrent System Specification. *Formal Aspects of Computing*, 4(3), pp. 239–272, 1992.

[GHR95]   D.M. Gabbay, C.J. Hogger and J.A. Robinson, eds: *Handbook of Logic in Artificial Intelligence and Logic Programming: Epistemic and Temporal Reasoning*, vol. 4, Oxford University Press, 1995.

[GB92]   J. Goguen and R. Burstall: Institutions: Abstract Model Theory for Specification and Programming. *Journal of the ACM*, 39(1), pp. 95–146, 1992.

[NW95]   M. Nielsen and G. Winskel: Models for Concurrency. Chapter in the *Handbook of Logic and the Foundations of Computer Science, vol 4.*, pp 1–148, Oxford University Press, 1995.

[Reg90]   G. Reggio: Entities: An Instituiton for Dynamic Systems. *Proc. 7th. Workshop on Abstract Data Types (ADT'90)*, LNCS 534, pp. 246–265, Springer, 1990.

[ACR99]   G. Reggio, E. Astesiano and C. Choppy: CASL-LTL : A CASL Extension for Dynamic Reactive Systems — Summary. Technical Report DISI-TR-99-34, DISI — Unievrsità di Genova, 1999.

[ST84]   D. Sannella and A. Tarlecki: Building specifications in an arbitrary institution. *Proc. Intl. Symp. on Semantics of Data Types, Sophia-Antipolis, France 1984*, LNCS 173, pp. 337–356, Springer, 1984.

[CSS98]   A. Sernadas, C. Sernadas and C. Caleiro: Denotational Semantics of Object Specification. *Acta Informatica*, 35(9), pp. 729–773, 1998.

[Tar99]   A. Tarlecki: Institutions: An Abstract Framework for Formal Specifications. Chapter 4 in E. Astesiano, H.-J. Kreowski, B. Krieg-Brückner eds.: *Algebraic Foundations of System Specification*, Springer, 1999.

Author Index

Lecture Notes in Computer Science

For information about Vols. 1–3331

please contact your bookseller or Springer